SARDAR SAROVAR PROJECT
ON
THE RIVER NARMADA

History of Rehabilitation and Implementation

Volume 2

About the Editors

R. Parthasarathy is a Professor, Faculty of Planning and Public Policy at CEPT University, Ahmedabad. He has about 25 years of teaching and research experience and has been a Visiting Scholar at the University of California, Berkeley; Professor and later Director of Gujarat Institute of Development Research, Ahmedabad. He has been a member of many government committees and has also been a consultant to DFID, ADB and other national and international organisations. He also works closely with NGOs. His research interests include, water management, large-scale irrigation systems, natural resources, infrastructure management and environmental economics. He has published 3 books and over 70 articles.

Ravindra H. Dholakia has been a Professor of Economics at IIM, Ahmedabad since 1985. He has about 34 years of experience in teaching Economics to different groups like students, executives, policy makers and senior government officials. He has practical experience in high-powered policy making and evaluation bodies in both state and central governments such as the Sixth Central Pay Commission in India (2006-8), State Public Finance Reforms Committee (1998-2001), Expert Committee to Restructure Public Sector Units (2004-7) and Committee to Restructure Public Debt (2003-5), appointed by the Government of Gujarat. Currently, he is a member of the High Level Expert Committee appointed by the Government of India to suggest measures for efficient management of public expenditure (2010-11). He has published 39 monographs, 17 books, and about 130 research papers and international cases on economic planning and policy reforms.

Sardar Sarovar Project
on
the River Narmada

History of Rehabilitation and Implementation

Volume 2

Editors

R. Parthasarathy

Ravindra H. Dholakia

Published for

CEPT University Press, Ahmedabad-380009

by

Concept Publishing Company Pvt. Ltd., New Delhi-110059

ISBN-13: 978-81-8069-795-1 (Set)
978-81-8069-796-8 (Vol. 1)
978-81-8069-797-5 (Vol. 2)
978-81-8069-798-2 (Vol. 3)

First Published 2011

CEPT UNIVERSITY
Kasturbhai Lalbhai Campus
University Road,
Ahmedabad-380009 (Gujarat) India.
Phone : 26302470, 26302740,
Fax : +91-079-26302075
Email: publications@cept.ac.in
Website: www.cept.ac.in

Printed and Published by

Concept Publishing Company Pvt. Ltd.
Regd. Office:
A/15-16, Commercial Block, Mohan Garden
New Delhi-110059 (India)
Phones : 25351460, 25351794, *Fax* : 091-11-25357109
Email : publishing@conceptpub.com
Website: www.conceptpub.com

Editorial Office:
H-13, Bali Nagar, New Delhi-110 015, India.

Cataloging in Publication Data--*Courtesy:* D.K. Agencies (P) Ltd. <docinfo@dkagencies.com>

Sardar Sarovar Project on the river Narmada / editors, R. Parthasarathy, Ravindra H. Dholakia.
3 v.. cm.
Includes bibliographical references (p.) and index.
Contents: v. 1. History of design, planning and appraisal -- v. 2. History of rehabilitation and implementation -- v. 3. Impacts so far and ways forward.
ISBN 13: 9788180697951 (set) ISBN 13: 9788180697975 (v. 2)
ISBN 13: 9788180697968 (v. 1) ISBN 13: 9788180697982 (v. 3)

1. Sardar Sarovar (Narmada) Project--Congresses. 2. Water resources development--Environmental aspects--India--Maharashtra--Congresses. 3. Water resources development--Environmental aspects--India--Gujarat--Congresses. 4. Dams--India--Maharashtra--Design and construction--Congresses. 5. Dams--India--Gujarat--Design and construction --Congresses. 6. Rehabilitation--India--Maharashtra--Congresses. 7. Rehabilitation--India--Gujarat--Congresses. 8. Irrigation--India--Maharashtra--Congresses. 9. Irrigation--India--Gujarat--Congresses. I. Parthasarathy, R., Dr. II. Dholakia, Ravindra H., 1953- III. Center for Environmental Planning and Technology University (Ahmadābād, India)

DDC 333.910095479 22

Shri Narendra Modi
Chief Minister, Gujarat

Foundation stone of the Sardar Sarovar Dam on the river Narmada was laid on 5th April 1961 hence this Project will also be completing its long journey of five decades as Gujarat State is celebrating its Golden Jubilee. Sardar Sarovar Project is not only an icon of water resources development but it is also an expression of the strong will of the people of this State. Against a portrayed perception of SSP as a controversial project, this compilation reveals its multi-faceted dimensions. Not only the dimensions and benefits of this project are awesome but it has also got many innovative features in planning, designing, implementation, financing and managing. It is one of the most studied water resources projects in the world and its implementation has been a long struggle against all odds, which could be possible mainly due to wide public support and participation from people of Gujarat. Successive Governments made positive contribution in this effort, and the will of the people of Gujarat has been unstinting, and therefore in a true sense SSP is Project of the People of Gujarat.

Last one decade has seen accelerated development in this Project as a result of which the dam height could be raised to the spillway crest level, two hydro power houses could be made operational, 458 kilometer long Narmada Main Canal could be completed and command area of more than 5 lakh hectare could be developed by constructing a canal network of 19,000 kilometer length. Increased and assured availability of Narmada water for irrigation has helped in achieving and sustaining an impressive agricultural growth rate of more than 9% in the State against the national average of about 3% during the same period.

SSP is no less than a *"living technological university"* and the need for proper documentation of its evolution and growth was increasingly felt. As one of the firsts, these three volumes have now documented history of many facets of this very important project and have captured the highlights of this unique endeavor, while also suggesting ways forward.

The articles written by pioneer engineers, involved activists, academicians and government officials present various aspects of the project and Editors too have done a commendable job in bringing the individual contributions to a comprehensive publication level. Articles in this volumes address several concerns raised by critics at various stages of the Project. In my opinion on this 'World Water Day' during the golden jubilee year of the State, there can not be a better tribute to the people of Gujarat who have wholeheartedly supported the Sardar Sarovar Project in this journey. Sardar Sarovar Narmada Nigam Limited also deserves compliments for having supported this invaluable commemorative effort that will help the future generations to understand the evolution and growth of Sardar Sarovar Project.

(Narendra Modi)

Preface

The Sardar Sarovar Project (SSP) on the River Narmada has a long and interesting, yet sometimes controversial history. Even before the formal proposal was made in 1959 for a dam at the Sardar Sarovar site at Navagam, there had been various attempts to harness the Narmada. While Independence solved the problems of political complexity associated with organisation among the princely states for construction of the dam, the reorganisation of states on linguistic basis in 1956 added a new dimension. When preliminary work on the dam began in 1961, the disagreements between the riparian states on the sharing of waters led to the project being suspended. In 1969 the Narmada Water Disputes Tribunal (NWDT) was set up to decide the inter-state allocation of water and the costs of the dams and other infrastructure needed to exploit the river. In 1979 the Tribunal Award was made and full-scale construction of Sardar Sarovar began in 1987.

In its long history, SSP has faced many hurdles. It was not until the year 2003-04 that many of the issues relating to construction of the dam were resolved. Meanwhile, the main canal was constructed and many minors and sub-minors are in the process of being built. Other important features of this project are the rehabilitation of the project affected persons and the planning of use of its waters. While many of the controversies have been well documented, many salient features associated with the project have not been. It is precisely the many dimensions of the setting up of this project that acted as the stimulus to document the evolution and realisation of the project during the Golden Jubilee year of the state of Gujarat. To this end, three main themes emerged, namely, (a) the History of Design,

Planning and Appraisal including the Environment; (b) the Resettlement and Rehabilitation Process during Implementation; and (c) the Future Outlook and Way Forward.

The idea of producing edited volumes on these themes was discussed in a half-day Workshop at CEPT University on August 6, 2010. The Workshop, titled 'Evolution and Growth of Sardar Sarovar: The Life Line of Gujarat was attended by Engineers, most of whom have superannuated, but who had been involved in the design and construction of the dam and the canals; academics, NGO activists, serving and retired officials of Sardar Sarovar Narmada Nigam Limited (SSNNL); practitioners and policy researchers. In many ways this was an exploratory workshop; some came with the notion of processes being important; some focused on the history, and yet others suggested that the many innovations during the project reflect the flexible and sensitive responses to the changing environments and challenges moving away from the conventional designs. Continued discussions with the participants and the editors evolved into a commitment to contribute to the volumes well-researched articles on the different themes. It was also resolved that this volume need not go into the many controversies but should focus on processes and innovative aspects of SSP.

Thus the important feature of these volumes is not only that most of its contributors were directly involved in the project at one time or the other, but also had intimate knowledge of the various interrelated issues. As discussed in the August Workshop, the contributed papers before being published were to go through varied and multiple levels of review. The draft papers were discussed at CEPT University in the Second Workshop held from October 29 to 31, 2010. During this Workshop special invited lectures by Professor Y.K. Alagh (former Union Minister of Planning and Programme Implementation, Science and Technology and Power, GoI, and, Vice-Chairman of the Narmada Planning Group); Dr. C.C. Patel (former Secretary, Ministry of Water Resources, GoI, and formerly CMD of SSNL); Mr. Jay Narayan Vyas, Minister of Health and Tourism, and formerly Minister of Narmada Development, GoG); and Mr. I.P. Gautam (Municipal Commissioner, Ahmedabad and formerly, Director Finance, SSNNL) were also arranged. In fact, we later persuaded Mr. Jay Narayan Vyas and Mr. I.P. Gautam to convert their lectures into papers for the volumes.

During the Workshop the peer discussions on the papers contributed immensely to further sharpening the focus. Some of the authors also presented the final version of their papers during the two days of January 3 and 4, 2011 in the final workshop held at CEPT University.

An effort of this magnitude would not have been possible but for the strong and committed response of the authors. The papers were subjected to a refereeing process and some of them were re-reviewed before acceptance as publishable material. For this publication, we also had an Advisory Board consisting of Mr. N.B. Desai, Mr. N. Ramaswami, Mr. G.L. Java, Mr. P.H. Rana, Dr. Anil Patel, Mr. Hasmukh Patel, Dr. Sanat Chavan, Dr. Tushaar Shah and Dr. M.B. Joshi. We are indeed thankful to the authors, the referees and the Advisory Committee Members for their invaluable support in bringing these volumes within the time constraints. Our special thanks go to Mr. Vivek Khadpekar for his commitment and involvement with the editors while carrying out a commendable and a very competent editing of the papers.

All the analysis, views and opinions contained in the papers are those of the respective authors and do not reflect the views of the organisations they belong. Similarly, while we have taken all possible care to check the facts and figures presented in the paper, the errors of omission and commission still rest with the authors of the papers.

We are very grateful to Mr. S. Jagadeesan, Mr. K. Srinivas and Mr. G.R. Aloria of SSNNL, Gujarat for generously sponsoring and participating in the various workshops during this year of Swarnim Gujarat celebration. The present volumes are the outcome of these workshops. Dr. M.B. Joshi, along with his colleagues at the SSNNL, Dr. K.D. Acharya and Dr. Vinod Rana, provided very crucial support at all stages during the work including providing data, documents, maps and many reference materials. Ms. Mahima Gupta, an irrigation engineer and a planner herself, provided able research support; Ms. Aparna, Mr. Harish Joshi and Ms. Soumini Raja were always willing to help us not only during the organisation of the three workshops but also during the various stages of bringing out these volumes.

At the CEPT University, Dr. R.N. Vakil was always very encouraging right from the time the plan to bring out these volumes was discussed. In fact he speeded up the setting up of the CEPT University Press and the Centre for Documentation and Publication at the University to facilitate the work. Ms. Nirmala Khadpekar, Associate Professor at CEPT University took charge of the publication of these volumes and provided able and unflinching support. Professors Utpal Sharma and Shivanand Swamy not only provided crucial support but, along with Professor Ashwani Kumar, also attended some of the workshops and gave valuable feedback. We are also thankful to the Faculty members for their encouragement. A work of this kind requires administrative and library support. The CEPT University administration, particularly the Registrar, Staff at Centre for Research, Development and Consultancy, the library staff and Jitubhai and Dhuljibhai provided this critical support. We deeply appreciate and thank all of them.

The management team of the Concept Publishing Company was extremely cooperative and efficient in bringing out these volumes in time.

Ahmedabad
March 2011

R. Parthasarathy
Ravindra H. Dholakia

Contents

Part 2: Planning for SSP

Part 3: Appraisal for SSP

Volume 2: History of Rehabilitation and Implementation

Part 4: Resettlement and Rehabilitation in SSP

Part 7: Ways Forward

Abbreviations

AIBP	Accelerated Irrigation Benefit Programme
AISLUS	All India Soil and Land User Survey
AMC	Ahmedabad Municipal Corporation
APMC	Agricultural Produce Marketing Cooperative
ARCH	Action Research in Community Health and Development
ARWSP	Accelerated Rural Water Supply Programme
ASS	Argillaceous Sand Stone
BARC	Bhabha Atomic Research Center
BCM	Billion Cubic Metres
BHEL	Bharat Heavy Electricals Limited
BIP	Broach Irrigation Project
BIS	Bureau of Indian Standards
BMC	Bombay Municipal Corporation
BOC	Board of Consultants
BOT	Build-Operate-Transfer
CA	Compensatory Afforestation
CAD	Catchment Area Development
CAPs	Canal-Affected-Persons
CARE	Credit Analysis and Research Ltd
CAT	Catchment Area Treatment
CBR	Costs-Benefit Ratio
CCA	Cultural Command Area
CDO	Central Designs Organisation, Gujarat
CEO	Chief Executive officer
CES	Consulting Engineering Services
CGWB	Central Ground Water Board
Ch.	Chainage
CHPH	Canal Head Power House
CI	Confidence Interval

CICFRI	Central Inland Capture Fisheries Institute
CMCE	Critical Maximum Credible Earthquake
CRISIL	Credit Rating and Information Services of India Ltd
CSS	Centre for Social Studies
CSWCR&TI	Central Soil and Water Conservation Research and Training Institute
Cumecs	Cubic Meter Per second
CVC	Constant Volume Canals
CW&PC	Central Water & Power Commission
CWC	Central Water Commission
CWINC	Central Waterways, Irrigation and Navigation Commission
CWPC	Central Water and Power Commission
CWPRS	Central Water and Power Research Station
D&R	Design and Research
DACNET	Directorate of Economics and Statistics
DAP	Diammonium Phosphate
DBE	Design Basis Earthquake
DCC	Data Collection Centre
DDB	Deep Discount Bonds
DDO	District Development Officer
DDRP	Dam Design Review Panel
DF	Design Floods
DMAs	District Metering Areas
DOC	Divisional Operation Centres
DoEF	Department of Environment and Forest
DSC	Development Support Centre
DSP	Dam Safety Panel
DST	Department of Science and Technology
EDF	Environment Defense Fund
EIA	Environmental Impact Assessment
EPC	Engineering, Procurement and Construction
EPGA	Effective Peak Ground Acceleration
ERR	Economic Rate of Return
ESG	Environment Subgroup
FAO	Food and Agriculture Organization
FCA	Forest Conservation Act
FCI	Food Corporation of India
FEM	Finite Element Material
FFDA	Fish Farmers Development Authority
FMG	Five Member Group

FRL	Full Reservoir Level
FSD	Full Supply Depth
FSL	Full Supply Level
GAD	General Administration Department
GCA	Gross Cropped Area
GCEL	Gujarat Communication &. Electrical Ltd
GCMMF	Gujarat Cooperative Milk Marketing Federation
GDP	Gross Domestic Product
GEB	Gujarat Electricity Board
GEER	Gujarat Ecological Education and Research Foundation
GERI	Gujarat Engineering Research Institute
GGRC	Gujarat Green Revolution Company
GIS	Geographical Information System
GNFC	Gujarat Narmada Fertilizer Company
GoG	Government of Gujarat
GoI	Government of India
GoM	Government of Maharashtra
GoMP	Government of Madhya Pradesh
GR	Government Resolution
GRA	Grievances Redressal Authority
GRC	Grievances Redressal Cell
GSDP	Gross State Domestic product
GSFC	Gujarat State Fertilizer Company
GSI	Geological Survey of India
GWIL	Gujarat Water Infrastructure Limited
GWRDC	Gujarat Water Resources Development Corporation
GWSSB	Gujarat Water Supply and Sewerage Board
Ha	Hectare
HB	Hybrid
HH	Household
HP	Hand Pump
HYS	High Yield Strength
HYV	High-Yielding Variety
IAS	Indian Administrative Services
IBM-IFSSR	Inter-State Board for Management of Integrated Fisheries in the Sardar Sarovar Reservoir
IBPT	Irrigation By Pass Tunnel
ICAR	Indian Council of Agricultural Research
ICICI	Industrial Credit and Investment Corporation of India
ICID	International Congress on Irrigation and Drainage

ICOLD	International Commission on Large Dams
ICOR	Incremental Capital-Output Ratio
ICS	Indian Civil Services
ICSSR	Indian Council of Social Science Research
ID	Irrigation Department
IDA	International Development Association
IDBI	Industrial Development Bank of India
IFPRI	International Food Policy Research Institute
IIM	Indian Institute of Management
IITM	Indian Institute of Tropical Meteorology
ILO	International Labour Organization
IMD	India Meteorological Department
IMT	Irrigation Management Transfer
INR	Indian Rupee
IPCC	Intergovernmental Panel on climate Change
IR	Independent Review
IRM	Independent Review Mission
IRMA	Institute of Rural Management Ananad
IRR	Internal Rate of Return
IS	Indira Sagar
ISP	Indira Sagar Project
ISRO	Indian Space Research Organization
IWMI	International Water Management Institute
J&K	Jammu and Kashmir
JNNURM	Jawaharlal Nehru National Urban Renewal Mission
JP	Jayaprakash Narayan
KBC	Kachchh Branch Canal
LA	Loan Agreement
LF	Load Factor
LIC	Life Insurance Corporation
LPC	Land Purchase Committee
LPCD	Litres Per Capita per Day
LVFO	Lake Victoria Fisheries Organization
M&I	Municipal and Industrial
MAF	Million Acre-Feet
MC	Municipal Corporation
MCC	Main Control Centre
MCE	Maximum Credible Earthquake
MCM	Million Cubic Meter
MDDL	Mean Draw Down Level

MDE	Maximum Design Earthquake
MEQ	Micro Earth Quake
MHa	Million Hectares
MHM	Million Hectare Meter
MHUPA	Ministry of Housing & Urban Poverty Alleviation
MIS	Management Information System
MLD	Million Litres per Day
MNP	Minimum Needs Programme
MOEF	Ministry of Environment and Forests
MOUD	Ministry of Urban Development
MOWR	Ministry of Water Resources
MP	Madhya Pradesh
MPISSR	Madhya Pradesh Institute of Social science Research
MSP	Minimum Support Price
MSU	Maharaja Sayajirao University
MW	Megawatt
MWL	Maximum Water Level
NABARD	National Bank for Agriculture and Rural Development
NBA	Narmada Bachao Andolan
NCA	Narmada Control Authority
NCSDP	National Committee on Seismic Design Parameters
NDDB	National Dairy Development Board
NDS	Narmada Dharangrast Samiti
NEERI	National Environmental Engineering Research Institute
NFHS	National Family Health Survey
NGNS	Narmada Ghati Navnirman Samiti
NGO	Non-Government Organisation
NGRI	National Geophysical Research Institute
NHPC	Narmada High Power Committee
NIFDA	Narmada Inter-State Fisheries Development Authority
NIH	National Institute of Hydrology
NMC	Narmada Main Canal
NOF	Non-Overflow
NPDDC	Narmada Project Dam Designs Circle
NPG	Narmada Planning Group
NPK	Nitrogen, Phosphorus and Potassium
NPSSM	Narmada Project Steady State Model
NPUSM	Narmada Project Unsteady State Model (NPUSM)
NRSA	National Remote Sensing Agency
NRSC	National Remote Sensing Centre

NRW	Non-Revenue Water
NSA	Net sown area
NSP	Narmada Sagar Project
NSSO	National Sample Survey Organisation
NSV	No Source Village
NTPC	National Thermal Power Corporation
NVDA	Narmada Valley Development Authority
NWDT	Narmada Water Disputes Tribunal
NWDTA	Narmada Water Dispute Tribunal Award
NWRDC	Narmada Water Resources Development Committee
O&M	Operations and Maintenance
OBC	Other Backward Castes
OBE	Operating Basis Earthquake
OECD	Organisation for Economic Cooperation and Development
OECF	Overseas Economic Co-operation Fund
ONGC	Oil and Natural Gas Corporation
ORG	Operation Research group
PAFs	Project-Affected Families
PAPs	Project-Affected Persons
PAS	Performance Assessment System
PCR	Project Completion Report
PF	Provident Fund
PFC	Power Finance Corporation
PGA	Peak Ground Acceleration
PIM	Participatory Irrigation Management
PINS	Pressurised Irrigation Network System
PMF	Probable Maximum Floods
PMP	Probable Maximum Precipitation
PPP	Public Private Partnership
PUCL	People's Union for Civil Liberties
QL	Quintal
R & D	Research and Development
R&E	Research and Evaluation
R&R	Resettlement and Rehabilitation
RBPH	River Bed Power House
RCC	Reinforced Cement Concrete
RCNCA	Review Committee of NCA
RDSO	Railway Designs and Standards Organisation
RIDF	Rural Infrastructure Development Fund
RIS	Reservoir Induced Seismicity

RL	Reduced Level
RO	Reverse Osmosis
RRWHSs	Rooftop Rain Water Harvesting Systems
RTU	Remote Terminal Unit
RWSS	Regional Water Supply Scheme
SAR	Staff Appraisal Report
SBC	Saurashtra Branch Canal
SC	Scheduled castes
SCADA	Supervisory and Control Data Acquisition
SCBA	Social Cost-Benefit Analysis
SCF	Standard Conversion Factor
SD	Standard Deviation
SHGs	Self help Groups
SJMMSVY	Swarna Jayanti Mukhya Mantri Shaheri Vikas Yojana
SOI	Survey of India
SPF	Standard Projected Flood
SPIESR	Sardar Patel Institute of Economic & Social Research
SPPWCP	Sardar Patel Sahbhagi Jal Sanchay Yojana
SPV	Special Purpose Vehicle
SRR	Social Rate of Return
SS	Sardar Sarovar
SSCAC	Sardar Sarovar Construction Advisory Committee
SSD	Sardar Sarovar Dam
SSL	Suspended Sediment Load
SSNNL	Sardar Sarovar Narmada Nigam Limited
SSP	Sardar Sarovar Project
SSPA	Sardar Sarovar Punarvas Agency
SSRA	Sardar Sarovar Rehabilitation Agency
ST	Scheduled Tribes
STPR	Social Time Preference Rate
SWRC	Social Work and Research Centre
TCIL	Telecommunication Consultants India Limited
TDO	Taluka Development Offcier
TDS	Total Dissolved Solids
TECS	Tata Economic Consultancy Services
TISS	Tata Institute of Social Sciences
TMC	Thousand Million Cubic
UIDSSMT	Urban Infrastructure Development Scheme For Small and Medium Towns
UIP	Ultimate Irrigation Potential

ULBs	Urban Local Bodies
UN	United Nations
UNCED	United Nations Conference on Environment and Development
UNDP	United Nations Development Programme
UNICEF	United Nations Children's Fund
UNIDO	United Nations Industrial Development Organisation
UP	Uttar Pradesh
USBR	United States Bureau of Reclamation
USSR	Union of Soviet Socialist Republics
VLW	Village Level Worker
VSA	Village Service Area
WASMO	Water and Sanitation Management Organisation
WB	World Bank
WCD	World Commission on Dams
WHO	World Health Organization
WIHG	Wadia Institute of Himalayan Geology
WUA	Water Users Association
WWF	World Wildlife Fund

List of Contributors

Acharya, K. D., Assistant Engineer, Sardar Sarovar Narmada Nigam Limited, Gandhinagar. acharya_kishore2005@yahoo.co.in

Bassi, Nitin, Institute for Resource Analysis and Policy (IRAP), Hyderabad.

Chandra, A., Former Student, IRMA, Anand.

Chavan, Sanat, Former Principal Chief Conservator of Forests, Gujarat, A-2/3, Kabir Housing Complex, Opp. Don Bosco School Makarepura Road, Vadodara.

Das, Biswaroop, Professor, Centre for Social Studies, University Campus, Udhana Magadalla Road, Surat. dbiswaroop@gmail.com

Datta, K. Samar, Professor, Indian Institute of Management, Ahmedabad. sdutta@iimahd.ernet.in

Desai N.B., Former Secretary (Narmada), Government of Gujarat, 11, Triveni Park, Opp. S.N.D.T. College, Akota Stadium Area, Vadodara.

Desai, Vaidya Prerana, Sneh-Setu Foundation, Vadodara, snehsetufdn@gmail.com

Dholakia, H Ravindra, Professor, Indian Institute of Management, Ahmedabad. rdholkia@iimahd.ernet.in

Garg, Amit, Professor, Indian Institute of Management, Ahmedabad. amitgarg@iimahd.ernet.in

Gautam, I.P., Municipal Commissioner, Ahmedabad and Former Executive Director, Resources, and Director of Finance, SSNNL, ipgautam@egovamc.com

Goswami, Subhrangsu, Assistant Professor, Faculty of Planning and Public Policy, CEPT University, Ahmedabad. s.goswami@cept.ac.in

Gupta, R.L., President, Jaypee Ventures Limited, NOIDA, Uttar Pradesh

Iyengar, Sudarshan, Vice Chancellor, Gujarat Vidyapith, Ahmedabad. sudarshan54@gmail.com

Jagadeesan, S., IAS, Managing Director, Sardar Sarovar Narmada Nigam Limited, Gandhinagar. sjagadeesan53@gmail.com

Java, G.L., Former Director (SSNL) and Secretary (Retired), Narmada Water Resources and Water Supply Department, Government of Gujarat, gl7java@yahoo.com

Jhala, Neha, Research Associate, Indian Institute of Management, Ahmedabad, nehajhala@ iimahd.ernet.in

Joshi, M.B., Superintending Engineer, Sardar Sarovar Narmada Nigam Limited, Gandhinagar. dr.mbjoshi@gmail.com

Kankal, Bhushan, Research Associate, Indian Institute of Management, Ahmedabad, bhushank@ iimahd.ernet.in

Krishnan, Sunderrajan, Director, CAREWATER, INREM Foundation, Anand. carewater@gmail.com

Kumar, M. Dinesh, Executive Director, Institute for Resource Analysis and Policy (IRAP), Hyderabad. dinesh@irapindia.org

Mehta, Ambrish, ARCH Vadodara, E-702, Samrajya Complex, Near Fatehgunj Post Office, Fatehgunj, Vadodara. m_ambrish@hotmail.com

Mehta, Dinesh, Professor Emeritus, Faculty of Planning and Public Policy, CEPT University, Ahmedabad. dineshmehta@cept.ac.in

Mehta, Meera, Professor Emeritus, Faculty of Planning and Public Policy, CEPT University, Ahmedabad. meeramehta@cept.ac.in

Parmar, B.J., Former Secretary, Water Resources Department, Government of Gujarat, bjpsays@rediffmail.com

Parthasarathy, R., Professor, Faculty of Planning and Public Policy, CEPT University, Ahmedabad, r.parthasarathy@yahoo.com

Patel, Anil, Arch Mangrol, Rajpipla, District Narmada

Pathak, Mahesh, Former Chairman, Gujarat State Agriculture Commission and Former Executive Member, Narmada Planning Group, Government of Gujarat, mhpathaks@yahoo.in

Pathak, Swati, Research Associate, Indian Institute of Management, Ahmedabad, swatip@ iimahd.ernet.in

Pullabhotia, Hemant Kumar, Research Analyst, International Food Policy Research Institute, NASC Complex, CG Block, Dev Prakash Shastri Road (Opp. Todapur) Pusa, New Delhi 110012.

Rathod, K.G., Chief Engineer and Director(Canals), Sardar Sarovar Narmada Nigam Ltd, Government of Gujarat, Gandhinagar, dire-canal-ssnnl@gujarat.gov.in

Sah, D.C., Professor & Director, Madhya Pradesh Institute of Social Science Research, 6 Bharatpuri Institutional Area, Ujjain, Madhya Pradesh. dc_sah@yahoo.co.uk

Shah, Tushaar, Senior Fellow, International Water Management Institute, Anand. tushaar@irma.ac.in

Sheth, Pravin, Former Professor and Head of Political Science Department at Gujarat University, Ahmedabad. pravinsheth@hotmail.com

Srinivas, K., IAS, Joint Managing Director, Sardar Sarovar Narmada Nigam Limited, Gandhinagar. ksrinivas_ias@yahoo.co.in

Sudhir, C., Former Student, IRMA, Anand

Tomar, Shubhra S., ICSSR Doctoral Fellow, MPISSR, 6 Bharatpuri Institutional Area, Ujjain, Madhya Pradesh

Verma, Shilp, UNESCO-IHE Institute for Water Education, PO Box 3015, 2601 DA Delft, The Netherlands.

Vyas, Jay Narayan, Minister of Health and Family Welfare, Government of Gujarat and Ex-Chairman, Sardar Sarovar Narmada Nigam Limited, min-health@gujarat.gov.in

11

Implementation of Sardar Sarovar Project: Overview and Impact

R. Parthasarathy and *Ravindra H. Dholakia*

Implementation Issues and Impacts

Innovation has been one of the keystones in the implementation process of SSP and it has not been limited to design and financing but also on the issue of people displaced by the project. This volume mainly discusses this facet. Evidently, implementation of a mega project like SSP on the river Narmada would necessarily face several challenges from various sources. Availability of highly skilled manpower like surveyors, designers, civil engineers, mechanical engineers, electrical engineers, information technologists, geologists, managers, accountants, and so on, can prove to be a very real constraint on the speed of implementation of the project of such a dimension. Similarly, semi-skilled workers like masons, carpenters, plumbers, miners and so on would also pose a problem of relative scarcity with the consequent demand for higher wages. The project of such large magnitude would certainly lead to massive problems of adequate supply of inputs of building materials, machinery and equipment and their availability at the right time. All this would also create unprecedented pressure on the existing infrastructure in and around the region.

These and related challenges in the implementation of SSP are essentially connected to the horizontal and vertical linkages, and sequencing of activities, physical and manpower planning. They represent economy-wide concerns and can be usually resolved by pursuing policies that are more liberal on imports, private sector participation, in-migration of skilled, semi-skilled and even unskilled workers, and would need an open mind to accept innovative and bold solutions to specific problems. Raising adequate amount of financial resources needed for the project with appropriate time schedule and ensuring timely repayments and servicing of liabilities also represent similar type of challenges in implementing such a long drawn out and mammoth project.

All such challenges were faced and successfully overcome by the state and national governments during the project implementation period. To some extent, those challenges paved the way to realisation of the constraints of the command economy and to early recognition of the need for economic policy reforms and liberalisation of imports, globalisation, and increased private sector participation. The state of Gujarat benefited particularly because the requirements of SSP implementation led to not only speedy development of several related infrastructural facilities but also ensured that the quality of such infrastructure are of certain standards. The liberal stance of Gujarat on immigration of workers, capital, technology, goods and services was further nurtured and strengthened by the SSP implementation process. This helped the state to take a lead in the matters of economic policy reforms and liberalisation to emerge as a growth leader among all states in the country (see, Ahluwalia, 2000).

Another set of serious challenges coming with the implementation of the SSP concerns extremely sensitive issues pertaining to displacement of humans, and maintaining the ecological balance. Large scale irrigation dam projects invariably involve a vast land mass getting inundated leading to permanent loss of agricultural, forest and residential lands to the society. If the land to be submerged in the dam reservoir area has human habitation, those affected families have to be resettled and rehabilitated in other parts of the country. In a democratic country with clearly defined individual private property rights for land, this poses a serious human rights issue. However, in many cases, the irrigation projects come up in the hills inhabited by tribals who often do not have clearly defined property rights. It thereby follows that without proper and satisfactory resolution of this issue, no major irrigation dam project, or for that matter, no substantial infrastructure development project should ever be implemented in a democratic society which otherwise has clear private property rights. In order to resolve such a potential deadlock and allow imperative developmental projects, the Constitution of India has an enabling provision for the state to acquire private land by paying "adequate" compensation for ensuring greatest good of the greatest number. However, the debate about what is "adequate" could be unending and could effectively derail implementation of the project as is evident from the case of SSP.

What is striking in the history of implementation of the SSP relating to the aspects of resettlement and rehabilitation of the project affected people (PAP), is the long drawn out elaborate social negotiation and bargaining process without it becoming violent and undemocratic. Considerable public opinion was mobilised on what is right and what is not. Not only the PAPs and the state government of Gujarat, but also activists, intelligentsia and media from all over the globe were involved in the grand debate and the social bargaining process for the Rehabilitation & Resettlement (R&R) in SSP. The Kaldor-Hicks compensation principle for potentially viable development project to compensate the adversely affected population was virtually put into practice to finally arrive at one of the most liberal and unique packages for R&R offered to the PAP of SSP. This package as offered to PAP in the three affected states of Gujarat, Madhya Pradesh, and Maharashtra has been appended to this chapter (Appendix I) to provide the basic reference point to the discussion contained in the rest of this volume.

The impact of such a liberal compensation package as offered to the PAP for the developmental processes and projects in the country and the state is both positive and negative. The negative effect of such a liberal package, particularly of providing cultivable irrigated land to every displaced adult is that most of the development infrastructural and irrigation projects of large dimension would become financially unviable and practically non-implementable given high scarcity of arable land in the country. It, therefore, poses a big question mark for the sustainability and operationality of a democratic development model based on market friendly private property rights regime. Alternatively, differing compensation principles need to be accepted rather than applying uniform ones for diverse development and infrastructural projects. It has the potential to lead to disappointments and discontentment.

The positive impact of the compensation package offered to the PAP of SSP in Gujarat was in terms of opening up of the rural land market and bringing several unrecorded informal scattered land transactions into the open market. The state government also realised that, after all, agricultural land is not as scarce as was made out to be if the right price is offered, and that the private sector and open market forces of demand and supply are the best ways to discover the right price. The lessons learnt from this long process of managing R&R for SSP came in very handy later while dealing with land acquisition for Special Economic Zones (SEZs) or for other infrastructural projects like roads, ports, and pipelines, or for institutions and industries. Land market liberalisation has given a special edge to Gujarat over other states in the country, for which a significant part of the credit should go to the implementation process of R&R under the SSP.

Overview

The essay by N.B. Desai in Chapter 12 documents the conception, planning and implementation of the Sardar Sarovar Project and narrates the constitution and involvement of several national and state level agencies over the last six decades. The history of the dam which dates back to 1946 eventually led to the acceptance by the Planning Commission for stage-l in 1960 with a dam at Navagam. He also traces the review of the proposals in 1964 by the Khosla Committee towards harnessing the Narmada waters and the Master Plan. He notes that the States, however, could not arrive at a mutually acceptable agreement resulting in the constitution of NWDT, which gave its final decision in 1979 on parameters like allocation of water and hydropower, height of dam, full reservoir and supply levels, etc. Importantly two significant policy developments changed the course of the project — the Forest Conservation Act, 1980 and the Environmental Protection Act of 1986. Environment impact studies through evolving guidelines for environmental clearances subsequently became mandatory for investments. Thereafter, with an aim to follow the highest standards, the Narmada Planning Group was formed with a multi-disciplinary team of experts in engineering, agriculture, economics, sociology, ecology & environment and planning & public policy.

It was not until 1983, after due technical studies that the Sardar Sarovar Development Plan was presented to the World Bank. It agreed in 1985 to provide credit assistance for certain components in stage I and later in 1990 offered more for rehabilitation, resettlement and environmental protection measures for the entire project. In 1986, the Overseas Economic Co-operation Fund of Japan also agreed to provide credit for electrical components of the river bed power house. What is important to remember here is that none of these assistance programmes worked out when the Government withdrew the SSP from the World Bank's lending portfolio and decided to complete the project without any external assistance.

Even as the major works of the main dam and power houses commenced in 1987-88, the NBA filed a comprehensive writ petition in the Supreme Court of India seeking stoppage of all construction activities of the SSP raising social and environmental issues. Converting this into an opportunity both the government and NGOs eventually set a precedent for an advanced rehabilitation package not just for the project affected landholders but also to their sons, landless labourers and 'encroachers.' In spite of a comprehensive R&R package being through by January 1998, and nearly 80 per cent of the work of main dam completed, there was virtually no progress for quite some time due to slow R&R processes.

Chapter 13 by Anil Patel has sought to answer the question as to what do the Narmada valley tribals want in order that the R&R process would pick up pace. The paper is a first-hand account of the experiences of the Civil Society Organisation ARCH-Vahini, which was instrumental in the formulation and implementation of the R&R policy for the SSP. The author recounts that the NWDT's decision regarding R&R was path-breaking in India for it had allotted agricultural land to a landholding family in lieu of the land lost, which is a clean break from the past when the landholders got only monetary compensation. The decision, however, fell short of taking into account the characteristics of tribal society, areas and their economy; the most important being that a majority of the tribals did not formally own land. The paper chronicles the successive debates and developments after the NWDT's decision in 1979 and the role played by the government bodies, international agencies and the non-governmental organisations in bringing the central issues of the project affected people to the fore. He recounts how the tribals prompted by ARCH-Vahini, moved the Gujarat High Court against the government and showed that the government was not obdurate. Importantly, the paper shows the positive role played by the World Bank in the process of evolving the R&R policy. In fact, the overarching principle of the Bank's guideline that the standard of living of the displaced people should be regained as soon as possible after the resettlement was attempted for the first time in India. Further, the landless were also resettled in agriculture thus gaining a stable means of livelihood. The encroachers' entitlement became a significant issue that the state governments could ignore no more. However, as the author points out it was not an easy and straightforward task. The many events during the process of evolving the R&R policy of which issues relating to an Independent Review is the subject of the following chapter.

The chapter also contains a captivating revelation of the intentions of a section of Civil Society, which impacted not just the course of the SSP but also delayed the formulation and implementation of the R&R policy in the other two states. In Maharashtra, the tribals always wanted to resettle in the denuded forest land in the same region but the MoEF was averse to it. Soon after the R&R policy of Gujarat in 1987, some organisations in Maharashtra and Madhya Pradesh raised doubts about the policy of Gujarat. International agencies also joined in and affirmed that the tribals in the valley were against the project, and opposition to the project spread. The project was hailed as environmentally disastrous and economically ruinous and was portrayed as a symbol of planned development disaster. What was curious is that even as the mechanisms were set in motion to render the R&R impossible, the tribals of Madhya Pradesh and Maharashtra did not have any inkling of these developments. The further step-by-step account of the tribals taking a stand for their cause shows the transition of people's empowerment because of access to information and the support of well-intentioned Civil Society. The travails of R&R however, did not end; in fact, soon came the GoG's resistance to concede the demand for private land based on the argument that such lands in the command of the project were not available. Again the people's power came to the forefront when the tribals gave a list of landowners in the command who were ready to sell land at the market price and even assured the availability of more of such lands for sale. Meanwhile, a major problem of interpretation of the Gujarat policy arose as to whether adult sons of the landholders were entitled to land compensation. The tribals confident of their organised power issued a public warning with the slogan, "No land, No dam".

The turning point of political will for R&R in Gujarat arrived when ARCH-Vahini's and the tribals' focus concentrated on identifying land; to have the prices of land fixed through the formally installed Land Purchase Committee (LPC); and, to make sure that land parcels of two hectares were allotted to each family. Anil Patel then describes how the tribals took major initiative to form groups to identify land and to press for the fixation of price and allotment. The right of choice in the R&R agreement worked out in two ways. The oustee chooses not only land, village and area but also the group the family wants to resettle with. This ensured an extremely high level of participation of oustee families apart from minimal hostile guest reception in the new village. Only in 1991 the tribals of Maharashtra and Madhya Pradesh became acquainted with the R&R policy of Gujarat and the manipulation of their demands and aspirations was shocking to them. The author narrates how not only the state government and the MoEF but also the activists had worked in tandem. In such an event, a number of tribal families from these states selected land in Gujarat and the GoG offered the same resettlement and rehabilitation package to them. Clearly the preference for good quality land, security of tenure, greater access to market and other infrastructure facilities in addition to the social and economic concerns of the R&R policy of Gujarat is what the requirements were. Accomplishing these even in a span of two decades was no mean achievement. Implementing R&R of Gujarat towards building the SSP has been an important innovation.

As mentioned, the role of international agencies including the World Bank had not always been helpful in setting up the SSP. In Chapter 14, Anil Patel and Ambrish Mehta describe the way in which large-scale interventions by international groups and enquiry commissions like that of the Independent Review (IR) headed by Bradford Morse had created controversy by raising questions about the scope and validity of the project. The authors had examined a few critical issues surrounding the environment and R&R, and showed that the anti-SSP critique was not substantiated. The authors meticulously analyse the reports to come up with two major findings. The first is that the IR suffers from grave methodological flaws that eventually compromised their conclusions, which were also based on an inadequate database, errors in reasoning and a failure to carry out logical analysis. The second type of inaccuracy includes selection and omission of vital information and distortion of facts that could have led to an entirely different set of conclusions than what the team had reached. The authors state that only a careful and dogged reading of the text reveals these questionable characteristics. What anguishes the authors is the fact that the report of the IR had seriously weakened the trust on which major international human rights campaigns were mounted. In fact they also state that the IR has repeatedly gone out of its way to "prove" or "establish" facts that would show the project in a negative light and moreover failing to recommend thoroughgoing reforms thereby wasting an opportunity to help many tribal oustees. Encroachers in Maharashtra and MP could have gained immense help from the IR recommendations; as could have the marooned villages in Maharashtra. Similarly, major sons in the tribal areas of MP and Maharashtra would have also secured decent means of livelihood had the IR team understood their legal situation. Instead of setting in motion constructive and creative reforms the team ended up recommending the Bank to step back from the project.

D.C. Sah and Shubhra Tomar also bring out the debates on the R&R issues in Chapter 15 and argue that these had fluctuated between the two extremes. Sometimes it had appeared to be discrediting the R&R efforts yet proclaiming it as the best. It is the attention and scrutiny received by the SSP that had eventually changed the rules of how projects were formulated in the context of socio-economic and environmental viability with minimum human displacement. The chapter also documents some of the controversies and important lessons for future development projects. Till SSP, the authors point out that though there was a provision for providing compensation for land, but there was none for other sources of livelihood loss such as from forest and common resources. An important institutional intervention as part of the SSP was to introduce accountability in land acquisition and relocation. The NWDTA definition of PAP is also an important innovation as the chapter highlights. The definition eventually included even landless labourers as oustees who would be affected by submergence. In defining the 'family' of an oustee also, the award had stipulated that every major son would constitute a separate family for the purpose of providing compensation and other benefits. In fact, while implementing the R&R policy the implementation plan was subdivided into a relocation plan and a resettlement plan. Barring a few exceptions, the state and the NGOs did a creditable job; PAPs did get what was due to them in terms of land and infrastructure.

Sah and Tomar's field analysis is focused on PAPs from 19 villages of Gujarat. They point out that the R&R policy was in fact put into practice even before it was accepted. Early re-settlers, however, did not immediately shift to the new sites: over 70 per cent of them took over 5 years to make new sites their home and during this period they either commuted or members of households split themselves between locations. Fragmentation of families, lack of access to fuel and fodder, relocation site outside the command area, fragmented or encumbered plots, lack of access to drinking water, hardship of tin-shed dwelling, and institutional isolation were some of the major first generation problems faced by early re-settlers which the state was quick to solve. More serious were the second generation rehabilitation problems that got dealt with once the Narmada water started flowing in the resettlement sites. The authors also describe how the traditional mutual exchange economy was replaced by markets and in coping with this, how the tribals needed support to get adjusted to the competitive markets especially in terms of resources and information to become active participants in the technologically dynamic agriculture. The process of integration with mainstream agriculture without extension services, irrigation, and credit support has been studied in the chapter which also shows how the PAPs unfamiliar with many of the intricacies of markets have responded to commercialisation in agriculture.

In Chapter 16, Biswaroop Das documents the agricultural and livestock rearing practices of 'resettled' oustees from parts of Maharashtra, Madhya Pradesh and Gujarat states as it had prevailed at their home villages and changes brought in at the new sites, based on a field research study of 2003. He finds that the traditional agricultural and livestock practices of the people in the submerging sites had some prominent characteristics. Rain-fed irrigation co-dependent on other factors had resulted in uncertain yields and livestock had become integral to agricultural practices. Land owned per household was much less than the actual area cultivated as a large number of households cultivated forest and other barren government land. It was mainly subsistence agriculture earlier and goat-rearing depending on forest vegetation and crop residuals was common. Some of the tribal groups residing in interior hilly regions did not even practice settled cultivation and cattle rearing. While, forests and rivers were important resources for sustenance the need for ready money was met through channels such as sale of MFP, illegal sale of logs, harvesting bamboo, making charcoal, migration and by selling goats. Importantly, none of these activities were dominant as primary source of earning.

Despite moving to the new site for around a decade by 2003, plots received by many of the Project Affected Families (PAF) only served as complementary farm sites as the people continued to live in their earlier villages. In fact, the social differentiation in terms of tribals and non-tribals too had continued to be shaped, resulting in exclusion of the former from institutions and market. Even partial crop failure made some PAFs vulnerable enough to mortgage parts of their land. There were distortions such as yield fluctuations at sites without irrigation, un-productive, fragmented or disputed land, and, lack of attention to agriculture as 'family centric work' during rehabilitation. Yet, the benefits of technology based intensive cultivation had been realised in the new sites; tractors had replaced ploughs and bullocks were used for levelling the plots. Crop loans could be raised and the Nigam supported procurement of hybrid seeds and buying of bullocks at subsidised rates as part of the rehabilitation package. A large majority of the families had done much better through cultivation. They could engage with institutional structures, market, and technology and could also strategise to counter the market fluctuations. The host settlements played a positive contributory role facilitating experimentation and adoption of newer innovations by the PAFs. Although, after initial years more of the oustees responded positively to the change, the author feels that in spite of agriculture having moved towards the centre stage of their lives, livelihood facilitation of access to irrigation and extension services were required then. With inputs, a larger number of milch and draught cattle were also raised. Both the activities being complementary were expected to catalyse development when irrigation entered the fields in a major way.

Sudarshan Iyengar, in Chapter 17 traces the policy reforms and their implementation and highlights issues for future. The Chapter is divided into three sections. In Section I, the evolution of the R&R policy in SSP is discussed. Section II reviews the implementation of the new policy for two time periods, 1988-1993 and 1993 onwards and Section III is on the problems and prospects in R&R in SSP. The R&R programme of the project was initially as bad as it had been in any development project involving involuntary displacement. The NWDT provision had serious lacunae and was sociologically ill informed as it had failed to take into account the characteristics of the tribal settlements and their economy. Signs of retrogression were also visible from the three State governments as none of them wanted to abide by the NWDT provisions. With the declaration of the NWDT Award in 1978, GoG began the project work as well as the process of rehabilitation, which the tribals living in the affected villages refused to accept. During this time, Iyengar recounts that members of ARCH-Vahini started frequenting the submergence villages. In the process of evolving the R&R, the World Bank and the International Non-Government Organisations had also played a pivotal role. They assessed the R&R situation in the SSP and started negotiating the Loan Agreement with GoI. Gradually, GoI agreed to release the forest land demanded by the displaced tribals. Further, the signing of the loan agreement with the Bank also bound GoI and the three state governments with commitments to the tribals in the Valley. GoG was the first State to respond by issuing GRs and showing a commitment toward an R&R package. The author then provides the details on the implementation process like the setting up of a List Committee and the Land Purchase Committee with NGO representation and the process of allotment of alternative agricultural land. As a result of these initiatives, the resettlement and relocation of the displaced population in the first phase, i.e. 1988-93 was largely successful. However, problems again overwhelmed SSP's R&R programme until the Grievance Redressal Cell was set up. The first was the government's understanding of R&R issues and the second one relates to the role of IRM, activists and NGOs in handling these. The government and the administration were either not aware or had decided to ignore the seriousness of both the problems. What was important is that the two problems together had the potential to affect the evolving effective and fruitful GO-NGO collaboration. In fact, he also argues that the exit of the World Bank had weakened and further isolated the already isolated NGOs like ARCH-Vahini which had provided critical support to the project and had collaborated to implement the policy.

In Chapter 18, K Srinivas documents all the major legal actions that took place throughout the development of Sardar Sarovar Project starting from initial plans in 1946 to the final clearances obtained from the NCA to raise the dam height up to 121.92 m. The chapter gives a record of all the successive plans, studies, surveys and reviews that had commenced after inspection of all potential sites by engineers and geologists. Subsequently, a tribunal declared its award and the Union of India and the three States filed references. The Tribunal heard these references and final orders were passed and published in a Gazette Extraordinary by the Government of India. The paper also summarises the salient features of the Award on various aspects of the project. In order to provide an institutional mechanism for the purpose of securing compliance with; implementation of the decision and directions of the Tribunal; and to ensure efficient, economical and early execution of the project, the Tribunal in its Award had directed that the Narmada Control Authority (NCA) and Sardar Sarovar Construction Advisory Committee (SSCAC) be constituted. A Review Committee was later formed to review the decisions of both the NCA and SSCAC. Further, to facilitate the monitoring and implementation of various environmental activities effectively, NCA created an independent Environment Sub-Group.

Financial closure of the World Bank funding after the IRM report was a big disappointment for the project. Rejecting this report outright, GoI decided to unilaterally withdraw from the funding agreement with the World Bank and to go ahead with the project by raising funds from its own resources and from the public. Narmada Bachao Andolan was the principal 'dissenter' to the project and its methods captured much media attention. The movement led to much realignment of policies, and improvement in the delivery methodologies and strategies. The resultant post R&R situation received widespread public approval for the entire SSP. NBA filed petitions before the Supreme Court of India pleading to restrain the construction of the dam. However, the Court, finding the plea impractical did not allow it. Lifting the stay on further raising the dam, the Apex Court gave a historic judgment in October 2000. It barred the NBA's plea to reopen questions such as the dam height, hydrology, environmental studies and seismicity and held its objections to be belated and procedurally poor. Finally, during 2000-06, stage-wise clearances were obtained from the NCA to raise the dam height up to 121.92 m. Even during this period, several attempts were made to suspend the construction work of the dam at different levels. Tremendous public support to SSP, however, succeeded in overcoming all these obstacles, which reflected the true value of this project. SSP has over time conquered various obstacles—legal, rehabilitation-related, environmental, financial and technological since inception and has reached a state where it enjoys a special place.

The Chapter 19 by Pravin Sheth first provides an account of the controversial history and development of the SSP from its inception. From the beginning, he states that the anti-project forces had been raising several issues related to the dam height, resettlement and rehabilitation, availability of land, ecological degradation and so on. These eventually contributed to the halting progress of construction of the dam. He also points out the not so helpful role of the Centre in facilitating Gujarat's case. The author describes the role of Dr. Anil Patel, leader of pro-tribal NGO, ARCH-Vahini in successfully fighting against the Gujarat government in the High Court and in the Supreme Court for the rights of the tribals and the displaced PAPs. Eventually when the state did come out with a progressive package, a unique mechanism was developed in which the state even included ARCH-Vahini as a partner in the R&R implementation process. Sheth's paper also gives a graphic account of how Medha Patkar and the NBA had succeeded in creating the powerful anti-project front and managed support from national and international celebrities and institutions. He also brings out how the state of MP was also not very cooperative during the various stages of SSP. The description of various demonstrations and skills with which the pro-dam and anti-dam activists steered their respective movements in different venues with varying intensities, is a spellbinding account. However, the world saw the people of Gujarat silently demonstrating and this had made the foreign and Indian media learn that the Dam was not only the concern of the state government as Medha Patkar and Baba Amte had made them believe but the cherished goal of the people of Gujarat.

When the Project was paralysed in the legal arena between1995-2000 and the construction work came to standstill, Chunibhai Vaidya and his Narmada Sangharsh Samiti legally assisted by Maganbhai Barot tried to push the procedure in the Supreme Court. It was during this time the Narmada Sangharsh Samiti's padyatra from Sardar Patel's residence in Karamsad to Gandhi Ashram in Ahmedabad served as an impetus to the SSP. Many scholars and officials linked to SSP played an important role in clearing many misconceptions. After the Court judgment in 2000, focused efforts of Chief Minister Narendra Modi had resulted in securing unanimity among the four states in 2006. He also meticulously handled the matter of speedily raising the Dam height first to 110 m. Reaching this height helped SSNL to start generating hydro-power for the first time which soothed the attitude of MP. In 2003, Narmada waters were sent to Bhuj by a pipeline network.

The chapter also talks about social and economic development benefits generated by the SSP including increase in agricultural production and cropping intensity. The author perceives that the SSP contributes to the nation's economic development. Widely misunderstood and not adequately supported by the Centre and national media, the project would indeed contribute to food security and energy self-sufficiency of the country. However, he points out that many areas that had been promised Narmada water are facing deprivation because of non-completion of the canal network. This is one of the major tasks before the state government.

Appendix I
PROVISIONS OF REHABILITATION POLICY 1[1]

Resettlement and Rehabilitation of Narmada Projects under the Aegis of NVDA

There are 29 Major, 135 Medium and 3000 small Projects of Narmada Valley which are in different stages of either completion or nearing completion or at planning stage. The Resettlement and Rehabilitation (R&R) of all these ongoing Intra-State Projects is being governed by the provisions of R&R Policy of the State formulated for Narmada Projects except for Sardar Sarovar Project in which the whole gamut of R&R is being governed by the provisions of Narmada Water Disputes Tribunal (NWDT) Award and R&R Policy of the State Government being an Inter-State Project.

Applicability of Different Provisions of R&R Policy of the State for Oustee of Narmada Project / NPRR - 2003 / NWDT Award

Sl. No.	*Name of the Project*	*Applicability of the provisions of the policy/ NWDT Award*	*Remarks*
1.	Indira Sagar, Omkareshwar, Man, Jobat, Upper Beda	Rehabilitation Policy for the Oustees of Narmada Projects as formulated by Government of Madhya Pradesh	This policy is being implemented to all ongoing projects of Narmada Valley.
2.	Sardar Sarovar Project	Provisions of NWDT Award and Policy of the State. All directives of Hon'ble Supreme Court regarding SSP.	Directives of Hon'ble Supreme Court in WP(C) 319/94 & 328/2002 being implemented.
3.	Upper Narmada, Halon, Lower Go, Hoshangabad, Handia and Boras Projects.	National Policy on R&R for Project Affected Families - 2003 (NPRR-2003)	To be implemented for all up coming projects of Narmada Valley.

1. *Source*: Narmada Control Authority.

R&R Policy of the State Government of Madhya Pradesh for Oustee of Narmada Project

Objectives and Principles of R&R Policy of the State for oustee of Narmada Project are as follows:

- Improve the standard of living or at least regain the standard of living the PAFs had been enjoying prior to their displacement.
- Be relocated as village units, village sections, or families in accordance with the oustees preference.
- Be fully integrated in the community in which they are resettled.
- Be provided with appropriate compensation and adequate social and physical rehabilitation infrastructure including community services and facilities.
- Ensure adequate participation by the PAFs in their R & R.

Provisions of the Policies Applicable on Different Ongoing Narmada Projects I.E. Indira Sagar / Omkareshwar/Man/ Jobat and Uper-Beda Projects

Provisions of the policy	*As per State Policy for Narmada Projects.*
Compensation of Agricultural Land	Agricultural land will be compensated as per Sec. 23 of Land Acquisition Act (market value of the land compensation of wells, pipelines and trees as per valuation report given by PHE, Forest and Horticulture Depts. + 30% solatium+12% interest (Interest payable from date of notification of section 4).
Compensation of Houses and Properties	Compensation of houses and other immovable properties will be paid as per section 23 of Land Acquisition Act (as per CSR) + 30% solatium + 12% interest (Interest payable from date of notification of section 4).
Rehabilitation Grant	(1) Agricultural landless labourers families, SC/ST/small and marginal farmers families = Rs. 18,700/- (2) Other families Rs 9350/-
Employment Resources Grant for Productive Assets to Landless Persons	(1) Rs. 49,300/- to landless agricultural labourers, landless SC&ST families (2) Rs 33,150/- to other landless families

Provisions of the policy	*As per State Policy for Narmada Projects.*
Other provisions Land for Land	Eligible families will be allotted land for land as far as possible (minimum 2 ha. and maximum 8 ha.). As per provisions of Para 3 & 5 of the policy, the oustee shall be paid 50% compensation of his acquired land and after adjusting the remaining 50% compensation and the amount of S.R.G., the cost of the land allotted shall be recovered in 20 interest free instalments. Till the repayment, the land shall be mortgaged in favour of the State Government.
Transportation Grant	Free transportation or Rs. 5,000/- for transportation of their house hold goods to new R&R site.
Residential Plots	90' × 60'= 5400 Sq. ft. plot or Rs 20000 to each family in rural area. For urban R&R site of New Harsood the plot size of 2400 sq. ft., 1500 sq. ft. & 540 sq. ft. or Rs. 20,000/- grant in lieu of plot.
Stamp/Registration fee	For purchase of Agriculture land or immovable property within M.P. up to the limit of amount paid to PAFs (excluding transportation grant).
Special Provisions	(i) Option of acquiring total agriculture land if 75% of land holdings are acquired. (ii) Compensation of encroached land if encroachment is prior to 13th April 1987. (iii) Major sons and major unmarried daughters will be treated as separate family if they were major on the date of Section 4 notification.

Provisions of the Policies Applicable in Sardar Sarovar Project

Provisions of the policy	*As per NWDT Award & State Policy for Narmada Projects.*
Compensation of Agricultural land	Agricultural land will be compensated as per sec. 23 of Land Acquisition Act (market value of the Land Compensation of wells, pipelines and trees as per valuation report given by PHE, forest and Horticulture Deptts. + 30% solatium+12% interest (Interest payable from date of notification of section 4).
Compensation of Houses and Properties.	Compensation of houses and other Immovable properties will be paid as per section 23 of Land Acquisition Act (as per CSR) + 30% solatium + 12% interest (Interest payable from date of notification of Section 4).
Rehabilitation Grant	(1) Agricultural landless labourers families, SC/ST/$mall and marginal farmers families = Rs.18,700/- (2) Other families Rs 9350/-

Provisions of the policy	*As per NWDT Award & State Policy for Narmada Projects.*
Employment Resources Grant for Productive Assets to Landless Persons	(1) Rs. 49,300/- to landless agricultural labourers, landless SC&ST families (2) Rs 33,150/- to other landless families
Other provisions- Land for Land	Eligible Families losing 25% or more Agricultural land will be allotted land for land (minimum 2 ha. and maximum 8 ha.). The major sons of such families will also be allotted minimum 2 ha. of land or amount of SRP for purchasing of land.
Transportation Grant	Free transportation or Rs. 5,000/- for transportation of their house hold goods to new R&R site.
Residential Plots	90' × 60'= 5400 Sq. ft. plot or Rs.50000/- to each family in rural area. For urban R&R site of Dharmpuri the plot size of 40' × 60'= 2400 Sq. ft. or Rs.50,000/- Joint Families refusing the residential plot will have to take at least 1 plot minimum.
Stamp/Registration fee	For purchase of Agriculture land or immovable property within M.P. up to the limits of amount paid to PAFs (excluding transportation grant).
Special Provisions	(i) Option of acquiring total agriculture land if 75% of land holdings are acquired. (ii) Compensation of encroached land if encroachment is prior to 13th April 1987. (iii) Major sons and major unmarried daughters will be treated as separate family if they were major on the date of Section 4 notification.

National Policy on R&R for Project Affected Families - 2003 (NPRR- 2003)—To be implemented for all upcoming Narmada Projects

The Government of India, Ministry of Rural Development (Gramin Vikas Mantralaya), Department of Land Resources (Bhumi Sansadhan Vibhag), has formulated a National Policy on Resettlement and Rehabilitation for Project Affected Families-2003. Narmada Control Board (NCB) in its 29th meeting held on 9m March 2006 took a decision to follow the provisions of NPRR-2003 for all upcoming projects (e.g. Halon, Upper Narmada, Lower Goi, Boras, Hoshangabad, Handia etc.) of Narmada Valley.

Objectives of NPRR-2003

The objectives of the policy are as follows:—

a. To minimise displacement and to identify non-displacing or least displacing alternatives;
b. To plan the R&R of PAFs including special needs of Tribals and vulnerable sections;
c. To provide better standard of living to PAFs; and
d. To facilitate harmonious relationship between the Requiring Body and PAFs through mutual cooperation.

R&R Benefits for Project Affected Families

- The resettlement and rehabilitation (R&R benefits shall be extended to all the PAF whether belonging to below poverty line (BPL) or non-BPL.
- Any PAF owning house and whose house has been acquired may be allotted free of cost house site to the extent of actual loss of area of the acquired house but not more than 150 sq. m. of land in rural areas and 75 sq. metre of land in urban areas.
- Each PAF of BPL category shall get a one-time financial assistance of Rs. 25000/- for house construction. Non-BPL families shall not be entitled to receive this assistance.
- Each PAF owning agricultural land in the affected zone and whose entire land has been acquired may be allotted agricultural land or cultivable waste land to the extent of actual land loss subject to a maximum of one hectare of irrigated land or two hectares of un-irrigated land/cultivable waste land subject to availability of Government land in the districts.
- Stamp duty and other fees payable for registration shall be borne by the requiring body.
- The land allotted shall be free from all encumbrances. The land allotted may be in the joint names of wife and husband of PAF.
- In case of allotment of wasteland/degraded land in lieu of

acquired land, each PAF shall get financial assistance of Rs. 10000/- per hectare for land development. In case of allotment of agricultural land, a one time financial assistance of Rs. 5000/- per PAF for agricultural production shall be given.

- Each PAF having cattle shall get financial assistance of Rs. 3000/- for construction of cattle shed.
- Each PAF shall get financial assistance of Rs. 5000/- as transportation cost for shifting of building materials, belongings and cattle etc. from the affected zone to the resettlement zone.
- Each PAF comprising rural artisan/small trader and self-employed person shall get one-time financial assistance of Rs. 10000/- for construction of working shed/shop.
- Each PAF owning agricultural land in the affected zone and whose entire land has been acquired shall get one-time financial assistance equivalent to 750 days minimum agricultural wages for "loss of livelihood" where neither agricultural land nor regular employment to one member of the PAF has been provided.
- Each PAF owning agricultural land in the affected zone and whose entire land has not been acquired and consequently he becomes a marginal farmer shall get one time financial assistance equivalent to 500 days minimum agricultural wages.
- Each PAF owning agriculture land in the affected zone and who consequently becomes a small farmer shall get one time financial assistance equivalent to 375 days minimum agricultural wages.
- Each PAF belonging to the category of 'agricultural labourer', or 'non-agricultural labourer' shall be provided a one time financial assistance equivalent to 625 days of the minimum agricultural wages.
- Each displaced PAF shall get a monthly subsistence allowance equivalent to 20 days of minimum agricultural wages per month for a period of one year up to 250 days of MAW.
- In the case of acquisition of land in emergent situation such as under Section 17 of the Land Acquisition Act 1894 or

similar provision of other Act in force, each PAF shall be provided with transit accommodation, pending resettlement and rehabilitation scheme. Such families shall also get R&R benefits as mentioned in earlier paragraphs under the Policy.

- Acquisition of Long Stretches of Land: In case of projects relating to Railway Lines, Highways, Transmission Lines and laying pipelines wherein only a narrow stretch of land extending over several kilometres is being acquired, the Project Affected Families will be offered an *ex-gratia* amount of Rs. 10000/- per family, and no. other Resettlement and Rehabilitation benefits shall be available to them.
- The Project Affected Families shall be provided necessary training facilities for development of entrepreneurship to take up self-employment projects at the resettlement zone as part of R&R benefits.
- The Project Affected Families, who were in possession of forest lands prior to 25th October, 1980 shall get all the benefits of R&R as given in earlier paragraphs under the Policy.
- The PAFs of Scheduled Caste category enjoying reservation benefits in the affected zone shall be entitled to get the reservation benefits at the resettlement zone.

R&R Benefits for Project Affected Families of Scheduled Tribes

- Each Project Affected Family of ST category shall be given preference in allotment of land.
- Each tribal PAF shall be entitled to get R&R benefits mentioned in earlier paragraphs under the Policy.
- Each Tribal PAF shall get additional financial assistance equivalent to 500 days minimum agriculture wages for loss of customary rights/usages of forest produce.
- Tribal PAFs will be re-settled close to their natural habitat in a compact block so that they can retain their ethnic, linguistic and cultural identity.
- Tribal PAFs shall get land free of cost for community and religious gathering.

- Tribal PAFs resettled out of the district/taluka and will get 25% higher R&R benefits in monetary terms.
- The Tribal Land Alienated in violation of the laws and regulations in force on the subject would be treated as null and void and the R&R benefits would be available only to the original tribal land owner.
- The Tribal families residing in the Project Affected Areas having fishing rights in the river/pond/dam shall be given fishing rights in the reservoir area.
- Tribal PAFs enjoying reservation benefits in the affected zone shall be entitled to get reservation benefits at the resettlement zone.

Basic Amenities to be Provided at Resettlement Zone

While shifting the population of the Affected Zone to the Resettlement Zone, the Administrator for R&R may as far as possible, ensure that:

(a) In case the entire population of the village/area to be shifted belongs to a particular community, such population/families may be resettled enmasse in a compact area so that socio-cultural relations (social harmony) amongst shifted families are not disturbed.

(b) In case of resettlement of Scheduled Castes PAFs, it may be ensured that they are resettled in sites close to the villages.

The Project Affected Families shall be provided the basic amenities and infrastructural facilities at the resettlement site as per norms specified by the Appropriate Government. It is desirable that provision of drinking water, electricity, schools, dispensaries and access to the resettlement sites amongst others be included in the resettlement plan formulated by the Administrator for R&R.

SARDAR SAROVAR PROJECT (SSP)
REHABILITATION POLICY AS PER NWDT AWARD AND STATES LIBERALISED POLICIES[2]

Sl. No	Item	NWDT Award	Madhya Pradesh	Gujarat	Maharashtra
1.	Definition of oustee	An oustee shall mean any person who, since at least one year prior to the date of publication of notification u/s 4 of the Land Acquisition Act, has been ordinarily residing or cultivating land or carrying on any trade, occupation or calling or working for gain in the area likely to be submerged permanently or temporarily.	Same as under NWDT Award Clause XI-1(2)	Same as under NWDT Award Clause XI-1(2)	Same as under NWDT Award Clause XI-1(2)
2.	Family	A family shall include husband, wife and minor children and other persons dependent on the head of the family, e.g. widowed mother.	A family shall mean and include husband, wife and minor children and other persons dependent on the head of the family, e.g. widowed mother, widowed sister, unmarried sister, unmarried daughter or old father.	Same as under NWDT Award Clause XI-1(3) (ii)	Same as under NWDT Award Clause XI-1(3) (ii)

2. *Source:* http://nca.gov.in/rnr_prov.htm

Sl. No	Item	NWDT Award	Madhya Pradesh	Gujarat	Maharashtra
3.	Land Allotment				
	(a) Landed Oustees	Every displaced family from whom more than 25% of its land holding is acquired shall be entitled to and be allotted irrigable land to the extent of land acquired form it subject to the prescribed ceiling in the State concerned and a minimum of 2 hectares (5 acres) per family, the irrigation facilities being provided by the State in whose territory the allotted land is situated. This land shall be transferred to the oustee family if it agrees to take it. The price charged for it would be as mutually agreed between Gujarat and the concerned State. Of the price to be paid for the land a sum equal to 50% of the compensation payable to the oustee family for the land acquired from it will be set off as an initial instalment of payment. The balance cost of the allotted land shall be recovered from the allotted in 20 yearly instalments free of interest.	1. Land equivalent to that acquired with a minimum of 2 ha and maximum of 8 ha and also Government assistance in providing irrigation by wells/ tube wells or any other method if land not irrigated. If irrigation is not possible, minimum 4 ha will be allotted. Development of dry land would be subsidised by the Government to the extent of 75% of the cost involved. (a) For families of all SC/ST and other categories with land holding up to 2 ha. grant-in-aid would be paid to cover the gap, if any, between the amount of	1. Land equivalent to that acquired with a minimum of 2 ha and maximum limited to State ceiling. Joint holders are also eligible for land equal to his share subject to a min. of 2 ha. Joint holders and major sons will be considered as co-sharer of acquired land. Irrigation facilities to be provided by the State. 2. Where agricultural land is purchased by the oustees' family through committee, the difference, if any, in price of	1. Land equivalent to that acquired with a minimum of 2 ha and maximum limited to State ceiling. Joint holders are also eligible for land equal to his share subject to a minimum of 2 ha. Irrigation facility to be provided by the State. 2. The land is allotted to the oustee free of cost.

Sl. No	Item	NWDT Award	Madhya Pradesh	Gujarat	Maharashtra
		Where land is allotted in Madhya Pradesh or Maharashtra, all recoveries for the allotted land shall be credited to Gujarat.	compensation and the cost of allotted land in full. For other owing land from 2 to 8 ha. will be eligible for an additional amount of Rs.2000/- per ha. of 50% of the difference in cost of allotted land and the compensation received, whichever is less. (b) Oustee has two choice : (i) He can opt for cash compensation for his submerged land. (ii) can opt for compensation in the form of land. He will be entitled for 50% amount of compensation in cash and remaining 50% amount will be adjusted against the cost of the allotted land. Remaining cost of allotted land will be recovered in 20 yearly	land so purchased and the amount of compensation shall be paid as an ex-gratia. 1. The landed oustee/ encroacher who have received compensation of land has an option to contribute 50% of his compensation as initial instalment towards the cost of allotted land and rest of the amount of the cost of land to be repaid in 20 years, yearly interest free instalments of contribute 100% compensation towards the cost of allotted land and the difference between the price	

Sl. No	Item	NWDT Award	Madhya Pradesh	Gujarat	Maharashtra
			instalments as loan due from third year. Loan will be interest free. (c) The GOMP allotted agricultural land to the eligible PAFs from the land bank. On refusal PAFs can avail the special Rehabilitation package (SRP) to purchase land of their choice. (d) The PAFs will have exemption from stamp duty and registration fee on purchase of land to the extent of the amount they received as compensation Rehabilitation Grant and SRP.	of the allotted land and compensation would be treated as *ex-gratia* towards occupancy price. In other categories, i.e. landless agricultural labourer, encroacher (who has not received compensation) and their major sons are given full ex-gratia amount towards the cost of allotted land.	
	(b) Encroacher oustees	No land allotment benefit provided.	(a) They will be treated as landed oustees subject to two conditions. (i) Encroachment must be	Encroachers prior to 1 year of Notification under Sec. 4 of the Act are entitled for 2 ha. of land and	2 ha of land and compensation as *ex-gratia* payment for the balance land encroached upon up to

Sl. No	Item	NWDT Award	Madhya Pradesh	Gujarat	Maharashtra
			on or before 13.4.87. (ii) Allotment of agricultural land will be 1 ha. or 2 ha. Only subject to the size of encroachment coming under submergence. Encroachers will be entitled to get compensation for land under submergence.	compensation for the balance encroached land as *ex-gratia*	31/3/78. Later encroachers will be treated as landless and will get 1 ha. Agriculture land
	(c) Landless oustees	No provision for land allotment	No land According to liberalised R&R policy of GOMP letter No. 12/1/27/2/98/1286 dated 4/12/2001 which in over & above NWDT award, all landless agricultural labourers and al SC/ST landless outstees will get Rs. 49300 each for productive assets while other landless will get Rs.33150.l	2 ha of land to landless Agricultural labourers only	1 ha of land if oustee moves with others.
	(d) Major sons of above all categories	Every major sons will be treated as separate family	Every major son/unmarried major	2 ha of land to each major sons of	Every major son/unmarried

Sl. No	Item	NWDT Award	Madhya Pradesh	Gujarat	Maharashtra
	of oustees		daughters will be treated as separate family. There eligibility will be 18 years of age one year prior to the date of notification under section-4 of LAQ.	all category of oustees with 1/1/87 as cut-off date for major sons.	major daughters will be treated as separate family. There eligibility will be 18 years of age one year prior to the date of notification under Sectin-4 of LAQ
		No provision for land allotment	They will be entitled to cash compensation according to the category to which they belong.		The GOM vide corrigendum dated 21/04/2007 on the GR dated 26/02/1992 the entitlement of major sons & major daughters for land increased to 2 ha from 1 ha allotted to them as per liberalised policy of 1992.
		Major sons of landed PAFs losing more than 25% land are entitled for 2 hectares of land as per the decision of Hon'ble Supreme	Major sons of landed PAFs will be allotted 2 hectares of land as per the decision of Hon'ble		Losing more than 25% land.

Sl. No	Item	NWDT Award	Madhya Pradesh	Gujarat	Maharashtra
		Court dated 15/03/2005.	Supreme Court dated 15/03/2005.		
4.	House plot	Free of cost allotment of house plot to each oustee family measuring 18.29 x 27.43 m (60' x 90') including major sons.	(a) Developed residential plot will be given to the oustee families and their major sons/unmarried daughter.	1. Residential plot of 502 sq.m. would be given free of charge to the oustee families and their major sons.	1. Residential plot of 502 sq.m. would be given free of charge to the oustee families and their major sons and unmarried major daughters.
			1. 502 sqm(60' x 90') for rural area coming under submergence.		
			2. 222.95 sqm (40' x 60') only for urban area coming under submergence.	2. Free readymade core house. Financial assistance of Rs.45000/- for construction of core house to the oustee families and their major sons.	2. Free Mangalore tiles and bamboos.
			(b) Cash Compensation is of house plots		

Sl. No	*Item*	*NWDT Award*	*Madhya Pradesh*	*Gujarat*	*Maharashtra*
			Rs.50,000/-		
5.	Rehabilitation Grant, Grant- in-aid, Subsidy	Resettlement/rehabilitation grant @ Rs.750/- per family inclusive of transportation charges. Grant-in-aid up to Rs.500/-	The GOMP has further liberalised the R&R policy dated 04/01/2001 which is over and above NWDT Award accordingly, the scale of grant-in-aid for rehabilitation will be as follows: *Rehabilitation Grant* All agricultural landless labourers/SC/ST labourers, small marginal farmer at the rate of Rs18700/- each. All other labourers & landless families at the rate of Rs. 9350/- each *Productive* Major son of each oustee family will be treated as landless. He will be categorised amongst landless oustee as per his occupation.	Subsistence Allowance Each family will be paid subsistence allowance at Rs.15/- per day for 25 days in a month for a period of one year after resettlement. Rs. 750/- per family with an increase of 8% per year worked out with January, 1980 as base, as resettlement grant. Grant-in-aid up to Rs. 500/- as per NWDT Award. Rs. 7000/- to purchase productive assets to all category of oustees and their major sons.	Subsistence Allowance Each family will be paid subsistence allowance at Rs.15/- per day for 25 days in a month for a period of one year after resettlement. Rs. 750/- per family with an increase of 8% per year worked out with January, 80 as base, as resettlement grant. Grant-in-aid up to Rs. 500/- as per NWDT Award. The above benefits at Sl. 2

Sl. No	Item	NWDT Award	Madhya Pradesh	Gujarat	Maharashtra
					will be provided to all categories of oustees and their major sons and unmarried major daughters.
6.	Transportation Grant	Transportation charges to be met out of resettlement/ rehabilitation grant of Rs.750/-.	Free transport will be provided by the project. In case this facility is not availed of, a lump sum relocation grant will be paid.	Free transport will be provided by State Government.	Free transport will be provided by State Government
7.	Acquisition of Private land/ Houses which get isolated or physically cut-off.	Not dealt with.	Would be acquired and owners treated as oustees.	Would be acquired and owners treated as oustees.	Would be acquired and owners treated as oustees.
8.	Compensation (a) Land	As per Land Acquisition Act in operation at the time of acquisition.	Compensation for land on the basis of price of similar land in adjacent command areas.	As per Land Acquisition Act as amended from time to time.	As per Land Acquisition Act as amended from time to time.
	(b) Houses	As per Land Acquisition Act.	Replacement value of House.	As per Land Acquisition Act.	As per Land Acquisition Act.
9.	Civic Amenities	1. One primary school (3 rooms) for every 100 families	1. Drinking water well/tube-well with trough.	As per NWDT Award.	1. One Primary School (3 rooms)

Sl. No	Item	NWDT Award	Madhya Pradesh	Gujarat	Maharashtra
					for 100 families
		2. One Panchayat Ghar for every 500 families.	2. Link and approach road and drains.		2. One Panchayat Ghar for 500 families.
		3. One dispensary for every 500 families.	3. Electricity		3. Samaj Mandir (cultural centre) for 500 families
		4. One seed store for every 500 families.	4. Dispensary		4. One health dispensary for 500 families.
		5. One children's park for every 500 families.	5. Primary School		5. One seed store for 500 families
		6. One village pond for every 500 families.	6. Panchayat Bhawan-cum-Community Hall.		6. One children's park for 500 families.
		7. Drinking water well with trough for every 50 families.	7. Playground/ Children's Park.		7. One village pond for 500 families
		8. One tree platform for every 50 families.	8. Cattle shed.		8. Drinking water well with trough for 50 families.
		9. One religious place of worship for every 500 families.	9. Place of Worship.		9. Approach and internal roads.

Sl. No	Item	NWDT Award	Madhya Pradesh	Gujarat	Maharashtra
		10. Each colony should be linked to main road by roads of appropriate standards.	10. Threshing ground.		10. One tree platform for every 50 families.
		11. Electrification, water supply, sanitary arrangements etc.	11. Seed Store.		11. Play ground for School (1 Acre for Primary and 2 Acre for secondary).
			12. Tree platform.		12. Electricity supply.
			13. Cremation and burial ground.		13. Open gutters.
			14. Pond, wherever feasible.		14. Public Latrines.
			15. Social amenities for each municipal town viz. water supply, sanitary arrangements etc.		15. Open place for collection of animals.
			16. Any other facility such as middle school which was existing in the affected village and its improvement. 17. Ear marking of pasture land and its		16. Khalwadi (Threshing platform) 17. S.T. Stand 18. Grazing Land 19. Open Place for Bazaar.

Sl. No	Item	NWDT Award	Madhya Pradesh	Gujarat	Maharashtra
			improvement.		20. Cremation /burial ground.
10.	Other facilities	Nil	1. Where the option of interest free loan is not availed, the family would be assisted to the extent of a grant-in-aid of Rs.1000/- per ha. per year for 2 years.	1. Special schemes for rehabilitating Dungri Bhils in their marriage circles.	Relocation would be given priority in public employment subject to meeting minimum qualifications and subject to age relaxation up to 3 years. Also attempt would be made to absorb as many oustees as possible in project works.
			2. Age relaxation of 2years in Class III posts in Government service.	2. Priority in allotment of tank bed land.	1. Priority in Class-III and IV posts. Reservation in project establishment.
			3. All welfare schemes being run at the site of displacement will be	3. A non-agriculturist family, i.e. Trader,	2. 50% reservation in ITI. 3. 5% reserved for

Sl. No	Item	NWDT Award	Madhya Pradesh	Gujarat	Maharashtra
			implemented at relocation sites also.	Shopkeeper, Artisan, are given financial assistance up to Rs.5000/- at the new site for resettling at the new location. He is also provided floor area equal to that lost at new site for carrying out the trade. The differential price is treated as ex-gratia. Preference in employment. 4. All ongoing welfare schemes will be implemented at relocation sites also.	PAP's children in Class-III and IV posts in all Government /Semi Govern-ment Organisations and local authority. 4. House building advance (i) Landholder-Rs.8000/- (ii) Landless labourers Rs. 4000/- 5. Priority in allotment of tank bed land as per existing state 6. All ongoing welfare schemes will be implemented at relocation sites also.

Part 4

Resettlement and Rehabilitation in SSP

12

Evolution and Implementation of Sardar Sarovar Dam

N.B. Desai

History of Investigation and Planning 1945-1946

The Government of India (GoI) constituted the Central Waterways, Irrigation and Navigation Commission (CWINC) under the chairmanship of A.N. Khosla in April 1945 to act *inter alia* as a central fact finding, planning and coordinating organisation with the authority to facilitate construction work. The then Governments of Central Provinces & Berar and of Bombay requested CWINC in 1946 to take up investigations on the Narmada river system for basin-wide development of the river including flood control, irrigation, power and extension of navigation. CWINC took up studies of the topography and hydrology of the basin in the beginning of the year 1947. These studies revealed excellent storage sites on the main river and some of its tributaries. Many sites were inspected by engineers and geologists who, among other things, recommended detailed investigations for a Broach weir below Gorge on the Narmada.

The GoI Ministry of Works, Mines and Power appointed in 1948 an *ad hoc* Committee headed by A.N. Khosla, Chairman, CWINC to scrutinise the estimates based on investigations, *inter alia* for the Broach Weir as advised and to recommend priorities. The *ad hoc* Committee observed that the Narmada Basin appeared to hold great potential of development and would have a far-reaching effect on the economic advancement of the country in general and the basin in particular. The committee recommended detailed investigations of four projects including the Broach Project. GoI approved the estimates of investigation in March 1949 and CWINC took up their investigations in the same year. Investigations of all the projects except Bargi were completed and the project reports prepared. CWINC was redesignated in 1951 as the Central Water and Power Commission (CWPC).

Broach Irrigation Project (BIP) 1956-1959

CWPC carried out preliminary geological and topographical surveys of all the candidate sites first proposed a site near Gora village for a weir with Full Reservoir Level (FRL) 160 ft. in the first stage, envisaging an annual irrigation of about 400,000 hectares (1,000,000 acres) through a right bank canal. The project report was first prepared in the beginning of the year 1956. The Member (Designs & Research), CWPC inspected the project site in February 1957 and suggested further investigations for the Gora site and also at Navagam, about 3 km upstream, where there was exposed rock in the riverbed affording high abutments that would allow raising the dam height. The Navagam site was finally chosen in consultation with the then Government of Bombay.

In 1959 CWPC modified the Gora weir project considering the Navagam site. Its implementation was envisaged in two stages – FRL 160 ft. in Stage I, with provision for a wider foundation to enable raising it to FRL 300 ft in Stage II.

The BIP Report prepared by CWPC was scrutinised and reviewed by the Bombay Government, which replied with its comments in July 1959. The main modifications it suggested related to the raising of the FRL of the dam to 320 ft. in stage II, provision of a power house in the river bed and a power house at the head of the low level canal.

However, less than a year later, on 1st May, 1960, Bombay was bifurcated into two states – Gujarat and Maharashtra. BIP stage I was cleared by the Planning Commission in August 1960, and its acceptance was communicated to the Government of Gujarat (GoG). The acceptance of stage I also provided for wider foundations and additional length for a masonry dam and building an earth dam to full cross-section corresponding to the dam with FRL 320 ft.

The dam at FRL 162 ft. in stage I provided for diverting water into a low level canal at Full Supply Level (FSL) 160 ft. to command 538,000 hectares in the Broach and Baroda districts and annual irrigation for 389,000 hectares. Stage-II envisaged the raising of the weir and providing crest gates for FRL 320 ft. Irrigation was proposed to be extended to an additional area of at least 364,000 hectares by means of a high level canal off-taking with FSL 295 ft. Power generation to the extent of 625 MW at 60 per cent load factor (LF) was also envisaged in the project. GoG gave administrative approval to Stage I in February 1961.

Feasibility of Construction of High Dam, 1963

The surveys of the command area and reservoir submergence area were done by the Survey of India. GoI indicated that a much larger area could be commanded by a +300 canal and that it was possible to store more water if the dam height could be raised beyond +320. Accordingly, after detailed engineering and geological investigations in 1963, the dam site was shifted 5.6 km upstream to Navagam dam site No. 3, which was suitable for construction of a high dam.

Navagam is the last in the series of dams of the Narmada Valley Project. Hence, its design must necessarily consider probable future planning and phasing of projects for the control and use of the waters upstream. With upstream consumptive use estimated by MP and the increased water requirement of Gujarat for areas to be commanded by the proposed Narmada high level canal, it was necessary to apportion the waters and, on the basis of such apportionment, to decide the optimum height of the Navagam Dam. Until this was done, it was not possible to proceed with its construction.

Proposals for High Navagam Dam, 1963

It appears that until the year 1960 engineers and planners believed that the gorge in the 113-km stretch between Harinphal and Navagam could not afford sizeable storage, even if the Navagam dam's height was to be raised. However, the available topographic survey sheets and other survey data by Gujarat Engineers indicated that the gorge between Navagam and Harinphal could store a very large amount of water if the Navagam dam was to be raised to the same height as Harinphal to provide a terminal storage reservoir. The proposition appeared so promising that GoG entrusted to the Survey of India the work of carrying out a contour survey by aerial photography for assessing the storage capacity of the proposed high Navagam dam. The advantages of the high dam are very large storage capacity, greater harnessing of Narmada waters, increased irrigation from the high level canal and an effective flood cushion to downstream areas.

A meeting of the Chief Ministers of Gujarat and MP was arranged at Bhopal with good offices of the Union Minister for Irrigation and Power, GoI in November 1963. It was agreed at the meeting that the Narmada should be developed in the best national interest and that FRL of Navagam dam should be 425 ft. The MP Government, however, did not ratify the Bhopal Agreement and did not actuate its terms.

Khosla Committee Report 1964-65

In September 1964, GoI with the concurrence of the concerned states appointed a the Narmada Water Resources Development Committee (NWRDC) under the chairmanship of Dr. A. N. Khosla, then Governor of Orissa.

The Khosla Committee held a number of sittings and also met a number of times with the representatives of all concerned state governments. The committee invited their viewpoints and proposals on the schemes for harnessing the Narmada Waters. Each submission from a state was referred by the committee to the other States, eliciting their comments on it. Comments and responses on almost all the documents initially submitted by the concerned states were received. The committee also made various independent studies and made on-the-spot inspection of the Navagam dam site and its commanded areas as well as the Punasa (Indira Sagar) site in MP.

In September 1965 the committee forwarded a unanimous report to GoI giving an outline of the Master Plan for the development of the Narmada valley. The Master Plan envisaged twelve major projects to be taken up in MP and one in Gujarat.

For Gujarat the committee recommended that the terminal dam should be located at Navagam, the optimum FRL of Navagam dam be 500 ft, the FSL of Navagam canal be 300 ft at its off-take and the installed capacity of the river bed power station and canal power station be 1000 MW and 240 MW respectively. Irrigation benefits from the Navagam dam would be shared by Gujarat and Rajasthan and power generation would be shared by MP, Maharashtra and Gujarat.

While GoG broadly endorsed the recommendations of the Khosla Committee, the Governments of Maharashtra and MP did not accept them. MP claimed absolute rights over the hydro-power which could be generated in its territory. It was not amenable to the submergence of any part of its territory by a dam in Gujarat and questioned the necessity of a high dam or a high level canal from this site. It also asserted that the Khosla Committee's assessment of water availability at different sites was on the high side and that each State has constitutional right to develop hydro-electric power which it could generate in its own territory. Thus it rejected almost all the recommendations of the Khosla Committee.

Maharashtra agreed generally with the basic considerations for drawing up the Master Plan by the Khosla Committee. It advocated restriction of the FRL of the Navagam dam to 210 ft. and a canal with FSL 185/190 to be taken from there for irrigation in Gujarat. It rejected almost all the recommendations of the Committee.

Efforts for Inter-state Understanding 1966-67

The Minister and the Minister of State for Irrigation and Power, GoI went around the concerned states in May/June 1966 and separately met the respective Chief Ministers to evolve a satisfactory basis for resolving the disputes. Their efforts proved futile.

The Union Ministry of Irrigation and Power arranged joint discussions at the official level, presided over by the Chairman, CWPC to consider the technical aspects of NWRD. These meetings, attended by the representatives of the four states and of the Planning Commission besides those of CWPC and the concerned Ministry, were held intermittently for twelve days between 19th July and 14th August 1966. Agreement was reached on quantum of utilisable flow at Navagam and its dependability, and the load factor to be adopted for hydro-electric projects. But on sharing of water, areas to be irrigated in each state and level of the Navagam dam, the differences remained unresolved.

On 22nd August, 1966 a meeting of the Chief Ministers of the four States was convened by the Union Minister for Irrigation and Power in New Delhi. It was also attended by the Union Minister of State for Irrigation and Power. At this meeting it was decided that the Chief Ministers of Gujarat and MP should meet as early as possible with a view to resolve the dispute amicably. Accordingly, in 1967, the two Chief Ministers held discussions at Pachmarhi on 23rd May and in New Delhi on 22nd June. Thereafter, the Union Minister of Irrigation and Power and the Chief Ministers of all the four states met in New Delhi on 18th December. None of these meetings could resolve the differences among the States.

Petition by Gujarat, 1968-69

In spite of GoI's earnest efforts over six years to settle the dispute by negotiations, no mutually acceptable agreement was reached on the distribution, control and use of Narmada waters and the height of the Navagam dam. The differences only widened. GoG therefore petitioned GoI in 1968 to appoint a Tribunal under the Interstate Water Disputes Act, 1956. GoI issued a notification constituting the Tribunal on 6th October, 1969 and on the same date referred to it GoG's petition dated 6th July, 1968. By their communication dated 16th October, 1969, GoI also referred certain issues raised by the state of Rajasthan to the Tribunal. In 1970 GoG decided to name the Navagam Reservoir as Sardar Sarovar and the Navagam Dam as Sardar Sarovar Dam.

Proceedings before the Tribunal 1970-79

The tribunal identified two preliminary issues – namely, whether Rajasthan could be a party to the dispute and whether Gujarat could construct in its territory a dam that would submerge the territories of Maharashtra and MP – which were required to be resolved before adjudication. The tribunal, taking into account the fact that Rajasthan could not, by virtue of its location, control the waters of Narmada, decided that Rajasthan could not be a party to the dispute. It also held that the construction of a dam in Gujarat territory, involving consequent submergence of portions of territories of Maharashtra and MP could be a subject of water dispute, and that the tribunal had jurisdiction to give appropriate directions to MP and Maharashtra to acquire land for making available to Gujarat.

After the tribunal had given its decision on the aforesaid preliminary legal issues, MP, Maharashtra and Gujarat agreed to the allocation of 500,000 acre feet of Narmada Water to Rajasthan for irrigation in Barmer and Jalore districts and 250,000 acre feet to Maharashtra. The tribunal accepted this agreement and gave its decision that Rajasthan and Maharashtra respectively were entitled to the specified share of Narmada waters. The tribunal also allowed Rajasthan to represent its case before it by virtue of agreement of the three basin states.

After considering all the material available to it, NWDT gave a final decision in December 1979. The main parameters of decision of the tribunal were as follows:

(a) Allocation of Narmada Water at 75 per cent dependable yield.

	Million acre feet
Madhya Pradesh	18.25
Gujarat	9.00
Maharashtra	0.25
Rajasthan	0.50
	28.00

(b) Height of Sardar Sarovar dam
  (i) Full Reservoir Level 455 feet (138.68 m)
  (ii) Maximum Water Level 460 feet (140.21 m)

(c) Full Supply Level of Narmada Main Canal 300 feet (91.44 m)

(d) Allocation of Hydro-power generated at Sardar Sarovar Dam

(i) Madhya Pradesh	57%
(ii) Maharashtra	27%
(iii) Gujarat	16%

The tribunal ordered setting up machinery to ensure implementation of its decision. Accordingly GoI constituted the Narmada Control Authority (NCA), Review Committee and Sardar Sarovar Construction Advisory Committee (SSCAC) to perform various functions to implement the decision of the tribunal.

Sardar Sarovar Project Report 1980

After the tribunal gave its decision on the Narmada Water disputes among the party states of MP, Maharashtra, Gujarat and Rajasthan in December 1979, GoG prepared, in accordance with the decision, the Sardar Sarovar Narmada Project Report (FRL 455 ft) in January 1980 and presented it to GoI for investment clearance by the Planning Commission and Technical scrutiny by CWC. The Project envisaged construction of a concrete gravity dam 1210 m in length and 128 m in height from the river bed, across the Narmada near Navagam village which is now Narmada District (carved out of Bharuch and Vadodara districts in 1997). The dam creates a reservoir with a gross storage capacity of 9.5 MCM and a live storage of 5.8 MCM, extending upstream for about 214 km of the Narmada valley and enveloping a surface area of about 37,000 ha. Its central spillway can pass a design discharge of 87,000 cumecs.

The project has two power houses, the river bed power house and canal head power house. The former is an underground power station located about 165 m downstream of the dam. It has six units of 200 MW Francis type reversible turbine generators. For creating the impoundment required for pumping operations a weir about 28 m high and 990 m long is proposed across the river near Garudeshwar. The canal head power house is located at the toe of the Vadgam Saddle dam. It is a surface power house with 5 units of 50 MW Kaplan Turbine Generators.

The Narmada Main Canal (MC) has a discharge capacity of 1133 cumecs at the head regulator, reducing to 70.8 cumecs at the Gujarat-Rajasthan Border. It will irrigate a cultivable command area of 2.124 million hectares with annual irrigation of 1.792 million hectares. The water for irrigation is conveyed to 8-hectare blocks through a 75,000 km network of branch canals, distributaries, minors and sub-minors. The Sardar Sarovar Project was estimated to cost Rs. 640.6 billion at 1986-87 prices.

Narmada Planning Group, 1981-2004

The Government of Gujarat was cognisant of the fact that the successful completion of SSP would require comprehensive and systematic efforts in several directions. It recognised that the project planners would have to ensure equitable and timely supply of the limited quantity of water allocated by the tribunal. The project was also expected to avoid problems of waterlogging and salinity.

The project authorities were expected to make satisfactory arrangements regarding resettlement and rehabilitation (R&R). The project was to be planned so as to ensure sustainable development by paying attention to issues of environmental concerns. Environmental planning was to be done as an integral part of total project planning.

GoG wanted SSP to be planned, designed, constructed and operated with the highest standards of organisation and management and with a vision of the requirements of agriculture in the 21st century. It was with this objective that GoG constituted the Narmada Planning Group (NPG) with a multi-disciplinary team of professional experts in engineering, agriculture, economics, sociology, ecology and environment and planning and public policy.

NPG commissioned about 25 technical studies covering aspects such as reservoir simulation, ground water investigation, supplies from *en-route* rivers crossed by NMC, canal losses and special problems of NMC. All NPG's policy initiatives were incorporated in the Sardar Sarovar Development Plan 1983 presented to the World Bank.

NPG has noted that three-fourths of the command area is drought prone. With irrigation from the project this area would turn green. The economy of the state would become more stable and achieve a higher rate of growth.

External Assistance from World Bank and OECF, 1983-1986

GoG got NPG to prepare the Sardar Sarovar Development Plan (1983) and presented it to the World Bank for its appraisal. For the Dam the Government set up a Dam Designs Review Panel (DDRP) in consultation with GoI and WB. DDRP carried out a total technical scrutiny with particular emphasis on safety of the dam against earthquakes and floods. The Design Organisation of Gujarat carried out all studies and presented them to DDRP, which endorsed the design. Similarly, other WB groups examined the design of the canal system.

As mentioned above, the project was estimated to cost Rs.640.6

billion at 1986-87 prices. The World Bank agreed to provide credit assistance totalling US $450 million in 1985 for certain components of works included in Stage I of the project. Out of this US $300 million was for the dam and civil works of power houses, while US $150 million was for the main canal up to the Mahi River and part of the distribution system under this segment of the canal. The Bank in 1990 offered credit assistance of US $200 million for R&R and environmental protection measures for the entire project.

The Overseas Economic Co-operation Fund (OECF) of Japan agreed to provide credit of 28.5 billion Yen, equivalent to Rs.1.5 billion at the prevailing the exchange rate in 1986, for electrical components of the river bed power house.

SSP could not avail of the credit assistance to the extent of US $180 million from the World Bank and almost the entire credit assistance from the OECF as explained later.

Forest and Environmental Clearance, 1987

Two significant developments took place after the NWDT gave its decision in December 1979. The Forest Conservation Act was passed in 1980 and the Environmental Protection Act in 1986. The passage of the FC Act, 1980, precluded any possibility of diversion of forest land to non-forest use as such use was expressly barred except through special dispensation of GoI. The submissions made before the tribunal had assumed that degraded forest lands would be made available for rehabilitation. Environment impact studies gradually came to be required on the basis of evolving guidelines, and environmental clearances became mandatory in the procedural progression of projects to investment clearance.

In April 1986, the Ministry of Environment and Forests (MoEF) came up with a paper on the Environmental Aspects of the Narmada Sagar and Sardar Sarovar Multi-Purpose Projects. The study argued that it would take another two or three years to complete environmental impact studies for these two projects. The Ministry of Water Resources (MoWR), GOG and the Government of Madhya Pradesh (GoMP) pressed for environmental clearance for both projects to avoid delays and to ensure *pari passu* actions for environmental protection with construction. MoEF granted conditional environmental clearance in

June 1987 for both NSP and SSP and forest clearance for SSP in September 1987.

The Planning Commission's Investment Clearance, 1988

The Planning Commission, after examining SSP, gave its clearance in October 1988 for an investment cost of Rs.640.6 billion, mentioned the need for compliance with conditionalities accompanying the environmental and forest clearance. GoG was required to give them priority. The programme of drainage and groundwater balance studies was to be completed for command areas beyond the Mahi and submitted for monitoring by the Planning Commission. Suitable water rates were to be devised to ensure annual revenues from the project sufficient to cover annual operating costs. Canal siltation studies were to be undertaken. Micro-level canal network plans and estimates were to be got ready within five years and the entire system planned to be completed in segments up to farm outlets so as to ensure early utilisation of available irrigation supplies. In according investment sanction, the Planning Commission also, therefore, laid down a number of environmental conditionalities.

Evolution of Rehabilitation Norms, 1987-88

NWDT laid out a rehabilitation package which, certainly as far as landholders were concerned, was a considerable advancement over previous practice. The package did not cover other land related categories such as agriculture labourers, encroachers and coparceners. The tribunal did not recognise encroachers on forest land or non-forest lands as entitled to land benefits. Landless labourers would lose their livelihoods once the lands on which they were employed were submerged. They too must, therefore, be provided land in recompense. There were some merits in these arguments and Gujarat had by 1988 accepted both propositions. Both encroachers and the landless would, therefore, receive 2 hectares of land, with encroachers being entitled to *ex-gratia* compensation for encroached lands in excess of this figure. Gujarat also granted major sons full entitlement to land benefits. The R&R packages of Maharashtra and MP have steadily improved and implementation has progressed on such basis.

Construction of Sardar Sarovar Dam, 1987-2000

The history of the project is long and chequered. Preliminary works like access roads and a colony at project sites were taken up in 1961. Additional colony, dam foundation, excavation and its treatment, rock fill dams, link channels, cofferdams, diversion of river flow, submersible bridges and experimental cross-drift for power house cavern could be taken up on large scale only in 1979 after the tribunal's decision.

The commencement of major works of the main dam and power houses could be taken up in the year 1987-88 when MoEF gave the project environmental and forest clearance in 1987 and the Planning Commission gave investment clearance in 1988.

The work for construction of the dam was awarded in April 1987 with a stipulated period of construction period of 110 months. Accordingly, the work was to be completed by June 1996. The period of construction was extended by 19 months to January 1998.

In the initial years of the contract work proceeded largely according to the programme. The progress started falling behind schedule from 1993-94, when 829,000 m^3 of concrete against the planned quantity of 1,136,000 m^3 could be placed in the dam. The work started lagging due to non-completion of R&R of affected families on time and after the Supreme Court order dated 5th May, 1995. However, by January 1998, nearly 80 per cent of the work of main dam was completed. Some significant work could be done in 1998-99 when the Supreme Court allowed raising of the dam from EL 80.3 m to 85 m. Finally the Supreme Court dismissed the writ petition and ordered on 18th October, 2000 that the work of the dam be completed expeditiously *pari passu* with R&R works and environmental protection measures.

Independent Review Mission 1992

The World Bank President, Barber Conable decided to appoint an Independent Review Mission (IRM) to review SSP. It started its work in September 1991. The IRM equated Narmada Valley Resettlement and Rehabilitation with human rights and cited the ILO convention 107 adopted in 1957 and ratified by India. The displacement in relation to the Narmada Valley Projects is not forced eviction in the sense of arbitrary ouster, but is in consonance with law, planned development and assisted relocation on generous terms, judicially determined, closely supervised and based on choice. Land acquisition is well established and is done in accordance with prescribed procedures.

The IRM claimed that environmental and social trade-offs had been made and continued to be made without a full understanding of the consequences. It observed that as a result of these trade-offs benefits tended to be overstated, while social and environmental costs were frequently understated. The IRM also claimed that the Sardar Sarovar Project would not work as intended because there was not enough water in the river to deliver the designed benefits. It expressed apprehensions about considerable inundation and displacement that might be caused by the back-water effect of delta formation at the upper end of the reservoir. Finally, it stated its opinion that the wisest course would be for the World Bank to step back from the projects and to consider them afresh.

The IRM did offer some useful insights and could have made recommendations to improve the projects and their implementation. That would have been a real contribution. Instead, from small premises it asserted that rehabilitation was impossible and the project environmentally unworkable. The World Bank was also surprised that its constructive efforts and genuine concerns with ensuring a positive outcome through a sensible incremental *pari passu* approach was turned against it and it was put in the dock.

Finally, GoI informed WB that it was withdrawing SSP from the Bank's lending portfolio, giving up an unspent $180 million. It instead decided to complete the projects without external assistance and agreed to provide them additional financial assistance in lieu of forgone amount of $180 million.

Narmada Bachao Andolan's Petition to the Supreme Court, 1994-2000

Narmada Bachao Andolan (NBA) filed a comprehensive writ petition seeking complete stoppage of all construction activities of SSP. It alleged that the project was not in national or public interest. It raised issues regarding R&R, environment and review of the Tribunal's award. It submitted that it was necessary for some independent judicial authority to review the entire project, examine the current best estimates of all costs, namely social, environmental and financial, benefits and alternatives in order to determine whether the project was required in its then existing form in the national interest or whether it needed to be restructured or modified. It also submitted that no work should be resumed until environmental impact assessment has been fully done and its implications for the project viability assessed in a transparent and participatory manner. This could best be done, it was submitted, as a part of a comprehensive review of the project.

After hearing NBA, the respondent states, namely MP, Maharashtra, Gujarat and Rajasthan, and the Union of India, and while issuing directions and disposing of the case, the Supreme Court observed that two conditions had to be kept in mind: (i) the completion of the project at the earliest and (ii) ensuring compliance with conditions on which clearance of the project was given, including completion of R&R work and taking ameliorative and compensatory measures for environmental protection in compliance with the scheme framed by the Government, thereby protecting Fundamental Rights under Article 21 of the Constitution.

The Supreme Court issued various directions to the states. A significant direction was that construction of the dam would continue as per the award of the tribunal. It also directed the states of MP, Maharashtra and Gujarat to implement the award and give relief and rehabilitation to the oustees in terms of the packages offered by them, and directed them to comply with any direction given in this regard by NCA, the Review Committee or the Grievance Redressal Authorities.

It also directed that if for any reason serious differences in implementation of the award arose and the same could not be resolved in the Review Committee, the committee could refer them to the Prime Minister whose decision in respect thereof would be final and binding on all concerned.

The Supreme Court further directed that every endeavour be made to see that the project was completed as expeditiously as possible.

Construction of the Sardar Sarovar Dam After 2000

After the judgment of the Supreme Court, the Narmada Control

Authority allowed construction of the dam from EL 85.0 m to EL 90.0 m during 2000-01. Thereafter, NCA in its 61st Meeting held on 17th November, 2000, decided important milestone dates for completion of the concrete dam. According to this programme the dam was programmed to be raised to 100 m, 110 m, 121 m and 138.68 m by June 2002, June 2003, June 2004 and June 2005 respectively. However, the actual progress could not meet the completion date of June 2005. The dam could be raised to spillway crest level of 121.92 m by December 2006. There is virtually no progress for last four years due to slow progress on R&R works and pending environmental clearances. Now MoEF has given environmental clearance in May 2010, but approval of the R&R Subgroup and that of NCA is awaited for raising the piers and installing gates. Once these approvals are given, it is planned to complete the work within a time span of three years.

A Retrospect of SSP: A Personal View on the Morse Report

A Prologue to Chapters 13 and 14

Anil Patel

The two Chapters 13 and 14, that follow were originally written about a year apart in the early 1990s. They are reproduced here, slightly modified, in a sequence opposite to that in which they were written - and first put in public domain. Thereby hangs a tale.

In November 1992, I wrote an early draft of the first paper (which comes to you as the second paper), The Independent Review: Was it a Search for Truth? I sent it for publication to William Fisher of Harvard University. He was then putting together, as editor, the book Toward Sustainable Development - Struggling over India's Narmada River, which was published in 1995.

My second paper (the first here), What Do the Narmada Valley Tribals Want? was written in March 1994. I presented it at the International Conference on Resettlement and Rehabilitation held in Montreal. A copy of this was sent to Fisher. He read it and was keen to incorporate it in the book he was editing. So it appeared as Chapter Eight of the book.

Fisher had been a little disappointed in 1992 when, in response to his request for a paper from ARCH-Vahini, we proposed a critique of the Bradford Morse Commission's report, which had come out in Jun 1992. He was not too keen that we write such a critique (I suspect his sympathies lay with the international campaign then raging against the SSP in the name of sustainable development, and he did not want the Morse Report to become our main focus). As a true academic concerned with both sides of the controversy, Fisher was interested in getting a version of the unfolding story of R&R from a less known agonist of the dispute, ARCH-Vahini. He made his preference clear, but I insisted on the Morse critique and he relented. He edited the paper, made several useful suggestions, asked for a number of clarifications and checked all the references. The process was long and took many months. My reasons for insisting on the Morse critique were complex. I did not fully explain them to him. But he accepted my word.

From the very beginning of its involvement in the SSP controversy in 1980, ARCH-Vahini had regularly and extensively written on the R&R process pertaining to the project. The press – particularly the English language press – was initially quite receptive and regularly took note of our analysis and concerns. This changed suddenly in August 1988.

In that month NGOs in Maharashtra and MP turned against the SSP. A new organisation, the Narmada Bachao Andolan (NBA) came into being, which declared from Bombay in August 1988 that the project was a planned social, economic and environmental disaster and the tribals in the submergence areas, facing displacement, would rather drown than tacitly approve it by agreeing to shift.

After this there were no takers for our side of the story. The English-language media and the Delhi, Bombay and Pune based intellectuals, who had supported us until then, distanced themselves. The so-called independent citizens' committees did no better. We could not quite comprehend this strange and disturbing development.

A major move was afoot in which 47 international NGOs submitted a joint memorandum to the President of the World Bank listing serious defects and violations in the planning and execution of the SSP. They held the Bank responsible for sins of commission and omission. They urged it to stop funding the project. An orchestrated protest was launched in major cities of Europe, America and Japan. Powerful Executive Directors of the Bank joined in. In the autumn of 1991, the Bank's resolve wilted and it appointed an Independent Review Commission under the chairmanship of Bradford Morse (ex-Administrator of UNDP). He was to be assisted by a young, dynamic ex- legislator and senior judge of Canada Mr. Thomas Berger.

About the same time I travelled to the UK, Europe and the USA. With a copy of the Joint Memorandum in my hand I met many well known NGOs and journalists in those places and also a member of the European Parliament and some prominent citizens. I tried to present a radically different version of the problems, and events from that presented in the Memorandum. They were courteous, kept their appointments and heard me out. Although they were all signatories to the Memorandum, not one raised a single pertinent question or tried to counter my facts. This was quite intriguing. In those days of raw idealism and sheer naïveté we had assumed that European NGOs and intellectuals would be more, open rational and enlightened than their counterparts in India!

Amidst our bewilderment, confusion and disappointment the announcement was made of the appointment of the Morse Commission. In November 1991, in Seattle, I met Thomas Berger and his assistant on R&R, Hugh Brody, an anthropologist from the UK. I extended them an advance welcome to Gujarat and assured them of our full cooperation. I told them about the indifferent and ambiguous reception I had received in Europe. The only thing I asked for on behalf of ARCH-Vahini was a patient and fair hearing of our case.

It was clear from the outset that Morse's indifferent health would not permit his active involvement in the enquiry. The actual investigation was to be done by two young assistants under Berger's guidance and directions.

We did get a fair hearing from Brody, who spent many days with us at Mangrol, made numerous field visits, discussed minute points arising from them, asked for relevant documents, studied and discussed them, asked for numerous clarifications and verifications The same cannot be said of the other assistant, Donald Gamble, an environmentalist who was looking at the environmental aspect of the project. In sharp contrast to Brody, Gamble's approach was perfunctory. He was not interested in the crucial details, nor was he interested in studying and discussing vital documents related to environmental clearance and others by MoEF (the Ministry of Environment & Forests) and other related documents which had formed the backbone of his critique in the Report.

He betrayed his bias against the SSP early on (Brody perhaps shared it but he never let it interfere with his objectivity and fairness of enquiry). He was visibly uncomfortable with the actual contents of the critical documents shown to him and eventually misinterpreted the facts. He refused to accept that MoEF had a prime role in maligning the project and in encouraging gross misuse of the documents by the anti-dam activists.

It was a deeply entrenched alliance between the key officers of the MoEF and anti-dam activists that we were trying to bring to his attention so that he could dig deeper and examine our accusations and interpretations. In the Report he did nothing of the kind, ignored our arguments and after several highly selective interpretations of information arrived at the same damning conclusions as the MoEF and the activists had earlier. He had not bothered to critically examine the issues we had raised and virtually accepted the other story.

Berger, the real moving force behind the Mission, fared no better. He spent a day and half with us at his Baroda hotel. He was yawning most of the time (which we attributed to jet-lag) and evinced little interest in our arguments and presentations. Except for some rare flashes of insight, the crucial documents stirred little interest in him. Somehow we trusted him then, and hoped he would bring Gamble round. The Baroda meeting and Gamble's indifferent handling created some apprehensions but still hope persisted. After all, we reasoned, he was a senior former judge of Canada, a country where rule of law prevailed.

The Morse Commission Report came out in June 1992. It was designed to deliver a mortal blow to the SSP. We were taken aback by the ferocity of its attack. Even the first reading convinced us that it had abandoned all norms of objectivity, fair enquiry and rationality. It was a no-holds-barred attack. The Bank was the target and the SSP was the means to batter it. It showed little respect for the sanctity of facts (except for some notable portions of Brody's report on Gujarat). Facts were omitted, distorted, and even invented; falsehood employed to destroy the project and the Bank's prestige. In a letter to the Bank's President which received widespread publicity, many damning conclusions were drawn without corroboration in the main body of the Report.

The Bank put up a feeble defence. The NGOs and the metropolitan intellectuals rose in unison to protest. Full-page paid advertisements appeared in practically all leading dailies of Europe and America. On the basis of the Morse report the project was condemned and the Bank was threatened that its donor governments would be forced to cancel funds. The Bank must have been rattled. All the EDs, mainly from the USA and other rich countries, had thrown their weight behind the Morse Report.

The Morse Commission took the unprecedented step demanding to meet the EDs (in its own words: the Review had no longer any formal existence). The meeting took place on 22 October 1992, with the EDs of the Northern nations (G-10). Not surprisingly, they were largely supportive of the Morse Report. The most outspoken supporter was Patrick Coady of the USA, who spoke with passion and eloquence. He said, "We have all read the Morse Commission report. ... I think it is fair to say that their concerns and their recommendations are substantive, detailed, clear and have withstood scrutiny... the Review has prepared a serious, objective and intellectually honest analysis... we recommend that the Bank step back and reappraise the project." The fate of the Project was sealed. The only face-saving device left for India and the Bank was to withdraw it.

In retrospect it is pertinent to recount an incident which occurred midway through the enquiry. Hugh Brody, who was investigating R&R, was initially inclined to believe the activists' version that R&R would be a disaster. As the enquiry proceeded he probably realised that the truth was not so stark and needed finer scrutiny. He proposed that after the enquiry was over, there should be a thorough private debate between selected members of NBA and ARCH-Vahini to thrash out all the outstanding issues. We promptly agreed, with a proviso that it should be moderated by him, without time constraints to thrash out each issue and that it should be a closed door meeting. While the NBA dithered and tried to wriggle out of the tight corner, Berger had the final say. He firmly told Brody not to get involved any further. "We will leave the country the day the enquiry is completed," he told Brody. Berger thus shot down a very useful idea to get at the bottom of things. If his demeanour in the Baroda meeting had left some room for doubt, it was conclusively dispelled by his curt advice to Brody. This was a clear indication, if one was needed, that the de-facto Commissioner had made up his mind. The enquiry was just a ritual that had to be gone through. The conclusion was preset.

For ARCH-Vahini, in those days of limited perspective, the discovery that the Morse Commission was neither a search for truth nor a helpful hand to the tribals of MP and Maharashtra, was a rude shock and a sort of betrayal. It was a phoney grandstand spectacle in the phantom fight for "sustainable and humane development." We knew that the chances of our crying foul and being heard were close to nil, but we had to bear witness. Fisher's invitation was an unexpected gift! So the second paper (Chapter 14 in this book), The Independent Review: Was it a Search for Truth? was written first.

During the year that Fisher was going though the article with a fine-tooth comb, raising searching questions and suggesting corrections and improvements I had an opportunity to examine and re-examine the Morse Report. It occurred to me that while our critique of the environmental aspect of the Report was accurate and adequate, an important aspect of the Report on R&R, especially as stated in the Letter to the President, was not adequately presented – that a major failure of the Bank's India Department was to totally neglect and ignore the Bank Guidelines for R&R (1980) and thereby neglect the vital interests of the tribals, especially the 'encroachers' who constituted the vast majority and had no entitlement in the eyes of the State.

The Morse Report repeatedly maintained that while the Bank had appraised the technical aspects of the SSP quite thoroughly it had not appraised the R&R prospects of the tribals; that it was a botched up job. The Morse Commission thus joined the fashionable chorus of branding the SSP as 'a mere technocratic Project.' Morse provided much needed legitimacy to the virulent campaign to malign the project.

The truth of course was quite the opposite. The R&R of the SSP was appraised systematically and thoroughly at least twice by an internationally renowned expert on R&R, the Caltech based anthropologist Thayer Scudder. This was well before the Loan Agreement (LA) signed in May 1985. Scudder had grasped the immense significance of the encroacher's non-entitlement status way back in the 1983 appraisal. He made sure they got an adequate package. This came in the form of minimum 5 acres of irrigable land of own choice, to each major landless oustee (effectively encroachers were treated as landless individuals). This formed the focal point of the LA's R&R package that went well beyond the Bank's 1980 guidelines for R&R.

This in fact made the Gujarat R&R package unprecedented and historic. The lengthy, elaborate process of appraisal and policy formation was very much on the record, but the Morse Mission chose to sidestep it and later lambasted the Bank for the sin it had not committed. This was cynicism at its worst. It also meant that the Mission had abandoned its grave responsibility of providing much needed help to the encroacher tribals of Maharashtra and MP because of a defective policy marked by a characteristic meanness of the State.

This great betrayal of the International Commission of Jurists and legislators was good enough for us to present a comprehensive account of the R&R process as it had unfolded since 1980. The story of the perfidy and meanness of the State before 1988; the serial betrayal of tribal interests by the valley's NGOs since August 1988; and the stamp of approval to it by the international NGOs, notably the US-based Environment Defence Fund (EDF) in the name of human rights and sustainable development, and finally the coup de grâce delivered to the tribals and the Project by the 'independent' Morse Mission, for the same cause had to be told.

Thus the second paper, What Do the Narmada Valley Tribals Want? (April 1994, Chapter 13 in this book), was written a year after the first, for an R&R conference in Montreal. I sent a copy of it to Fisher at Harvard for his comments. He jumped with excitement and chided me on the phone that this was precisely what he had wanted for the book he was editing. He asked if he could also include this paper as well in his book. I readily agreed. As time was running out he skipped the meticulous checks and corrections that he would have preferred. He made some editorial improvements and sent the paper to the publisher. Since there was no time to include references he implicitly trusted the veracity of the facts that I had stated.

For the present publication I have taken the liberty of reversing the order of the papers to put the narration in a historical perspective that better contextualises critique of the Morse Report.

A few months ago, when I was reading for this volume, I also read a paper by Lori Udall of the EDF and others. Before 1988 she, with John Clark of Oxfam, UK among others, had played an important role in lobbying with key officers, including WB's Executive Director,

UK to help formulate WB's historical R&R policy. (The Morse mission had no time for this important process at the Bank and other places. In 1992-93, I had seen Lori Udall in a new avatar. She and EDF were then active in the campaign against SSP. Reforms were no more their agenda. Demolition was.

But reading her paper in Fisher's book a few weeks ago, I realised for the first time that she and EDF were the prime movers of the international campaign. They conceived the grand strategy and coordinated the orchestrated campaign in the rich countries. Until then I had believed that Patkar and her colleagues/intellectuals in NBA had initiated and effected the dramatic U-turn in August 1988, and that EDF and others had followed the NBA lead, revising the agenda from one of reform to one of demolishing SSP.

Now, after 15 years, reading Udall's paper and re-reading some old material of 80s I saw that at least four months before the August 1988, declaration against SSP from Bombay, EDF's Bruce Rich, in a US Congressional hearing, had stated that the Narmada Valley had risen in revolt and tribals everywhere were opposing the project. I now have no doubt that it was EDF that was actually calling the tune and others like the NBA in August 1988, and the Morse Mission in 1992, were following its lead.

The role of the Supreme Court in the years following the exit of the World Bank from the Project also needs to be looked at in a similar perspective.

In those days the picture was still hazy, but the contours of an emerging new anti-development ideology were discernible. The Berlin Wall had fallen, the Soviet Union and its empire had vanished, and "Communist" China had become a ferociously capitalist country in all but name. India too had embarked on liberal economic reforms, bidding goodbye to Nehruvian Socialism. Marxism was no longer a live discourse and socialism was bewildered and in retreat.

A new ideology was struggling to be born, drawing its sustenance and vocabulary from Post Modernism which idolised the debasement of the language, ridiculed "facts," made intellectual dishonesty respectable provided the idiom was right and, above all, occupied a high moral perch. In the last 20 years this dangerous ideology has become more entrenched. The Morse Report gave us a foretaste of things to come.

13

What Do the Narmada Valley Tribals Want?

Anil Patel

Background

The Narmada is the largest westward-flowing river of peninsular India. It is rain-fed and carries an annual mean discharge of about 36 MAF, or about 4.43 MHM. Through most of its total course of over 1300 km it traverses between the mountain ranges of the Vindhyachal in the north and the Satpura in the south, before it enters the plains of coastal Gujarat to run its final 160 km to the Arabian Sea. Because of such a difficult and varied terrain, it remains one of the least developed rivers of India. In the hills of these ranges live tribal people. They are mainly of three types – Bhils (Vasavas), Rathwas (Bilalas, Pavras) and Tadvis.

In the 1960s, Gujarat and Madhya Pradesh (MP) the two main riparian states, started water resources development plans for the valley (Maharashtra has a very small stake). Gujarat wanted to build a large terminal dam to divert the Narmada water across several river basins to the drought-prone areas of North Gujarat and parts of Saurashtra and Kachchh to irrigate water-scare fields and for domestic consumption. A dispute inevitably arose between the two states as to how much water each could claim and for what purpose. Since the dam proposed by Gujarat led to the submergence of areas in the two adjacent states of MP and Maharashtra, the resettlement and rehabilitation (R&R) of the oustees also became an issue. To resolve it, the Government of India (GOI) constituted the Narmada Water

Dispute Tribunal, 1956 (NWDT). After exhaustive deliberations over ten years, the Tribunal gave its final decisions in 1979. These are binding on all the three states and GOI, and above appeal.

The Tribunal, its Decisions, and the Tribal People

The Tribunal's decision regarding resettlement and rehabilitation (R&R) was path-breaking in India. It explicitly provided for allotment of agricultural land to a landholding family (minimum 2 hectares) *in lieu* of the land it lost. This was a clean break from past practice, when landholders got only monetary compensation under the Land Acquisition Act. The brunt of the displacement would be borne by the tribals. The decision, however, failed to take into account the characteristics of tribal interests and their economy. The states complained about the displacement in their states for a project to be built in Gujarat, and Gujarat, which eventually was to bear the cost of R&R, did not bring to the notice of the Tribunal that the majority of the tribals do not formally own the land. They eke out their living by cultivating denuded forest land and are considered illegal encroachers by the State. The future displacement of these people by developmental projects nearly brought to the surface a serious and embarrassing problem of the unacknowledged tribals' right of access to the resources of livelihood in their ancestral land. The roots of this problem lay deep in history, when the British rulers arbitrarily extinguished tribal rights by taking control of the forests in 1860 and then imposing the Indian Forest Act in 1927. With some modifications in 1952, the government of independent India retained the Act and carried on with deep-seated, dubious legality. The tribals continued to bear the legal stigma of encroachers. The legal tangle is so complex that the states neatly skirted the whole issue of entitlement to the so-called encroachers. The Tribunal too remained silent on this issue.

Who Are the Tribals?

Tied up with the above analysis is the issue of the place of tribal people in Indian society. Are they "indigenous" in the sense the term is used in North America, Australia, New Zealand, and so on? Are they distinct races conquered by invaders who came to India over the past four millennia? The terms of debate about the resettlement of tribals because of the Sardar Sarovar Project (SSP) in Gujarat, or, for that matter, the issue of tribal welfare in relation to the development strategy in India, has not so far involved such questions, at least not before the Bradford Morse Commission (Independent Review – IR) appointed by the World Bank (WB) raised it in their report on SSP in June 1992. Although the Morse Report is not forthright and categorical in its assertion, it does assert enough to convey that they are indigenous people, and that WB had ignored its own guidelines on indigenous people in the case of SSP (Operational Manual Statement No. 2.34, February 1982). This triggered a heated debate about the place of tribals, and therefore, on the nature of their entitlement. A close reading of the Morse Report shows that, although it tilts in favour of the tribals being indigenous, its analysis is far from convincing and so is the prescription that they should have reserved areas to themselves as indigenous people in the United States and Canada do. On the other hand, some protagonists in the SSP controversy, like Amita Baviskar, whom the report extensively quotes with approval, have concluded that the tribals are not indigenous people. Indian anthropologists and sociologists are nearly unanimous in stating that such a distinction is not maintainable in India.

Economic Development and the Tribal Interests

This takes us to a deeper level. The founding fathers of the Constitution of India, while not considering the category of "indigenous" for tribals, did clearly acknowledge that they were among the most neglected subpopulations of the subcontinent and provided for explicit guarantees and protection to them in the Constitution adopted in 1950. Notwithstanding these provisions, the Tribunal and the contending states in their deliberations did not protect the tribals' rights. The then reigning concept probably dominated the Tribunal, that for the development of many, a few might have to sacrifice. Hence, the Tribunal did not give more refined consideration to the plight of mute tribals, who did not even know that their fate was being decided by those who claimed to represent the interests of all the people who had elected them to power. This formally correct mechanism hid serious anomalies and contradictions between constitutional provisions and the statutory law enacted by the British that preceded the Constitution. Given the limited scope of the inquiry, the Tribunal too could not have raised such a fundamental issue as a conflict between Constitutional provision and the Act. This obviously had to be done by the State and by the Supreme Court. The failure of a democratic State to resolve a serious problem of tribal rights, and its turning a blind eye to the problem, has led to loss of its face and has undermined its claim to represent the tribals' interests. This lapse brings, to the fore the central issues of what the tribals want. Who represents their real interests? And what are their aspirations *vis-à-vis* economic developments? Inevitably the answers to these questions are not straightforward and they should not be sought in quick trips to the area and in short interviews. They are to be found in the process that followed the Tribunal's decision in 1979.

PHASE I: ARCH-Vahini in the Sardar Sarovar Project and the Helpless, Passive Tribals

After the Tribunal's final decision in December 1979, the foundation work of the dam began in earnest in early 1980. Almost simultaneously, we began our work in the first group of tribal villages that were going to be affected in the following six months. WB evinced keen interest in the project and decided to enter in a bid to help modernise India's irrigation sector. The now-famous February 1980 WB guideline on R&R was already in place. A reputed consultancy firm from Bombay was asked to carry out a cost-benefit analysis and, in February 1983, the project was sent to the Ministry of Environment and Forests (MoEF) for environmental clearance.

The tribals knew nothing about the project and their imminent displacement. The landowners were served land acquisition notices, but they knew little about their entitlements. They knew little about the place where they were going to be resettled. The administrative setup for R&R was in a pitiable condition. The administration was informed by a smug belief that, as in all major irrigation schemes in Gujarat and elsewhere in India, the project-affected persons (PAPs) would just have to move, accepting whatever was offered to them. This indifference and apathy, no doubt, derived its inspiration and strength from the top echelons of the administration and the political leadership. A minister-in-charge was quoted as saying that the government didn't have to move a finger to resettle the tribals; the tribals would leave their habitat like rats from their holes when the water rose. The elected representatives of the people were nowhere in sight. People dared not approach them with their woes. Everywhere in the submergence zone in Gujarat and the other two states, nobody thought it necessary to properly inform the people about the displacement, the reasons for it, and their entitlements. Almost always they came to know about the project and the future inundation of their land and houses through surveyors who were pitching the boundary stones of the reservoir. Fear of displacement and an uncertain future was visible on their faces. The panic was palpable.

When we first approached them, even in the middle of panic and desperation they tried to size us up and wondered, "It is futile. What can these few young, inexperienced persons get for us from the clutches of the mighty government?" But then we were their only choice and hope. In their robust and practical sense they also decided, "no harm trying them". The first tenuous contacts were thus established in July 1980. We were aware of our inexperience and powerlessness. Needless to say, we knew nothing about the maze of land laws, about the Forest Act, about the Tribunal's decisions on R&R. We did not even know of the WB guideline; and even if we had known, what good could that do? The World Bank, even to us, was distant, remote, and powerful – a mere abstraction. If we had then known of these intricacies and the legal tangles, we would have been repelled by the sheer immensity of the task. Happily, we did not know this and got engaged in what can be called, even in retrospect, an unrealistic and uneven battle with the State. To be sure, there were sceptical well-wishers who repeatedly warned us about the futility of it all. We persevered because that was the right thing to do and because, strangely enough, the tribals persevered too. The news media gave us big coverage. The government machinery was forced to take some notice of our activity. The local politicians descended on the scene and started chiding people in one breath and wooing them in another: "Who are these rootless, city-based young individuals? Don't you see that only the government can deliver anything substantial? And don't you know we are the government? And who are you, after all, but tribals? We are your leaders and only we can get things for you." Even the then minister-in-charge of the Narmada Project – himself a tribal – used this argument in a series of meetings in the submergence villages in 1984, when the strength of the movement had reached a new high. People nodded their heads in wisdom. They even shared these feelings. But they also noticed with amusement "where were these leaders when we were voiceless. It is only because of these young people that they are here." The alliance grew, their confidence grew, and so did our intimacy with them.

By 1983, the first group of villages had accepted the R&R package, which fell far short of even the Tribunal Award. The encroachers had no choice. They were repeatedly called thieves of the government by responsible officers. Even the legitimate landowners, whose names were not on the government records, because of no fault of theirs, had no land entitlement. The R&R policy was deeply repressive, and it kept out the majority of the tribal families. We were reduced to monitoring the implementation of this bleak policy. The government slipped up and committed a silly mistake of retracting a written promise an officer had made. For the first time the tribals accepted our advice to move to the Gujarat High Court, which earlier they had always rejected because they did not believe that mighty government could be forced. Highly symbolic though this gesture was, it was taken out of sheer desperation. The case was a clear breach of promise – of promissory estoppels. The Gujarat High Court would have almost upheld the tribals' complaint. The government was forced to compromise. The effect was dramatic. Through their practical experience the tribals had come to know that the government was after all not invincible, and that the law could be used for their benefit as well. Organising the tribals of the most remote interior villages of Gujarat, for a higher scale of demand, became possible for the first time. This was the beginning of 1984.

Scudder, the World Bank, and the Encroachers

In August 1983, we had already written to the World Bank arguing for the encroachers and for all the eighteen-year-old and older adult males who were legitimate owners of the ancestral land through the Indian Succession Act. Earlier we had received information that the World Bank was not merely going to fund the project, but was also looking into its R&R aspects. Our letter was partly a blind shot, but it was a bombshell in the World Bank as we discovered about ten years later.

Apparently, the Bank was fed only part of the information and knew nothing about the encroachers. For the Government of Gujarat (GoG), for the other two state governments and for GoI, the encroachers' entitlement was a non-issue. The Tribunal knew about this and did not pass any order. The tension between the constitutional provisions for the tribals on the one hand, and the Indian Forest Act of 1927 and the Forest Conservation Act of 1980 on the other, was at best an irritating abstraction for the governments concerned. The Ministry of Environment and Forests (MoEF), then and later, was in no mood to even acknowledge the existence of the tension, and continued to emphasise the power it possessed through the two Acts.

WB, however, took no recourse to such legal technicalities. Professor Thayer Scudder of the California Institute of Technology, a well-known international expert on R&R, was asked to lead a Bank mission to India. The mission arrived towards the end of 1983 and confirmed all the facts we had made complaints about. His November 1983 report revealed a keen insight into the critical aspects of the tribal situation. Among the important recommendations he made, one was related to the encroachers' entitlement and another was related to the release of the denuded forest land for resettlement should the tribals insist on that. He had noticed the resistant stand taken by the Forest Department for the release, but felt it was necessary to make a recommendation. Its significance was to prove momentous, as we shall see. The report is the most accurate expression of the then prevailing World Bank guideline of February 1980, and in fact goes well beyond it to reveal other major elements that were to be incorporated in the June 1990 guidelines of the Bank. The encroachers' entitlement became a substantial issue that the state governments could no longer ignore, irrespective of the Indian Forest Act and the Forest Conservation Act.

PHASE II: The Encroachers Demand Land

In village meeting after village meeting, the tribals assembled in large numbers. They spoke with feeling and anger: "The government provides no jobs, no facilities, no education and no transport. We are left to fend for ourselves and still they call us thieves. The only livelihood option we have is agriculture. We cannot be thrown out of our land, encroached upon or not, and be deprived of our source of livelihood." The encroachers were in the majority. Their voices became more strident. They began rejecting the label "thieves" given to them by the officers, which they had at first meekly accepted and internalised. They now understood how they were so branded in the first place. This change from apology to assertion of their right was slow, painful, and faltering, but it was unmistakable. The bolder and more imaginative among them spoke out for all: "We cannot be treated like monkeys on trees, who will simply climb up to a higher level when the water rises. We demand land – for all the families. It is no use merely talking in these mountains, in the middle of nowhere. We will not be heard. We must organise, go out and let the world know that we are also human and must also get our due before you flood our lands, our homes."

Unsure of what all this implied, we decided nonetheless to organise a protest march of all the tribals of Gujarat and the adjacent villages of Maharashtra on March 8, 1981. Even at this stage, we knew little of the developments in the World Bank, its guidelines, the impact our letter had made there, and the impact of Scudder's November 1983 report. The rally's main demand concerned the encroachers. Such a rally in the area was unprecedented. The encroachers had written an agenda in complete ignorance of the developments in the World Bank. Newspapers flashed front-page stories with photographs. A friend in the state capital wrote, "Your arrow has hit the bull's eye." GoG was rattled. We did not know the reason that we now know. The last thing GoG wanted, at that juncture of delicate negotiations with the Bank, was such an organised show of strength by tribals refusing to accept whatever crumbs the mighty government was throwing them. Unsteady on their feet, they now had an agenda and were pushing it high. From this position, they never looked back.

John Clark of Oxfam (UK), the then development policy adviser, took up the issue of R&R with the World Bank and mounted an intensive, sustained campaign in close collaboration with ARCH-Vahini. He mobilised support of many international NGOs who became aware for the first time of the tribal displacement due to SSP. The R&R of the tribals of the SSP became a cause célèbre in the Bank. Oxfam (UK) became a strong link between the concerned officers and executive directors of the Bank and the tribals of the

SSP. In the hottest summer months of 1984, a month after the protest rally, the then Minister of Irrigation, himself a tribal, held a series of meetings in the submergence villages and promised that all the encroachers, and all the sons eighteen years or older of the landholders, would get a minimum of two hectares of land. This was the first major public announcement of this kind.

Towards the end of 1984, WB had entered final negotiations with GoI and the three state governments. R&R was the critical piece of negotiation. Scudder was playing a key role. He came to India once again to get a final feel of the situation in September 1984. We met for four hours. He urged us to tell him everything. He wanted all the details and facts, so that when a final R&R agreement was worked out, all the major issues related to tribals were taken care of. The R&R plan of the loan agreement bears all too clearly the imprint of Scudder's assessment and efforts. WB's R&R guidelines could not have been better translated into practice.

From Scudder's November 1983 report and from his letters to us and to Clark it becomes clear that the encroacher issue was in all probability the chief bone of contention. GoI and the state governments put up a stiff resistance to the Bank's proposals that encroachers' cultivations be regularised and they be given generous entitlements of land. This was the only way that the overarching principle of the Bank's guidelines, that the standard of living of the displaced people be regained as soon as possible after the resettlement, could be achieved. We do not know the details of what transpired between this and the final formulation of the terms of the loan agreement, but the agreement (May 1985) refers to landless "oustees" – a broader term that includes encroachers. The landless were to be resettled in agricultural or non-agricultural sectors and were entitled to stable means of livelihood so that their standard of living would not deteriorate. We shall presently see what this meant in actual practice, but the encroachers had scored a major point. The states at last had to concede their right to entitlements which until then had been stubbornly denied.

ARCH-Vahini and the Tribals in the Supreme Court of India

In Gujarat, after waiting for months for the new policy promised by the minister in April 1984, the tribals and ARCH-Vahini concluded that there was not going to be a promised policy. The government was clearly biding time. A massive protest was staged at the dam site in January 1985. We decided not to wait any longer and moved the Gujarat High Court. It issued an interim injunction staying displacement before a satisfactory R&R policy was worked out. But the writ was soon rejected on perfunctory grounds. However, the Supreme Court of India admitted our writ petition and issued an interim injunction not to displace, which was violated with impunity. GoG filed a series of affidavits during this time and bluntly rejected any claim of the encroachers for entitlement to means of livelihood. The encroachers were again termed as thieves. The provisions of the Indian Forest Act were invoked to support the state's decision. We were stunned by the statements, because by now we knew that GoG was speaking a different language with the Bank.

A Breakthrough for the Encroachers in the Loan Agreement

It is true that, when we first saw the terms of R&R agreement, we were disappointed by the vagueness of the provision for the encroachers. But our disappointment was soon dispelled when Oxfam sent us a copy of the letter that their then Executive Director (UK), Tim Lancaster, had written to John Clark on March 26, 1987. He was categorical: "With regard to the availability of land for oustees, the Gujarat Resolution of May and November 1985 (giving land entitlements to encroachers and landed oustees respectively) were designed to enhance, not to abrogate, the legal agreement with the World Bank. This was firmly established by the World Bank prior to declaration of project effectiveness in January 1986." Land entitlements to encroachers (one to two hectares) were already spelled out. Encroachers had finally achieved their tangible due through active intervention of the Bank. An informal alliance of Oxfam, Thayer Scudder, and a few key members of the World Bank had achieved a landmark decision for encroachers, granting rights that the states were denying them and that the Tribunal had overlooked.

The Supreme Court never really came to grips with the core issue of encroachers' entitlement. The petition was initially accorded urgency, but then was put at the bottom of the pile for unknown reasons. The World Bank obviously could not and did not change the laws of the country, but effectively managed to shield the encroachers from the adverse effect of the laws the Forest Department was all too eager to use.

The Government of Gujarat Exposed—Birth of a New Resettlement and Rehabilitation Policy in December 1987

The exposure of the double game of the Government of Gujarat was only a matter of time. This was done rather dramatically by the Bank mission in April 1987 to inquire, among other things, into the allegations of doublespeak that we had made against GoG. The tribals were getting weary because of the long wait and suspense in the Supreme Court case. They were losing heart and asking anxiously whether the land entitlements for the encroachers and for the adult sons were indeed coming through. The head of the April 1987 mission made a dramatic and uncharacteristic announcement at the meeting of one of the most backward and interior tribal villages in the presence of GoG officials. He said that whatever the government officers might tell them in the village, the government was legally bound to give a minimum of two hectares of land to each encroacher and each adult son. After this, GoG could put up little resistance. A wave of enthusiasm swept through the tribal villages. The formal victory came on December 27, 1987, when the minister, who had dithered for a long time and presided over the doublespeak of his officers, made a formal announcement of the policy. This was a direct result of the terms of the R&R package in the loan agreement, which the Tribals and ARCH-Vahini had consistently demanded since March 1984.

Tribals in Maharashtra and Madhya Pradesh: Before 1987

The process of organising tribals affected by SSP in Maharashtra and MP was begun in early 1986 by two organisations. Not surprisingly, the tribals there had the same type of problems as those in Gujarat. There too, encroachers were an issue and the tribals also wanted good quality agricultural land after being displaced. Their aspirations and demands were reflected in the memoranda these organisations were submitting to their state authorities.

In Maharashtra, the tribals had always wanted to resettle on the denuded forest land in the same region. The Maharashtra organisation – Narmada Dharangrast Samiti – made almost identical demands for resettlement as we had made in Gujarat. In addition they invoked the World Bank Loan Agreement to demand the denuded forest land identified by them. These demands were made village-wise and collectively. They were repeated many times in the subsequent two years. Repeatedly, the rigid attitude and intransigence of MoEF in failing to release the denuded forest land, thereby violating the loan agreement, were sharply criticised. MoEF persisted with this untenable position for five more years, and in fact castigated GoM for making such a request in the first place.

The Narmada Tribunal had ordered that the tribal oustees of Maharashtra and MP be resettled in the command area of Gujarat, but if the oustees chose to be resettled in their own states, they would have the right to do so. The respective states would arrange to resettle them with the same entitlements in their own states. GoG would have to bear all the R&R costs. In disregard of this decision, MoEF went further and started goading GoM to push all the oustees to Gujarat. The tribals in Maharashtra were vigorously protesting against this double violation. This was toward the end of 1987, about the same time that GoG succumbed to a combined pressure of tribals, ARCH-Vahini, the international alliance, and the World Bank.

PHASE III: 1988 - The Rise of the Anti-Sardar Sarovar Project Movement

The story of water resources development in the tribal area, wherein the tribals were denied their rightful entitlements by unfair use of legal instruments – by guile, fatiguing delays, and doubletalk – is straightforward in comparison to what followed next. Soon after the December 1987 R&R policy of Gujarat state was hailed as being revolutionary by all concerned, the organisations working in Maharashtra and MP stopped looking for ways to achieve similar policies in their states and started raising doubts about the Gujarat policy, stating that it was destined to be merely a piece of paper. There wasn't going to be enough land available, they concluded. Before long city-based intellectuals jumped into the arena, arguing vigorously that this was a charade, the tribals were bound to be cheated, and they must not trust the government. A US-based environmental organisation, Environment Defense Fund (EDF), which until this time had lobbied with the World Bank in concert with Oxfam (UK) and had supported the R&R demands of the tribals, suddenly changed course. In a submission to the US Senate sub-committee in June I988, it declared that the tribals in the valley were against the project, and the opposition to the project was widespread. The World Bank must not go ahead with the project without a reappraisal.

This was interesting, because the Maharashtra and MP organisations still had not declared their opposition to the project. When they did, in August 1988, it was as sudden as it was total. They declared that the project was environmentally disastrous and economically ruinous. They declared that the "tribals would prefer to be drowned by the rising waters of the dam, rather than [agree] to this destructive scheme by agreeing to shift." SSP became a symbol of planned development disaster for a developing economy like India.

We were completely taken aback, although we had sensed some rumbling against the project in the metropolises since 1985. We also knew that something dramatic from them was afoot. But we didn't believe that the organisations leading the tribals of the other parts of the valley for better R&R policy could even contemplate joining such a move, let alone leading it. We knew the tribals in Gujarat, although apprehensive and suspicious about the government's true intentions, were rearing to get the new policy implemented, that they did not want to join the battle against the dam announced from Bombay. This was natural. We were almost certain that the tribals in Maharashtra and MP would not want to abandon their quest and struggle for an R&R policy similar to that of Gujarat. And yet the battle against SSP was joined in the name of the tribals of the valley. Bewildered and anxious, we asked the organisations if they had informed the tribals, who had relied on them, about the revolutionary contents of the Gujarat policy, and if they then had asked for their informed consent. We also raised doubts about the strength of the anti-dam case they had advanced, given the fact that only a few months ago these very organisations had acknowledged that the case against SSP had to be developed, and had decided to develop such a critique of the SSP for the record, even when it was too late to stop its construction. In this context, we asked, "Was it a responsible activism to keep the tribal in front of the fight against the project?"

The Movement Grows, R&R is Impossible in Principle – 1989-90

In the flush of enthusiasm of their new strategy of a complete U-turn from the pursuit of a better R&R policy, and encouragement from the NGOs like EDF, our objections and arguments were easily brushed aside.

By 1989, fresh forces joined the fracas. The battle cry now was not only that the dam was a planned disaster, but that R&R was impossible in principle. A multisided comprehensive case was made out for the first time since the anti-dam movement began in 1988: the project was environmentally and economically ill-conceived; the R&R was impossible in principle; and the poor tribals and their cultures were being sacrificed so that a few rich farmers and industrialists in the plains of Central Gujarat might reap great benefits. A most potent argument was advanced for the first time that the command area of SSP was not predominantly drought-prone as claimed by GoG. The critique, wholly wrong on the substantive issues, but not refuted at the time because of the ineptness of the project authority and mounting hostility towards the project in the English press, gave a strong impetus to the notion of the poor tribals being made victims of a developmental strategy adopted essentially for the benefit of the rich and the powerful. In a matter of a few months, environmental and human rights groups in the United States, Western Europe, and Japan (led by EDF in the US) joined the battle against the development-gone-berserk in the Third World countries. Mechanisms were set in motion to render the R&R impossible. The anti-dam forces set out to fulfil their prophecy that the R&R was impossible.

The Ministry of Environment and Forests in a New Role

The R&R of tribals of Maharashtra and MP had taken a turn for the worse. Paradoxically, their plight remained in focus, but now the defenders of their interests were no longer struggling to wrest meaningful and substantial reforms from the unwilling and recalcitrant state. In the din and the dust of the battle then raging in the national and international metropolis, hardly anyone asked, "What do these tribals want?" For most, whose imagination was captured by the anti-dam movement, the question was unnecessary because the answer was self-evident: "How could it be different from what the activists in Maharashtra and MP had said it was?" Historically proven low credibility and the ineptness of the state along with strong support of the press only fortified this understanding.

In such a dramatically shifting context, MoEF assumed a very different role. It was no more portrayed as a dogmatic and conservative force denying the tribals their due by wielding the legal instruments of forest acts. It now donned a radically different mantle and with that the Bank's image was transformed. The Bank was made out to be a foreign interest out to destabilise the ecology of the country and its tribal culture in support of a disastrous project. MoEF now was tacitly perceived to be fighting valiantly the onslaught led by the World Bank on the ecology of the country. Its stance towards crucial tribal demands was pushed out of view.

In the mid-1980s, Scudder had fought hard battles with MoEF to win an important right for the tribals to resettle in the denuded forest land. This was hailed by the core organisations of the anti-dam movement – NBA in 1986-87 and even EDF – and was used to vehemently criticise MoEF as a violator of the loan agreement as well as the human rights of the tribals. Unmindful of this criticism, MoEF resisted the demand to release forest land in gross violation of the loan agreement and went even further to violate the Tribunal's directives, as described above. In the changed circumstances post-1988, these activists found the Ministry's intransigence very useful in the strategy of rendering R&R impossible. No protest march was ever organised, nor was any memorandum submitted to the authorities by the activist organisations of Maharashtra and MP to pressure MoEF to comply with the terms of the loan agreement and the Tribunal Award. GoM's efforts to get the denuded forest land released were successfully stalled by MoEF. In fact it was emboldened to tell off GoM for putting up such a demand. Scudder came in again in April 1989 after five years to review the situation as a part of the Bank mission. This was the early part of the implementation of the new R&R policy in Gujarat. We argued, for a change, that we were more assured than ever before that the political climate for R&R in Gujarat had taken a decisive turn for the better, that the tribals were satisfied about the correct interpretation of the policy and the will to implement it. He was sceptical. We agreed to disagree. Scudder, however, saw clearly that MoEF was stalling on a crucial point and so were the governments of Maharashtra and MP on other points, despite the provisions of the loan agreement. The governments of Maharashtra and MP had still not announced an R&R policy consistent with the loan agreement, giving land entitlement to encroachers, and MoEF had not released the degraded forest land. Swamped as he was by the NBA's critique, he did not lose sight of the objective: "R&R of the tribals of Sardar Sarovar Project." The mission made strong recommendations to amend these defects in a time-bound way.

The game was up for MoEF. In June 1990 it had to release the denuded forest land in Taloda, one of the sites demanded by the Maharashtra tribals in 1986-87. The organisations representing the interests of the tribals, now part of NBA, were furious. The World Bank office in New Delhi was the focus of their fury. They accused the Bank of pressuring MoEF, which was doing its best to preserve the fragile ecology of the forest. This was done in the name of the tribals. Not only this, NBA activists presented a memorandum to GoM on July 31, 1990, claiming to represent the views of 33 tribal villages of Maharashtra and stating that they rejected the Taloda forest land. The memorandum criticised the release of the forest land and declared, "The clearance is ... tantamount to selling India's ecological and social future at the behest of a foreign agency." EDF and other Northern NGOs were not to remain too far behind. Led by EDF, 47 NGOs, in a memorandum submitted to the World Bank in March 1991, criticised it for forcing on the tribals of Maharashtra villages the land they did not want in violation of the terms of the loan agreement.

NBA, together with EDF and other Northern NGOs, in this fateful act staked the most dramatic claim until then, that the crucial demands of the tribals had changed radically since December 1987 and that their true aspirations and demands were represented by these organisations. If they were right, the divergence of the interests and aspirations of the tribals of the same valley and of the same stock, of Gujarat on the one hand and of Maharashtra and MP on the other, could not have been greater.

Resettlement and Rehabilitation in Gujarat—Continuing Tribal Activism

Since the Government of Gujarat had conceded the demands of the tribals after nerve-wracking suspense, fatiguing delays, and stunning duplicity, neither the tribals nor ARCH-Vahini trusted the intention of the Government of Gujarat. The Government of Gujarat had done all in its power to stall a crucial clause in the R&R agreement, which the Bank had insisted on. This related to a right of land of preference to oustees, which in Gujarat virtually meant one thing – private agricultural land. GoG, in those fateful days of negotiations with Scudder before the May 1985 loan agreement was finalised, had resisted the demand for private land on the ground that such land in the command of the project was not available. This was a very plausible argument and even Scudder paused. However, when the tribals were informed about the GoG tactics, they responded in a most unusual way. In a matter of a few weeks, they gave us a list of landowners in the command who were ready to sell about 2,800 hectares of land at the right price. They further assured us that many more times this land was available for sale. The tribal initiative at a critical juncture was again evident. We passed on the list to Scudder, who produced it in the wrap-up meeting in Delhi to clinch the issue that the land of choice necessarily included private land.

Inevitably, the implementation of the policy got off on the wrong note. A major problem of interpretation arose as to whether adult sons of the landholders were entitled to a minimum two hectares. GoG officers insisted that they were not. The tribals, confident in the strength of their organised power and the justice of their demand, staged yet another convention on May 14, 1988 and issued a public warning that if there was going to be backtracking of land entitlement for the sons, the work on the dam would be stopped. For the first time, and also for the last time, the tribals of Gujarat raised, the slogan, "No land, no dam." GoG retreated in a hurry. In a crucial meeting on May 19, 1988 with the newly appointed chairman of SSNNL in the state capital, we were assured that there would be no misinterpretation inimical to tribal interests as perceived by the tribals and by us. The meeting barely lasted for fifteen minutes. As usual, we accepted the assurance. And that turned out to be a qualitative break with the past.

After this meeting we never had any conflict regarding the interpretation of the R&R policy. A long time passed before apprehensions subsided. Thereafter, the central role of ARCH-Vahini and the tribals has been to identify land, to have land prices fixed through the formally installed Land Purchase Committee (LPC) of which we are a permanent member, and to make sure that land parcels of two hectares each are allotted to each family. The tribals have taken the initiative to form their own groups to identify land and press for fixation of price and allotment.

Exercise of Choice

The right of choice formalised in the R&R agreement has in practice worked out as two kinds of choices – the oustee chooses the group he wants to resettle with, which in turn chooses the area, village, and land it wants to resettle in. This has not only ensured an extremely high level of participation of every oustee family, it has also taken care of other crucial requirements of the World Bank guidelines, that the oustees shall be resettled in groups of their choice and in the way that will minimise host-guest conflicts. Since the tribals identify a willing landowner in the village they want to settle in, the basic issue of hostile reception does not arise. Since the groups are formed with the utmost freedom, they have retained strong internal coherence, a sense of common purpose, even enterprise and identity, which we believe goes a long way to retain their own identity once they resettle at a new site. The combination of a strong material asset base with a coherence of the group is a much better guarantee that a painful and difficult transition from dislocation-induced resettlement to an enduring self-sustaining rehabilitation takes place within a minimum time.

From 1987 to 1992, Gujarat thus resettled 4,300 of 4,500 families. There have been moments of anxiety, heartache, doubt, and so on, but eventually the outcome has been in the right direction. The tribal groups have continued to take strong initiatives. They have demanded and received further improvements in policy like advancing the cut-off date for sons from the original 1980 to January 1987. Families getting physically or socially cut off due to the reservoir or displacement of neighbours have also earned a right to be resettled, and have been firm in demanding amenities like drinking water, roads, electricity, and schools. The list is long.

The Tribal Divergence from Narmada Bachao Andolan: Aims and Objectives

As resettlement work in Gujarat was climaxing and work in the first nine villages in Maharashtra was picking up momentum, we started receiving accounts in early 1991 that tribals in Maharashtra and MP were totally in the dark about the Gujarat's new R&R policy and its true magnitude of implementation. NBA was vigorously propagating in these villages information that was at complete variance with reality. From the days of the first opposition to the dam, the tribals were given to understand that the opposition was mainly to get a better R&R deal, and that the plight of the displaced in Gujarat was pitiable. But this wall of disinformation and misinformation could not be maintained after the beginning of 1991. The real information slowly but steadily began to trickle in. The real discovery that shocked the interior tribal villages of Maharashtra was made when they first contacted us in April 1992, when we informed them that the Taloda forest land was released as far back as June 1990. The NBA had kept them completely in the dark about the release and, as observed above, actually had made represented in writing to the authorities on their behalf that the tribals did not want the land. The tribals were stunned. The manipulation of their demands and aspirations was as shocking as it was revealing. In a public meeting on April 3, 1992, they protested in anger to the Chief Minister of Maharashtra that his government could accept such a false claim. How could it allow such a serious mischief undermining their vital interests? The NBA, in its effort to stop the dam, had now joined hands with MoEF, once its strong adversary, to deny the tribals the means of livelihood they had demanded. Their claim to represent the tribal interests was decisively shattered that day.

NBA, with ample support from supposedly independent institutions like TISS, Bombay and even the Morse Commission, has continued to raise problems about the Taloda forest land. The tribals in Maharashtra have learned the hard way, through actual experience, that obstacles to their R&R have been created not only by the state government or MoEF but also by activists in tandem with others who continue to claim that they represent the true tribal interests and aspirations.

Tribals and Resettlement and Rehabilitation in Maharashtra and MP

The story of R&R of the tribals of Maharashtra and particularly MP got immensely complicated in June 1988 because pursuit of R&R was dropped. The central task of encroacher entitlement had been more or less achieved in the R & R agreement, but its formalisation into actual R&R policy was missing at least two years after Gujarat announced its policy. This added to the anxiety of the tribals and considerably helped the anti-dam movement. The situation changed after the Bank's mission of April 1988 and Scudder's recommendations. Some 2,700 hectares of Taloda degraded forest land in Maharashtra was released. R&R policies for both the states were adopted. The tribals realised that the game plan of NBA was different from what they wanted. There were many problems in the policy documents, but no pressure from NGOs. They persisted in the state and got further policy changes. These, however, came about slowly and with difficulty. By February 1992, GoM had mended its policy on tribal areas. MP, which had always dragged its feet in R&R and had always seriously objected to Gujarat getting one-third of the river water, proved the most recalcitrant.

This provided a great advantage to NBA to fight its anti-dam battle from MP. As long as WB was in the picture, at least until March 1993, GoMP had to improve its policies in many areas, especially with regard to landless agricultural labourers in non-tribal villages and post-1987 encroachers in the tribal areas. After that GoI withdrew the project from the World Bank, realising that the Morse Commission had induced an onslaught on the Bank and on the project with the decisive backing of executive directors of some key industrial countries, notably the US. Those like the United States Executive Directors, who were unsparing in their criticism of the Bank's handling of R&R, accepting the Morse Commission's lead, did not achieve anything from the Bank's exit except further deterioration of R&R prospects of the tribals in Maharashtra and MP, because neither

NBA nor MoEF nor GoMP would think much about R&R in their common pursuit of blocking the dam.

One relieving feature for the tribals of MP has been GoG's willingness and ability to offer land of choice in the SSP command. Thus, in the face of the strong opposition by NBA, encouraged and often backed by GoMP, a number of tribal families have selected land in Gujarat. This constitutes, in our reckoning, a sizeable proportion of tribal families from MP. We know that many of these families and many more of those who still remain in the reservoir area would have preferred denuded forest land in MP; but in the present circumstances, with the World Bank out and the key executive directors' poor grasp of reality, the tribals have little chance of getting lands released from the clutches of MoEF. Violations of the loan agreement and the Tribunal's decision go on unabated. The hurdles are insurmountable. Unknown numbers of tribals had to make a second-best choice. In Maharashtra, fortunately, the displaced families do not have to make such forced choice despite the MoEF. There, the tribals still face an indifferent NGO movement and an ever-so-obstinate and scheming MoEF. The second instalment of 3000 hectares of degraded forest land, which WB persuaded GoI to release before its exit in March 1993, was effectively stalled by MoEF until the very end. And this late release, when it came through in February 1994 because of joint pressure from the Chief Ministers of Maharashtra and Gujarat on the Minister of Environment and Forests in the presence of the Prime Minister of India, was half-hearted. MoEF released only 4500 hectares. This caused a critical delay in the allotment of alternative land. Now MoEF, which is responsible for the delay and dislocation of orderly R&R, has turned into a judge – and has sent out a fact-finding team in the area to study the status of R&R. In spite of these hurdles, out of 3,000 tribal families 650 have selected land in Gujarat and 800 in Maharashtra. GoG continues to offer the same R&R package to the Maharashtra tribal families as is offered to those from Gujarat and MP.

Resettlement and Rehabilitation: Tribal Aspiration and the Morse Mission

The World Bank appointed Commission of Bradford Morse and Thomas Berger fared no better. During the course of their inquiry, we provided the documentary evidence to Berger as to how the tribals of Maharashtra were denied their rights by MoEF, how the NBA had made a U-turn, turning from a strong adversary of MoEF into a strong ally, how the tribals were cheated all along and how a systematic effort was being made to render R&R impossible. Berger heard the whole story, and vaguely concurred with our analysis, but in the report, the story was ignored and its implications for the fair R&R of tribals were forgotten. The Morse Report went off on a tangent and ended up criticising the World Bank for the sins it had not committed – the "original sin" of non-appraisal of R&R of SSP as pithily described by the Morse Commission. This "original sin," as we have shown, has no basis in reality.

Human Rights – The Forest Act and the Morse Report

For all their eloquence about tribal culture, the tribal economy being rooted in the culture itself and the forest wealth with which they live and from which they draw sustenance, the Morse Report has failed to draw correct conclusions from the pertinent observations they have themselves made.

The report candidly records that in the submergence zone of the valley, the forest is already degraded with low biodiversity and productivity. It also records that encroachment is illegal, that the Tribunal Award made no provision for the encroachers and ignored their customary rights as tribal people. Sidestepping the relevant analysis of the illegality of the encroachment as rooted in MoEF's negative conservatism, the Morse Commission has tried to hold the Bank responsible for not getting GoI to accept encroachers as landed oustees. It never occurred to them that it was an impossible task for the Bank. The Bank did its best under the circumstances to shield the encroachers from MoEF's obstinate stand. The fact that the Forest Act holds sway in the tribal areas is not acknowledged by the report. Given these two facts, the existing life of the tribals is itself highly precarious. What can one say of the future when forest deterioration is bound to continue and MoEF is bound to deny the tribals any economic rights in the forest area? The tribals will be forced to cultivate the vacant land and the Ministry will continue to invoke the Forest Act, treat them as encroachers and try to evict them. The description in the report of the vibrant forest-based tribal economy is appealing, but is devoid of reality.

Those who speak like the Morse Commission of tribal cultural economy in the middle of the rich forests have not faced the questions: Is there a reality to this choice? Is it open to them given the MoEF's stance? To be sure, in June 1987 a strong presentation of a similar kind of idyllic life style of the tribals of SSP was made to Tim Lancaster, an executive director of the World Bank. He carefully refuted the argument in September 1987 by saying, "Villagers in virtually every village were aware of their rights and recognised that a forest-based mode of production cannot be sustained with or without the project and expressed their desire to be relocated as plains cultivators."

Michael Cernea, a senior sociologist in the World Bank writing to John Clark in May 1987, argued in similar vein to say that on the basis of facts he had assessed in the field, and corroborated by others, he had reached the conclusion that land entitlement was in the short- and long-term interests of the tribals.

What Do the Tribals Want?

The tribals facing the reality in their habitat have spoken with their action. Tribals of all the states have shown clear preference for good quality land, security of tenure, and greater access to market and other infrastructure like roads, transport, electricity, schools, health services and so forth. That their access to markets and their integration with the market economy is limited is true enough. But the real questions are: Are they averse to free access to markets given the solid assets of land they have uniformly and consistently demanded? Are they effectively handling the market? Will this destroy their culture? Since "power" in the objectionable sense of the word – that is, the capacity to direct the energy and resources of others to the services of values that those others do not share – has a strong tendency to be misused, the "final question" and the dilemma of the tribals is how to know what they want. Who is to decide this?

14

The Independent Review: Was It a Search for the Truth?

Anil Patel and *Ambrish Mehta*

Introduction

Ever since 1980, when ARCH-Vahini began a single-handed battle to secure a just Resettlement and Rehabilitation (R&R) Policy for the tribals affected by the Sardar Sarovar Project (SSP), we hoped for the objective and sensitive intervention of the Non-Government Organisations (NGOs) of the North to help further the human rights of the tribals and environmental concerns without causing an outcry against "foreign hands" or fears of "interference" with the sensitive issue of "sovereignty". Until 1987, interventions by such groups as Oxfam (UK) or reports by individuals like Thayer Scudder were fact based, sensitive, and promoted the tribals' human rights. From 1988, however, this changed. International groups rejected the need for objectivity and sensitivity, and their large-scale interventions created a mega-controversy. The intervention by enquiry commissions like that of the Independent Review (IR) headed by Bradford Morse continued the pattern, raising questions about the very scope and validity of such interventions in developing countries. The tribals of the SSP deserve better.

In 1992, the SSP controversy had entered its sixth year, if we disregard the mild rumblings against the Narmada valley projects before 1988. The controversy, though widely perceived as a battle between grassroots and resource-poor activists on the one hand and a mighty government on the other, is complex and multilayered. The involvement from the beginning of powerful interests, including parts of the Indian government, the English-language media, and some of the most influential international NGOs has remained largely hidden from view. The controversy has reached new heights with the publication of the Report of the IR (1992). This report, contrary to widely held expectations that it would clarify issues, has only added more twists to the controversy.

When the attack on the SSP began, it was as a classic case study of the way things go wrong with large dams. A general critique built up against large dams by Goldsmith and Hildyard in their new-classic *The Social and Environmental Effects of Large Dams* (1984) was applied to the SSP point by point and critics claimed that it fitted extremely well. The SSP continues to be subject to wide criticism along the lines laid down in the Goldsmith and Hildyard paradigm. However, it is not unusual—in fact it is quite common—to come across criticisms of the SSP that take the form "Isn't the SSP a large dam? And aren't all the large dams, bad dams?" This argument ignores the fact that SSP has had to provide detailed empirical evidence to corroborate the general critique of large dams. The truth of the general theory cannot be assumed and then invoked in the specific case of SSP. But while the logical error in the argument that all large darns are bad dams and, therefore, SSP is a bad dam is obvious, the wide acceptance of the argument goes almost unnoticed.

The critics of the project have persistently pointed at the mistakes, lapses, and inadequacies that were found at one stage in the evolution of the project as evidence to prove their point, and they have conveniently ignored information about later corrective developments. Even the Report of the IR has not been able to avoid such a serious methodological flaw.

Having followed the critique through time, we are struck by the fact that, while at one time, the project might have been found wanting on various counts, damning facts have been marshalled to reach the conclusion that the project is a "planned disaster". And yet, with the passage of time, the original "facts" – which allowed the critics of SSP to reach negative conclusions – have been quietly dropped without any explanation or even acknowledgement, but the conclusion that the project is a planned disaster remains. Now new facts are brought to the fore to support this conclusion. The IR report follows this pattern. It has faulted the project on issues of R&R and on its negative environmental impact, as have national and international critiques of the project, but then facts and reasons in many instances are of an entirely different order. For instance, while their conclusions support those of the Narmada Bachao Andolan (NBA), the IR did not incorporate the facts and criticism cited by NBA in their report. This in itself is not an argument to refute the conclusions of the IR or, for that matter, any conclusions. What is needed then is a critical examination of the facts, a sifting of the evidence, and an analysis of these conclusions. We believe this critical evaluation of the controversy will demonstrate that it is being kept alive and given strength by various forces, without regard to the sanctity of facts and reason.

The issue is no longer merely whether R&R in the SSP is possible, nor whether SSP is an environmentally and economically sound project. Instead, the debate is being broadened to include, among other things, alternative lifestyles and developmental strategies on the basis of the SSP critique developed by the anti-dam movement, NGOs, and the IR. It is based on the assumption that the basic critique of the SSP is valid. We feel that the SSP is incorrectly treated as a paradigmatic case study of the way things go wrong with large dams and how development should not take place. To place the SSP into the same category with proven environmental problems such as CO_2 build up, ozone depletion, and the rain forest crisis, as some prestigious publications have tried to do, is a mistake.

Other Criticisms

One potent argument invoked on a different plane from the Goldsmith-Hildyard paradigm by the critics of the SSP is that, because of government corruption and inefficiency, no programmes (including those of R&R, afforestation, catchment area treatment, and the construction and maintenance of water delivery systems) can be properly implemented despite good design and policy on paper. They also argue that if past performance by the government has been so dismal, there is no guarantee that things will change for the better.

This apprehension has strong merit, but it should be noted that, in fact, the SSP has been different from other projects from the start.

The NGO ARCH-Vahini moved into the project area before the actual construction began, organised the oustees, and waged a seven-year struggle for a better R&R policy. The case went to the highest court in the country, and it was lobbied intensely with the World Bank. Finally, ARCH-Vahini and the oustees of Gujarat were awarded almost verbatim the policy that had been demanded. If a greatly improved policy had not been agreed upon, the project would have come to a standstill in 1988. Such a process had never been followed before. The new R&R policy of December 1987, with its many subsequent improvements, is a historic breakthrough, unprecedented in India. It provides for the allotment of a minimum of two hectares of irrigable land of choice to each oustee family, and, by early 1994, nearly 4,200 out of 4,500 oustee families in Gujarat had been allotted land. Provision of land for the remaining families was in progress.

Also unprecedented is the design of the SSP canal system and delivery system which is planned in an entirely different way so as to remove or minimise those very problems of land degradation in the command area and the inefficient use of water that the large dam critique predicts. But despite these innovative design features, the critics have argued that the problems have not been anticipated or that studies will not be done as planned to ensure corrective steps. Based on these critiques, opponents of the SSP say it was no different from other big irrigation projects and that it would face the same problems as in the past. The IR report took note of these novel features of design and planning, but concluded that they were at best intentions, and they went on to assert that no studies were being conducted that would help to bring the design features into reality.

Another important counterargument to the pessimistic position that "the future will replicate the past" concerns the tribal oustees of Gujarat. Historical experience notwithstanding, when told of the new R&R policy in December 1987, they opted to brace themselves for its implementation. So much was at stake for them that they could not afford the luxury of pessimism. As we will see, the same applies to the tribal oustees of Maharashtra and Madhya Pradesh (MP).

We at ARCH-Vahini find ourselves caught between the extreme pessimism of the most radical anti-dam activists and the optimistic predictions of the government. While we argue against pessimism, we also retain a strong measure of caution regarding the government's working style and the problems that can be created by bureaucracy. In the final analysis, our conviction is that as difficult as these problems are, they must be solved. Otherwise, what is the alternative?

SSP has been seriously criticised on another ground, namely that water from it will only make richer those who are already rich. Critics argue that irrigation benefits will go to a command area in which three-fourths of the land is already richly endowed with water resources, and of which only one-fourth is drought-prone. The project has been portrayed as a conspiracy of rich farmers, industrialists, corrupt bureaucrats, and politicians against poor, illiterate tribals. Activists have quoted the Sardar Sarovar Narmada Nigam's (SSNNL) and the Government of Gujarat's (GoG) documents to prove this point, creating the impression that tribal people, their culture, and their economic base were being destroyed to sustain and improve rich farmers' life styles. Our opinion is different. When we thoroughly examined these documents, we found either that they simply did not contain the clinching evidence purported to be there, or that they were misquoted and misinterpreted. Moreover, other more relevant documents that showed the true extent to which the command area is drought-prone are ignored by the critics.

Environmental Issues: A Critique

Specific criticism of the environmental aspects of the SSP surfaced in the second half of 1987. On June 24, 1987, the MoEF of the Government of India (GoI) issued a joint clearance letter (often referred to as the conditional clearance letter) on both the SSP and the Narmada Sagar Project (NSP) in Madhya Pradesh. This letter, together with other documents from and involving the MoEF, provide the core of the environmental critique of SSP.

While these documents are always referred to in the footnotes of the critiques, the form, contents, and context of these documents have never been the subject of close scrutiny, and the accuracy of the basic facts and analysis contained therein has remained unchecked. Until this is done, it is not possible to assess and answer the pointed questions the critics have been asking. For example, is it true that even though the SSP was not ready for environmental clearance in June 1987, it was given clearance because of political considerations?

Is it true that while there had been interaction with the SSP authorities for "almost three years ... the absence and inadequacy of data on some important environmental aspects still persisted"? Is it true that environmental clearance lapsed in September 1990 because conditions were violated? Is it true that the Narmada Water Disputes Tribunal (NWDT) merely assumed the quantum of water available in the Narmada? Is the economic viability of the SSP seriously jeopardised because the environmental cost of submerging forests alone turns out to be Rs. 80 billion, according to the MoEF's computation?

The IR team, conducting its inquiry in late 1991 and early 1992, was shown these documents and the following analysis. To our surprise, they have apparently not considered our reasoning and analysis and have only referred to these documents in the footnotes of their report, following the pattern set by other critics of the SSP.

The MoEFs' Clearance

The June 24, 1987, joint clearance letter for both SSP and NSP is a brief, two-page document (MoEF, 1987a). This letter states that both projects were examined in detail and that more details were sought from central and state authorities. There follows a list of eight broad topics: Catchment Area Treatment (CAT), Command Area Development (CAD), Compensatory Afforestation (CA), Resettlement and Rehabilitation (R&R), Flora and Fauna, Carrying Capacity of the surrounding area, Seismicity and Health Problems.

The authorities had applied for environmental clearance for SSP in February 1983, followed by that of the NSP in October 1984. The SSP application remained pending for more than four years, during which time, the letter states, it was examined in detail. The NSP application went through a similar process for three years. After such a long deliberation, if the MoEF still had questions about the above-named major environmental issues, and consequently wanted specific studies related to these questions to be done as a condition of issuing clearance, it does not seem unreasonable to expect that studies asked for in the letter would have reflected this by highly differentiated treatment of the issues. Moreover, since the NSP and the SSP would have obviously differed so much in details, it would seem logical that separate letters of clearance (or at any rate, separate sections for each) would have been necessary. The joint clearance letter states only general and vague conditions and lacks any kind of specificity.

The anti-dam movement, in its budding stages in large cities like Delhi and Bombay, seized upon this document and, disregarding its vague conditionalities and lack of specificity, held it as clinching evidence that the SSP was a "planned environmental disaster". Unfortunately, this document has thereafter retained its central place in various critiques published in India and abroad. It also remains a pivotal piece in the IR report. Careful scrutiny of this vitally important document, with all its striking and strange features, is called for. Criticism of the MoEF's non-specificity, however, would have remained only tangential had it not been for another document – a note to the Prime Minister prepared by the MoEF only about a month earlier, in May 1987.

MoEF's Note to the Prime Minister

Around May 1987 just prior to the environmental clearance of the SSP and the NSP, the MoEF sent a note to the Prime Minister of India (see Alvares and Billorey, 1988: Appendix 3, pp 107-117). It consisted of separate fairly detailed analyses of the two projects, addressing major environmental concerns like CAT, CA, Flora and Fauna and R&R. In each of these important areas except CA, the note had specific and fairly serious criticisms to make of the NSP but offered no criticism of the SSP except a moderate one concerning R&R.

Specifically, the note stated that the proposals submitted for compensatory afforestation in both projects were detailed and satisfactory. Indeed, not only was the environmental assessment of both projects different and differentiated, but, according to this note, the SSP turned out to be better prepared on these counts. Sections viii and ix of the note repeatedly stated that the SSP's preparedness was quite good while the NSP was not ready for clearance. When referring to this note, the anti-dam movement has dropped the words "the NSP" from the last sentence, substituting "the Project" (meaning the SSP) to make it read, "the Project is not ready for clearance".[1]

1. Medha Patkar in *Bhoomipuira*, a Gujarat periodical on Gandhian and Vinoba thoughts, February 1991 issue.

Similarly, a statement in the note about the "absence and inadequacy of data on some important environmental aspects" applies in its total context to the NSP, and yet the IR report has applied this to the SSP (224).[2] A comparison between the MoEF's note to the Prime Minister and its conditional clearance letter reveals a number of significant inconsistencies. These were ignored by activists of the anti-dam movement, who had access to the MoEF's note to the Prime Minister at the same time as they were highlighting the MoEF's conditional clearance letter. Ironically, while state governments did not get copies of the note to the prime minister, the letter was widely circulated among the activists and published only by them (Alvares, 1988: 107-117). This politics has been hidden from view.

In February 1992, during a day-long meeting with Thomas Berger, Vice-Chairman of the Independent Review, we showed him and two of his senior advisers these and other documents. Berger lost no time in grasping the central deficiency of the conditional clearance letter, remarking, "This is only a list of topics without any substance. . . . This is no conditional clearance." His conclusion could only have been strengthened when he saw the relevant parts of the note to the Prime Minister. If he had any reservation about the meaning of the two documents, he did not reveal it then, nor is it revealed in the Report of the IR. What is described in the report is very different. It points out that the 1983 application for SSP clearance was supported by a short-term benchmark survey by the Maharaja Sayajirao University (MSU) of Baroda. It holds that the information provided was not enough to meet the environmental guideline requirements so that environmental clearance was not granted (222)[3] . The guideline

2. Ministry of Environment and Forests, Government of India. "Environmental Aspects of Narmada Sagar and Sardar Sarovar Multi-Purpose Projects." Note to the Prime Minister. (In Alvares, 1988: Appendix 3. 107-17). The Report of the Independent Review has also quoted in this portion of the note, applying it to the Sardar Sarovar Project (IR, 1992: 24).
3. One of the senior advisers to the Independent Review, Don Gamble, during the course of leisurely discussions with us, had argued along the line of anti-dam activists that the MS University Report is not based on primary data but only on secondary data. We showed him that the Report was based on the primary data collected from the field in most of the investigations.

it is referring to is mentioned very vaguely in the Report as "first prepared in the late 1970s" (216).[4]

Clearly, there is a need to describe in more detail the ways in which the MS University study fell short of these guideline requirements. Moreover, the relevance of this pointed observation in a crucial letter to the World Bank president (xxi) needs to be explained in view of the fact (acknowledged by the independent Review) that in a letter to the Prime Minister, MoEF had described the SSP as in an "advanced stage of readiness". By more or less following the line of the MoEF and the anti-dam activists, and by failing to discuss the inconsistencies in the MoEF's documents, the IR report failed to shed light on the tangled issue of conditional clearance and the circumstances surrounding it. Thus, the entire issue still needs to be untangled.

Obviously, the MoEF, the clearing agency, was itself not fully prepared to perform its task. That its own understanding of environmental impact issues was undergoing evolution is made clear by the succession of guidelines that emerged both during and after the clearance application was being processed from the late 1970s through 1985 and 1989. Standards were changing and the circumstances for making ground rules for clearance themselves were in flux. A description is needed of the salient features of the guidelines of the late seventies, which according to the IR report compelled the MoEF to withhold the clearance, and of the 1985 guidelines which became operational while the clearance was pending.

In fact, a rapid turn of events was taking place in the early 1980s. The NWDT had given its decision in 1979, ending the decade-long uncertainty about the use of Narmada waters. By 1980, preparation started in earnest when the World Bank got involved to help modernise India's irrigation sector by assisting with the planning of the SSP. Tata Economic Consultancy Services (TECS) did a cost-benefit analysis in the early 1980s. In 1983 when the project went to the MoEF for clearance, the Ministry was itself struggling with its guidelines on environmental clearance. What it possessed was a

4. It is unfortunate that they do not give the exact date year of the guidelines. Updated versions came in 1985 and 1989, at least two years after the application for clearance was made by the Sardar Sarovar Project authorities.

sketchy guideline dating from the late 1970s, which would change again in 1985 and later in 1989. Similarly, WB environmental guidelines went through a series of changes in the 1980s. Thus the rapidly changing situation dictated that environmental studies and programme implementation had to be *pari passu* with the project works.

Ignoring this evolving process of creating guidelines, the IR mounted a furious attack on the very idea of *pari passu* provision in the conditions (xxiv: 335). They failed to recognise its necessity and made the provision one of the major highlights of their critique, raising a storm of controversy about it.

Hydrology

The MoEF's note to the Prime Minister lay behind other major environmental controversies. For example, in section vii the note states that the 75 per cent dependable runoff in the Narmada has been estimated to be only 2.829 MHM instead of the 3.346 MHM assumed by the NWDT. But while tacitly casting aspersions on the competence of the Tribunal, the note does not cite any authentic source to back this new estimate. One of the issues the Tribunal had to adjudicate was the quantum of water in the Narmada. Gujarat and MP each presented different estimates of the runoff. While the Tribunal was deliberating, experts from the three states of Gujarat, Maharashtra, and MP and from the GoI's Central Water Commission (CWC) met and agreed to adopt a series used by MP to measure the runoff. Gujarat accepted the MP estimates; the dispute was resolved. This figure was then incorporated into an agreement between the four chief ministers of the concerned states (the three named above and Rajasthan). The Tribunal accepted all parts of this agreement, including the agreed assessment of the Narmada's annual runoff (NWDT, 1979: Vol. 1, pp. 38-39).

Why, then, does the MoEF's note consider the Tribunal's estimate of the Narmada runoff an assumption? If an agreement was reached based on facts that were later put before the Tribunal, why should the runoff estimate be dubbed "assumed" and called into question? If any assumption of the runoff was made, it was made by the MoEFs, which failed to give any source for the runoff figure it cited. The note to the Prime Minister went so far as to suggest design changes in the project that related to water quantity. These issues were later picked up and amplified by the anti-dam movement. Publication after publication claimed that the water available in the Narmada was less than that assumed by the NWDT, each publication quoting different sources. But none of these sources were adequately substantiated. Nevertheless, the viability of the SSP was called into question; it was widely portrayed as over-designed and capable of causing avoidable adverse environmental impacts.

When the IR started its inquiry, the issue of water availability in Narmada already had many twists. The team also felt very strongly that hydrology was central to the analysis of environmental impacts.[5] In the chapter "Hydrology and Water Management" of the Report of the Independent Review, the whole critique rests on two planks.

One plank is that the NSP, a major project to be located in MP, 350 km upstream of SSP, will not be built, thus depriving the SSP's reservoirs of a critical supplement (17.8%) of water release. The NSP is a kingpin of the big three-dam hydropower complex in MP, with relatively few irrigation benefits to MP. The IR concludes that the SSP will not work as planned without the NSP.[6]

Moreover, the argument that the SSP is not viable without the NSP would seem to be refuted by the relevant chapter of the Tribunal Award, which shows that the quantum of water allotted to Gujarat-Rajasthan is not likely to be affected adversely if the NSP is completely dropped (NWDT, 1978: Vol. 2, Chapter 15). The second plank is the available quantum of water.

The Report of the IR never resolves the issue of hydrology. At different places, its authors note that this is beyond their responsibility and purpose (246, 249, 252). Nevertheless, the Report does not hesitate to declare that the assumption of the Tribunal Award may not hold in the future (250), and that "many previous assumptions now appear to be unfounded or are questionable" (251). These decla-

5. "But first it is necessary to review the overarching issues related to hydrology and water management" (IR, 1992: 234).
6. Towards the end of the chapter, however, the authors confront the possibility that the Narmada Sagar Project may go up after all, a situation that would throw all their arguments related to the Narmada Sagar Project component out of gear.

rations are made with only a vague reference to an independent hydrologist who was asked to "assess…contentious issues" (253). But neither the critical findings of this independent hydrologist, the report, nor the name of the hydrologist are provided, nor are the findings discussed critically *vis-à-vis* the findings reported by the country's apex body – the CWC. In his letter to the WB president, Morse continues to argue that they had found discrepancies in basic hydrological information and had therefore examined the stream flow data and completed their own analysis, finding that the SSP will not perform as planned (IR, 1992: xi-xxv). He further argues that the effects of Sardar Sarovar on upstream, downstream, and command areas will be different from what has been assumed (1992, xxii). Unfortunately, the details and analysis are not found in the text of the report, which serves as the basis for these conclusions.

Meanwhile, the anti-dam movement has seized upon this and other parts of the IR report as the substantive evidence confirming what they have been arguing all along. If the Independent Review had come up with strong and hitherto unknown arguments solidly supported by facts and sound analysis, the boost they gave to the anti-dam movement would have been justifiable even if not liked by the governments. In actuality, what one finds in this report are not good arguments backed by solid facts but opinions expressed with a sense of finality. When opinions are supported by men of stature and experience, like those of the Independent Review, who enjoyed rare and unfettered access to all the documents on all the sides, they carry considerable weight. It is most unfortunate that, without being substantiated by fact, they further muddy rather than clear the waters.

Environmental Costs

The MoEF's note to the Prime Minister also contained another dubious claim: that the environmental cost of submerging forest lands for the SSP alone was about Rs. 80 billion (the total cost of the SSP is Rs. 64 billion). The anti-dam movement lost no time in adopting this additional cost statistic, attacking the SSP's economic viability and, between 1988 and the middle of 1990, this argument enjoyed significant publicity in the media. The Environmental Defence Fund (EDF) presented this argument to the US Senate in June 1988 in an effort to prove that the SSP was economically nonviable (Rich, 1988: II).

Yet the environmental cost of the SSP was erroneously based on a rule of thumb devised by the Forest Research Institute (FRI), Dehradun, which was adopted by the MoEF as a guideline (MoEF 1986, Annexure 2, item 5). This guideline, issued in January 1986, three years after the SSP application reached the MoEFs, contained the MoEF's environmental cost computations. The GoG followed these guidelines to work out both the environmental costs of the SSP and its benefits. An economist connected with the NSP, S. Pant, however, found no economic rationale for the guidelines. He also found out that there was no economist on the staff of either FRI or MoEF.[7] Vijay Paranjapye, a noted activist and economist, has also stated that the MoEF has provided no justification or rationale for its figures (Paranjapye, 1990: 225).

Cost-benefit analyses based on the MoEF's guidelines were done by the GoG in 1986. The questions that still need to be asked are whether or not this was a serious exercise in environmental monitoring, and whether these considerations stalled clearance. How was the MoEF's exercise, which absorbed considerable time and energy, related to environmental studies and safeguards for SSP? The way MoEF has conducted itself on the issue of environmental clearance at various stages demands investigation. Although the Terms of Reference of the IR asked for an assessment that took into account, among other things, the environmental clearance decision of the MoEF, the IR team have maintained a discrete silence on MoEF's role.

Compliance

Controversy over compliance also persists, based on the information in the conditional clearance letter and associated documents. The IR reports difficulties in assessing compliance because of disputes between NCA and the MoEF over who was to assess compliance and who to define progress. As the IR report observes, "There are differences between the documentation available from NCA and that from MoEF" (229).

---

7. "Calculation of Benefit-Cost Ratio: Interview with Dr. S. P. Pant," Part 2 (Alvares, 1988: 90-100).

The Report of the IR made a major issue of the fact that studies that were to be completed before 1989 were not completed within the established time limit. According to the Report, "The 1987 clearance required 'complete details' in eight areas by 1989. The Narmada Control Authority maintains that, despite some slippage, all are on track and will be completed *pari passu* with the project. The Ministry of Environment and Forests is less optimistic" (229). A nine-point summary of MoEF's position as of December 1991 follows without comment. Clearly it is necessary, in light of the conflict between NCA and MoEF, to check with the NCA and ask for their version, as well as to analyse these findings within the framework of the *pari passu* provision. Apparently the IR did not do so.

On one other point, the arguments of the Independent Review invite scrutiny. Citing the minutes of the Narmada Control Authority meeting in early 1991 (NCA, 1991: 18), the Report states "the Chairman began with an inquiry about the scope of *pari passu* implementation. He noted that 'some of the environmental safeguard measures are ongoing as per schedule and some others (a major part) are not'" (231). We have checked the minutes of the 38th meeting of the NCA, which they have cited, but cannot find this quotation. The chairman – always the secretary to the Ministry of Water Resources (MoWR) – did not say, even remotely, what is attributed to him by the IR report. Although the exact quotation is not found in the document, something similar to what is quoted was said by the secretary to the MoEF (though even he did not make any inquiry about the scope of the *pari passu* implementation). The minutes record that D. C. Debnath, Executive Member of NCA, informed the Secretary of MoEF that "such *pari passu* action was already underway on most of the major items of work required to be undertaken in the environmental context." The minutes similarly did not record what the IR report claims they did – that "the Narmada Control Authority had not applied for extension of time for compliance with the conditions of clearance as had been decided at the meeting with the Chief Ministers held on September 18, 1990" (231). These discrepancies between what the report alleges the document says and what is actually in the document distort the meaning of what was said by the chairman of the NCA.

The MoEF's nine-point summary of December 1991 and the minutes of the NCA meeting of March 1991 constitute the two most important pieces of factual evidence reported by the IR to support their conclusion that compliance with *pari passu* conditions is lacking. We might also wonder why the IR did not cite the most recent sources when making their report. The IR team made inquiries at least up to March 1992 in India, and thereafter they continued to ask for and to get information from various sources. They had access to most authentic documents, like those of the Environment Subgroup (ESG), specifically set up to monitor compliance with conditional clearance after June 1987, and in numerous places they have used information from this source. Why, then, did they not consult the minutes of the latest meetings of ESG to check and report the state of compliance? Their reliance on the rather dated information from 1991 and from only one source – MoEF – distorts their analysis of compliance. Further, the clearance letter should be criticised for its very general and vague conditions, as well as its lack of specific criteria to assess compliance. Because of these shortcomings, assessment becomes highly subjective, arbitrary, and prone to abuse. Here, too, the IR report fails to further our understanding, and a detailed assessment is still needed.

The Agenda Notes of the August 1990 ESG meeting have been used by the anti-dam activists to support their claim that, according to the MoEF, the conditional clearance for SSP had lapsed. In our view, the minutes of the September 1990 meeting tell a different story: a lapse of the clearance is not even mentioned. The minutes record the comments of N B Lohani of MP, who noted that due to physical, financial, cost sharing, and other constraints, the catchment area treatment works were lagging. He was referring to an ongoing dispute between MoEF and MoWR as to which parts of the catchment area should be treated by the project authorities – directly draining, as claimed by MoWR, or freely draining, as claimed by the MoEF. The minutes further record that, except for CAT, other studies necessary for implementation of the action plans were going on normally. The anti-dam activists conveniently ignore the minutes of this meeting. Similarly, EDF disregarded the minutes of this meeting when they repeated the claim that the clearance had lapsed in the memorandum they submitted to the WB president on March 20, 1991. The anti-dam movement still continued to argue that the conditional clearance has lapsed on the basis of the agenda notes.

The Report of the IR refers to both the agenda and the minutes of the August-September 1990 ESG meeting, but quotes, without clearly stating so, only from the agenda notes prepared by the MoEF itself, to the effect that the MoEF's conditional approval "must be deemed to have lapsed." The IR report thus notes only that "the project authorities should seek renewal of environmental and forestry clearance" (230). In other words, the report refers to both agenda and minutes but quotes only the agenda, thereby creating the appearance that these quotations pertain to both the agenda notes and the minutes. What actually transpired in the meeting as recorded in the minutes is held back.

We believe the Report of the Independent Review erred again in the chapter "Upstream Environment". Here they argue:

> In India early considerations about the flora and fauna in the upstream area were based on general concerns of Sardar Sarovar and the Narmada Sagar Projects taken together. The 1987 briefing note to the Prime Minister by the Ministry of Environment and Forests deals with the flora and fauna for both projects and states the need for a Master Plan showing just not the present status but also the likely scenarios after the project is implemented. (IR, 1992: 275)

Since the two projects are now considered separately, the report seems to imply, the flora and fauna survey for both projects and the common Master Plan has fallen by the side. It appears to us that in the relevant portion of the note to the Prime Minister (section v), which the Report quotes, the entire section unambiguously refers to the NSP only. Surprisingly, the IR has taken what applies only to NSP as if it were applicable to SSP as well. Yet based on such an untenable assessment, the IR reaches the startling conclusion that with respect to MoEF's clearance requirements, WB is guilty of "gross delinquency" (233-34).

In the chapter "Findings and Recommendations" the IR report argues:

> We have been at pains in the sections of this chapter summarising our findings to demonstrate how these problems of human and environmental impact encompass all aspects of the Projects, including the uncertainties of hydrology, the upstream questions, the impact downstream, the command area issues, the health risks, the deficiencies in resettlement policy and implementation in each of the three states as well as the canal. None of these issues can be ignored. (355-56)

We fully agree that none of these issues can be ignored. But the central challenge has yet to be taken up: proving the quantum of water in the Narmada. Is it less than what was decided by the NWDT? We believe that the NWDT's assessment has yet to be proven wrong. Thus the MoEF's mischief in the clearance issue and how it contributed to the argument that *pari passu* compliance with the project works had failed must also be looked into. Also the IR's assessment of other important issues suffers from the same methodological flaws, selectivity of documents, and serious errors of facts and of reasoning.

Resettlement and Rehabilitation

Resettlement and Rehabilitation (R&R) in the SSP has a complex history. In 1980, when the foundation work on the project began, the states' poor R&R performance had become an issue. In the case of large development projects, especially in water development projects, oustees had been given a raw deal and the prevalent ideology held that in the process of development and modernisation, this was inevitable – for the benefit of many, a few had to sacrifice. We at ARCH-Vahini found this logic inhumane and unacceptable. We accepted the fact that Narmada waters would go to a predominantly water-scarce and economically depressed command area, but the implied rationale, derived from a vision of an inevitable march towards modernisation that considered heavy human costs inevitable, was unacceptable to us – then and now.

Early in 1986, a struggle began in Maharashtra and MP for a better R&R policy. The principal demand of the oustees there was also for minimum land entitlements as we have argued elsewhere (ARCH-Vahini, 1991). In December 1987, facing the first real threat against the SSP, GoG was forced to adopt the new R&R policy we demanded. This policy conceded all our demands, the core of which was that a minimum of two hectares of land be awarded to each oustee family (defined as an eighteen-year old son – married or unmarried), regardless of the nature or absence of the land title held by the family. Moreover, a family had a clear right to land of its choice. ARCH-Vahini and the oustees of Gujarat accepted the policy package.

It is interesting to note that until this time, no grassroots organisations active in the three states of the valley had raised any issue concerning the viability or desirability of the SSP on environmental or economic considerations. The oustees in each state were focused on their demand for land as just compensation, even when they were extremely doubtful that such rights would be conceded. Within a few months of the major policy breakthrough in Gujarat, in the middle of 1988, the grassroots activists in Maharashtra and MP and the intellectuals in the metropolitan centres declared their opposition to the SSP on the grounds that it was a planned disaster – social, economic, and environmental. Just resettlement and compensation was thought to be impossible because of the unavailability of land, not because of issues at that time considered secondary, such as tribal life style, destruction of tribal culture, and dependence on the river. Even the activists in MP conceded in 1988 that, while they were opposed to the project "for all the standard reasons that are too well known to require repetition," the tribals in MP would readily accept resettlement if they were given good quality land. Their opposition to the dam was rooted in their fear that the land would not be given to them (SWRC, 1988: 19).

When in April 1991 we questioned the rationale for the apparently sudden shift from a "just resettlement" to the more radical "no dam" position, the anti-dam movement replied that we did not understand the process and experience that the organisations in Maharashtra and MP had gone through (NBA, 1992). Our arguments rested on the published material of the anti-dam movement at that time. This does not provide a rational and coherent explanation for the sudden public shift in their position.

For three years after opposition to the dam was declared in the middle of 1988, the idea was spread that the valley's tribals were opposed to the project despite the quality of the R&R package. But by the middle of 1991, the ground-level reality was irrefutable both in Maharashtra and MP: the tribals' prime objective remained a better R&R policy. In Maharashtra, a majority of the tribals had broken away from the anti-dam movement. A group composed of the deepest interior tribal villages set up an independent organisation to press for policy reforms that had been abandoned by activist organisations. They held a rally in April 1992 and gave a memorandum of demands for better R&R to the Chief Minister of Maharashtra. Today a similar movement is on the rise among tribals in MP. Yet this process has not been analysed, nor have its implications for R&R been examined with the thoroughness they deserve. The IR report also categorically declared that R&R was impossible, though their detailed reasoning is substantially different from that offered by the anti-dam movement. The IR and the anti-dam activists also seem to agree with regard to their very estimates of the number of affected families in MP who ought to have land entitlement, which is in one way or another denied to them.

Gujarat

The anti-dam movement raised a host of questions regarding the possibility of R&R in Gujarat. The activists argued that the burden of finding land fell upon the oustees; because not enough land was available, this was an almost impossible task and tribals were rapidly losing hope; only a handful of all of the oustees would actually get land, while others would be left "high and dry"; only a few relocated villages in Gujarat were prospering and those villages were serving as showcases; the Land Purchase Committee (LPC), a key mechanism for the purchase of private land for R&R, was subject to corruption and exorbitantly high land prices were being offered to big landowners; the oustee villages were being split into small fragments, breaking apart the tribal community; landless labourers working on the acquired lands would become jobless, creating what was called "secondary displacement": and, finally, the oustees who were allotted land were discovering that the land was no good and were beginning to return to their original homes – a trickle of return which, it was claimed, would soon turn into a flood.

The facts, however, were different. When the review began in

Gujarat, more than 3,500 out of 4,500 families had been given land they had chosen themselves; allotment was not confined to a few select families. Furthermore, after thorough examination of all these issues, the IR raised no questions in their report about the quality of land offered to oustees and pointedly ignored the anti-dam movement's prediction that the trickle of return would turn into a flood. The Report freely praised the LPC for its energy and thoroughness and quietly ignored charges of corruption. Having examined the issue of secondary displacement, it found no basis to it and ignored it. The IR also accepted the fact that the most vulnerable tribal community, the Bhils, about whose welfare there was a legitimate and widespread concern, was not fragmented because of resettlement, although they expressed concern that the oustees were being scattered among a great number of sites where they had been given land (106). The Report concedes that this effect is in part a result of the choices made by the oustees, and that the "scatter effect" was observed especially in case of the rock-filled dyke villages.[8]

ARCH-Vahini contends that the tribal communities have not been more scattered than they were in their original places. We argue that the number of sites where the land is allotted is not a good criterion by which to evaluate the scatter effect. A more sensible criterion is a comparison of the distances individuals and their groups must travel to their fields, friends, and relatives in their original villages both before and after resettlement. The effect of the choice exercised by the oustees in choosing land on the distances now existent between groups and individuals needs to be acknowledges. In each village, even in each hamlet of each village, there are families and groups who deliberately opted to move away from those with whom they did not want to live as closely as they had in their original villages.

8. These five villages were displaced in the early 1980s because of the construction of rock-filled dykes and had been resettled in the vicinity of the dam site according to the old resettlement and rehabilitation policy of June 1979. The new policy of December 1987 offers them an enlarged package of land entitlements, but the additional lands in the surrounding area are not available. Major sons have, therefore, to move farther away, causing the split in the extended families. This is reprehensible but inevitable, given the recalcitrance of the government in the early 1980s.

Without understanding this process of choice, one fails to capture a realistic picture of social dynamics at work.

The Report of the Independent Review on "Broken Promises"

The Report of the Independent Review discusses other problems in Gujarat to which we now turn. First, of the three tribal groups facing relocation, the Tadvis and the Rathwas are generally considered to be the most enterprising. The Report has noted that these groups look forward to new opportunities offered by good quality land, and that they can cope with the new economic challenges and opportunities with which they will be presented. It anticipated that the Bhils, whom the IR report considers the least enterprising and the least integrated into mainstream life, will have immense problems. But these concerns for the Bhils remain unsubstantiated. None of the four detailed case studies they cite to show that rehabilitation will not work out pertain to the Bhil community (109-115). Furthermore, the four case studies cited suffer from inability to distinguish the short-term problems of transition (for example, fodder and fuelwood shortages and the inconveniences of temporary and transit tin sheds) from the long-term problems the oustees may face.

A second point revolves around "broken promises," a constantly recurring theme not only for R&R in Gujarat but for all three states. Yet when one tries to look for these broken promises, it is difficult to locate them. Only three promises stand out: of *pukka* houses, of irrigable land, and of in-house electricity. In fact, no policy decision has ever been taken to give *pukka* houses to the oustees. It may be that a field-level officer made unfounded promises to some oustees, but if this were true, it should be recorded and the government should be criticised for not taking the appropriate disciplinary action. Unfortunately, the IR report fails to shed light on this issue: it neither provides information about the proportion of project-affected persons who complain of broken promises nor records any instances where field officers made inappropriate promises.

As for the promise of irrigable land, it should be noted that most of the rehabilitation situations cited, especially by the IR report, were in the first phase of the SSP command area where the irrigation would begin by 1994-95. This was easy to find out. But the Review speculated that even "the most optimistic schedule will not deliver water for irrigation until 1995, and possibly not until 1998" (114). Finally, in regard to the confusion over electricity, the oustees cited in the IR report were living in transit sheds. Electrification, when it took place, would be in the permanent houses that the oustees would be building thereafter. Though aware that "visitors can easily become an opportunity for the airing of problems, and therefore, can cause bias towards negative statements" (113), the IR team apparently failed to check and cross-check the information they got from some oustees.

The necessity for this sort of cross-checking is illustrated by the IR's report on a large family of five brothers from Panchmuli (a rock-filled dyke village) which was resettled at Gadkoi in the early 1980s. The family reported that in Panchmuli they had 200 buffaloes, 200 goats, ten pairs of bullocks, and ninety cows and calves. In Gadkoi, however, they had only seven or eight buffaloes, eight pairs of bullocks, 25 goats and 22 cows and calves (99). In fact a relevant volume of the socio-economic baseline survey done by the Centre for Social Studies, Surat, just before the village was displaced, indicates otherwise: that in Panchmuli, out of 45 families who owned buffaloes, only ten had more than five buffaloes and cows, and the total number of buffaloes in the entire village fell considerably short of 200.

Similarly, to back their claim that as many as 50 per cent of the oustees recognised to have land entitlements (based on the 1987 policy and a major amendment to it in June 1990) from five rock-filled dyke villages had not been allotted agricultural land, the IR report cites SSNNL's February 1991 figures, according to which only 466 of 929 families had received land. More up-to-date figures from SSNNL, which had been submitted to the IR at their request, showed that by January 1992, 593 families from these villages had been allotted land. Several major sons had also become eligible for land only after June 1990 – another point the Report failed to take into account.

The IR team repeatedly expressed anxiety that socio-cultural factors were neglected in the resettlement of Gujarat oustees. In support of this conclusion, its report considers this another major undoing of the R&R process in Gujarat, citing the March 1991 Report No. 12 of CSS, Surat (106). Table 2-20 of the CSS report, from which the IR report draws its conclusion, describes different reasons for land selection by Bhils and Rathwas in the four most interior villages of Gujarat. While 402 out of 637 total oustee families had selected their land up to that point, only two cited social reasons as the sole criterion for their choice of land. Aside from thirty-three families for whom no reason was known, all of the rest had selected their land for economic and socio-economic reasons.

Since there was no coercion of oustees to select any particular parcel of land and since the IR report acknowledged that they had exercised freedom of choice, it should have also acknowledged the greater importance of economic and socio-economic grounds rather than insist, as it did, that tribal economic considerations were divorced from and in opposition to their socio-cultural aspirations. This type of conceptual rigidity has led the IR to miss the significance of the almost single-minded demand of the tribals in each of the three states for land entitlements in the memoranda and the charters of demands the tribals have submitted to the authorities.[9]

Maharashtra

The availability of suitable land was a central problem of R&R in Maharashtra too until 1991. From the beginning of 1986, oustees of 33 Maharashtra villages began to organise themselves under the banner Narmada Dharangrast Samiti (NDS), which later became a key part of the much bigger Narmada Bachao Andolan (NBA). During 1986-87, the NDS and the oustees formulated their demands for a better R&R policy in Maharashtra. Simultaneously, a similar set of demands was articulated by another organisation in MP, the Narmada Ghati Navnirman Samiti (NGNS). The main focus of these demands was the provision of land for oustee families – almost identical to those we were making in Gujarat. The main difference between the demands of Gujarat oustees and those of Maharashtra was that the latter were not demanding private lands like their counterparts in Gujarat, but were demanding degraded forest land instead. The Taloda forest land was one area in which they had expressed interest. In a memorandum submitted to the Government of Maharashtra (GoM) in April 1986, the NDS specifically demanded:

9. See the memoranda by Narmada Dharangrast Samiti and Narmada Ghati Navniram Samiti, 1986, 1987.

> As per the principle agreed to in the World Bank Agreement, forest land available in the nearby Talukas – Akkalkuva, Taloda and Shahada – should be released for the rehabilitation purpose. They should be released, in spite of Forest Conservation Act, 1980, as is agreed to in World Bank Agreement.[10]

In June 1987, Medha Patkar of the NDS read a paper at a seminar on the Narmada valley projects in Bombay, in which she stated that the only alternative was "to release large chunks of denuded forest land which the Centre is reluctant to do, taking a rigid view of Forest Conservation Act, 1980. The tribal-forest linkage, too, is not viewed seriously or not considered in planning alternatives."[11] As late as November 1987, just a month before Gujarat's new R&R policy was

---

10. Before 1980, the forest was under state jurisdiction. In 1980, parliament passed the Forest Conservation Act (1980), whereby the forest was placed in the Concurrent list. The Act gave power to the Ministry of Environment and Forests to decide whether the forest land could be diverted for non-forest use. In May 1985, the Government of India, together with the governments of Gujarat, Maharashtra, and Madhya Pradesh, entered into an agreement with the World Bank. In part related to resettlement and rehabilitation, the agreement stipulated that, if the tribal oustees insisted upon resettling on the degraded forest land, the land would be released by the Ministry of Environment and Forests. This condition was specifically inserted at the insistence of Professor Thayer Scudder, then consultant to the World Bank on R & R, because Scudder had found that at least some tribals would demand degraded forest land. In April 1986, Narmada Dharangrast Samiti (later part of the Narmada Bachao Andolan) formed by Medha Patkar, submitted a memorandum to the District Collector of Dhule (April 8, 1986) in which they demanded that the degraded forest lands of Taloda, Shahada, and Akkalkuwa should be released for the purpose of rehabilitation of the Maharashtra oustees and that the Maharashtra government should take up this issue with the Central Government.
11. In June 1987, Medha Patkar, still working for resettlement and rehabilitation, read a paper, "Sardar Sarovar (Narmada) Project: A Case of Environmental and Human Destruction," in a seminar on water utilization and the Narmada Valley Project, organised by the Forum for People Oriented Water Utilization (FOWL), in Bombay (June 27-28, 1947) in which she reiterated her arguments of April 1986.

announced. NDS and NGNS had submitted to NCA a list of 38 demands related to R&R, one of the most crucial of which was the release of denuded forest land for resettlement.

In June 1988, Bruce Rich of EDF, the American NGO spearheading the anti-SSP movement in North America and Europe, made a submission to the US Senate arguing that in Maharashtra, even the government had reached the conclusion that "no land for resettlement was available except denuded forest land which the Forest Department had not agreed to release. The fact that the land has not been released is another violation of WB Loan Agreement."[12] Soon after the EDF submission to the US Senate, organisations in Maharashtra and MP declared their total opposition to the project and claimed that the oustees were also opposing the project. The goal had shifted from a "better R&R policy" to a "no dam" position by claiming that R&R was no longer the attainable goal it had been in 1986-87. It was now impossible not only in practice but also in principle. In Gujarat, the movement anticipated that the required private land would not be available. In MP, they argued, the total number of oustee families who would need land would be five to six times larger than in Gujarat and that there would not be enough land for the more than 22,000 families needing to be resettled. In Maharashtra, NDS cited the refusal of the MoEF to release the degraded Taloda forest land despite a legal agreement that GoI and the three states had made with the World Bank.

We have discussed above the MoEF's role on environmental issues. Now, we draw attention to its unfortunate role with regard to R&R. As we have seen, in June 1987 the MoEF gave a joint conditional clearance to both the NSP and the SSP. In September 1987, in a separate letter, it allowed forest land to be diverted for the construction of the SSP (MoEF, 1987b). A similar letter for the NSP followed in October 1987 (MoEF, 1987c). The SSP letter of September 1987 contains a clause stating that no forest land would be diverted for the purpose of R&R. The NSP letter of October 1987 does not carry such a clause. But forest lands were an issue only in

12. On June 14, 1988, Bruce Rich of the Environmental Defence Fund, U.S.A. repeated the same line of argument before the Sub-Committee: on Foreign Operations of the Committee on Appropriations of the U.S. Senate.

Maharashtra, where oustees had been clamouring for Taloda and other forest land from the very beginning. The World Bank had been aware of this reality and had therefore encouraged the insertion of the clause concerning the release of degraded forest land in the May 1985 Loan Agreement. The MoEF, of course, was well aware of the agreement the GoI had entered into and yet, unhesitatingly, inserted this clause in the letter of September 1987.

The MoEF's actions were ambivalent. It continued to ignore the WB agreement and continued to resist the release of the Taloda forest land. At one point it suggested that the Maharashtra government was creating a crisis situation so as to resolve the issue as a special case. It also insisted that it had not categorically told the oustees that forest land would not be available (RSG, 1989: 3-4). It even went so far as to suggest to GoM that they actively encourage the oustees to go to Gujarat (MoEF, 1990). Such a suggestion is a violation of the oustees' legal right to choose the state within which they wanted to settle as embodied in the NWD Tribunal Award and WB Agreement.[13]

Under WB pressure, MoEF had to release the Taloda land in June 1990. On July 31, 1990, NDS submitted a memorandum to GoM on behalf of the oustees of its 33 villages, stating that they rejected the Taloda forest land (NBA, 1990: 6). It went on to say, "It is reprehensible that the Central Government agreed to release forest land in Maharashtra for rehabilitation in clear contravention of the conditional clearance given to the SSP which clearly stated that no forest land can be used for resettlement purposes." It called this clearance "tantamount to selling India's ecological and social future at the behest of a foreign agency." In a memorandum submitted to WB in March 1991, EDF, together with 47 NGOs in Europe, Japan, and Australia, severely criticised WB for forcing oustees to accept the Taloda forest lands that they had explicitly rejected, thus violating the terms of the Loan Agreement.

In April 1992, the oustees of Maharashtra's ten most remote interior villages of Akrani Taluka, incensed by the manipulation committed in their names by the NDS and its supporters overseas,

---

13. Though these documents were submitted to the Independent Review, the report makes only pasting reference to them on p. 225.

submitted a memorandum to the Chief Minister of Maharashtra stating that they had been cheated. It stated that they had never been asked their opinion nor informed of the release of the Taloda forest land, which they had always demanded and which they still wanted. In fact, they argued that the 2700 hectares of released forest land might not be enough for their resettlement and more should be released. This is the best documented case to date in which tribals have been manipulated, and they were manipulated by the anti-dam activists. Even Amita Baviskar, an avowed supporter of the anti-dam movement, has written that the NBA makes its decisions without consulting the tribals and that the tribals are just the passive recipients of decisions made at the top of the pyramid (Baviskar, 1991: 91-97).

The IR made a different case concerning the impossibility of R&R in Maharashtra. It argued that as many as 80 per cent of the project-affected persons were "faced with some degree of dispossession as a result of existing resettlement policies" (IR, 1992: 137). It also asserted that none of the oustees of the 24 villages of the Akrani submergence communities in Maharashtra, identified as forest villages, qualified for two hectares of land (165, xix), but would instead be treated as landless and eligible to apply for only one acre (0.4 hectare) (165). It further declares that "neither Madhya Pradesh nor Maharashtra acknowledge any right of encroachers to adequate land on resettlement" (xvi) and "that encroachers must be treated as landless oustees with no entitlement of adequate land for cultivation on resettlement" (xvii). In the chapter on Maharashtra, the IR provided an analysis of two relevant policy documents of the Government of Maharashtra – government resolutions dated June 29, 1989, and February 26, 1992. The June 1989 resolution specifically provided that any oustee who had encroached upon the forest land prior to March 1978, as per the survey carried out in 1985, would be eligible for two hectares of land only. The resolution excluded major sons of these encroachers from any land entitlement (i.e. they were not to be treated even as landless with entitlement of one acre of land). All post-1978 encroachers were treated as landless with entitlement of one acre (0.4 hectare) of land. Thus the June 1989 resolution created various categories of oustees – those with land, regularised encroachers, major sons of landholders and regularised encroachers, landless agricultural labourers, joint holders, and so on – each with its own land entitlement. Naturally, these categories created a lot of confusion. But, more important, they excluded many families of major sons of landed oustees and of regularised encroachers from any land entitlements.

The February 26, 1992 resolution sought to streamline the categories and also to correct deficiencies related to major sons and others. The category of "landed" was expanded to include two categories of the June 1989 resolution – landowning oustees and pre-March 1978 encroachers. A new all-embracing category of "landless" was created that included major sons and major unmarried daughters of landed oustees (i.e. landowning oustees and regularised encroachers and post- 1978 encroachers). The February 1992 resolution gives the explicit right of irrigable agricultural land to landless oustees. It also gives the regularised encroachers the status of landed oustees, which entitles them to irrigable agricultural land of an area equal to that of land acquired, subject to the ceiling and minimum two hectares of such land. The previous resolution gave them only two hectares of land and did not say specifically whether it would be irrigable or not. The net effect of this resolution was to give a status of landed oustees to a majority of oustees in the 24 villages of Akrani Taluka who were treated as encroachers by the state government. A historical injustice to these tribals was finally set right. Out of 1,485 oustee families of these 24 villages, 855 thus became landed oustees. Their major sons were now given a land entitlement of one acre of irrigable land. Similarly, another resolution of February 1990 pertaining to the acquisition of marooned *(tapu)* land gave entitlements of landed oustees to 192 more families in Akrani, making up a total of 1,047 (70.5%) out of 1,485 families (70.5%) entitled to a minimum of two hectares of irrigable land. The IR not only ignored this resolution and its beneficiaries but even raises a false issue of their plight in its Report (146-47). On page 165 the Report claims, "Since all of the Akrani submergence communities are identified as 'forest villages,' none of their oustees qualify for two hectares of land at Taloda." This statement stands in stark contrast to the hard fact that 70.5 per cent of them were treated as landed oustees with commensurate land entitlements. And yet this dramatic statement is incorporated in the letter to the then

WB President, Lewis Preston, which gives the strong impression that things are very bad on the R&R front.[14]

A number of disturbing questions arise from the discussion of the Taloda forest. The IR team, in the course of their visit to the area, met with the local people, who said that they were not unduly alarmed by the prospects of a resettlement site near their village (165). The IR report notes, "These findings appear to contradict the conclusion arrived at by the Tata Institute research team. However, their work is more extensive and took them to all 10 of the villages directly at risk as a result of use of Taloda Forest for resettlement sites" (165). The Report cites both the first TISS report from May 1991 on the Taloda forest land and the second report from February 1992, which is in fact their submission to the IR and apparently not based on "additional field work" as reported by the IR (165). Quoting the second report, the IR states that the ten villages in the Taloda forest area depend upon 800 hectares of forest for grazing and firewood. Then it states that a large number of encroachers were already displaced to make way for resettlement. The February 1992 TISS report also says that the host population is completely against the proposed rehabilitation (165-67).

The May 1991 report of TISS sharply differs from its February 1992 report. TISS had carried out a first comprehensive study of the area in May 1991, barely nine months before the second report was submitted to the Independent Review. The May 1991 report, while discussing all the relevant points, does not talk of the dependence of the 10 villages on 800 hectares of forest for grazing and firewood. Lastly, it does not state (as the February 1992 report does), that the host community is completely opposed to the resettlement. The IR team had to contend not only with what they were told by the people of the host villages during their visit to Taloda some time in February 1992, but also, effectively, with TISS's own May 1991 report. It is very difficult to understand why the two TISS reports, separated by an interval of nearly nine months, should contain such different information. It is also pertinent to note that the second report is not available for public scrutiny. Clearly the situation needs to be

---

14. The entitlement to landless major sons was liberalised further in December 1993 to one hectare each. Government of Maharashtra Revenue and Forest Department Resolution No. SSP-3I92/R & R/ CR-4 R-5, December 14, 1993.

investigated more thoroughly to settle these questions. We would like to know why the Report of the Independent Review relies on the 1992 report and implicitly rejects the findings of the 1991 report.

Madhya Pradesh

The anti-dam movement has always argued that submergence is by far the greatest in Madhya Pradesh (MP). Of the 245 villages in three states to be affected by the reservoir, 193 are in MP. By simple extrapolation, the activists argue that this means that roughly 80 per cent of the displacement will take place in MP. The anti-dam movement thus questions how, if the GoG had difficulty in finding land for 10 per cent of the oustees, there is any hope of finding land for the 23,000 oustees from MP. In 1991 we argued that this pessimism was misplaced. GoMP had done its first village surveys in 1987 and its findings were reported in January 1989 at the 30th meeting of the NCA.[15]

These findings brought out for the first time the fact that of the 193 villages to be affected in Madhya Pradesh, only in 65 would twenty hectares or more agricultural land be submerged. Submergence in the remaining 128 villages would be less than 20 hectares each. Analysing the MP government's published data; we discovered that whereas only 13.7 per cent of the total agricultural land (6,855 out of 49,937 hectares) would be submerged, many more houses – 73 per cent (15,552 out of 2l, 231) – would be inundated. Really heavy submergence, we found, would to be confined to 28 out of 193 villages. This analysis was criticised by the NBA, which questioned the reliability of our data. GoMP subsequently carried out a sixty-four-point survey in all except the twenty-eight villages where opposition was very intense. Even though this survey confirmed the same pattern of submergence as the 1981 survey, the NBA questioned the reliability of this data as well (NVDA, 1992).

The IR report basically adopted the NBA position on the two most critical issues concerning R&R in MP. It accepted not only the NBA's basic contention that the quality of the sixty-four-point survey

15. Vijay Paranjapye has quoted these findings (Paranjapye, 1990: 193-94).

by the MP government was unreliable, but also seemed to accept the NBA's figure of 73,000 families as the number displaced that would need alternative agricultural land.

The IR also accepted NBA's claim that the encroachers in MP would be reduced to 'landless' status with no land entitlement. In their letter to the WB president they wrote, "Neither Madhya Pradesh nor Maharashtra acknowledges any rights of encroachers to adequate land on resettlement" (IR, 1992: xvi). Further Madhya Pradesh and Maharashtra say that encroachers must be treated as landless oustees with no entitlement to adequate land for cultivation upon resettlement. The dispute here is whether tribal people, holding their land by customary usage, are entitled to be treated as landed oustees. Madhya Pradesh and Maharashtra say they are not, that they are illegal occupiers.

The result is that, in Madhya Pradesh and Maharashtra thousands of tribal families who are classified as landless but who are in fact cultivating land may not receive any or adequate land upon resettlement. . . . We estimate that, under the states' view, at least 60 per cent of tribal oustees engaged in cultivating land in Madhya Pradesh and Maharashtra will not receive adequate land on resettlement, (xvii).

In fact, the same point about encroachers is repeated in the letter at least six times, which is bound to leave a deep impact on the mind of the reader. What evidence is presented in the report to support these conclusions? In the first few pages of the text on MP, the authors observe that encroachers represent a much smaller proportion of oustees than in Gujarat and Maharashtra and that in the isolated tribal areas, where some 40 MP submergence villages are situated, encroachment is integral to people's economic life (176). The import of this insight is that the encroachment problem is mainly confined to this area. As we shall see presently, out of 40 tribal villages, 27 villages with their 2,000 families are badly hit by submergence (that is, the total land submergence will exceed 25 per cent of the land). Encroachers among these 2,000 families who may fail to be regularised under the existing state policy form the core problem in MP. The upper limit of the problem is more apt to be in the hundreds than in the thousands that the Morse letter would lead us to believe.

The IR report recognised that, according to the MP government's policy, encroachers prior to 1987 would receive compensation for land acquired from them and would be allotted a minimum of one and a maximum of two hectares of land (xvii, 175). The puzzle, given the policy, is how the Report reached the dramatic conclusion that thousands of encroacher families would be treated as landless. A clue lies in the advice they were given that proof of encroachment – that is, receipts of fines paid to the forest guards – are quite unlikely to be preserved by the illiterate tribals (176). This caused the IR to conclude that encroachers would face the bleak prospect of landlessness.

Their reasoning, however, was faulty. Before accepting the argument that fine receipts are unlikely to be preserved by tribals, the team ought to have spot-checked among oustees to determine if a large number of the encroachers could not produce receipts. More importantly, however, the team I ought to have considered another source of records: the forest department also has records of encroachers who pay fines. Indeed, GoMP claimed that 76 per cent of the tribal population in the affected area is formally recorded as cultivators in official records, securing for them the land entitlements to which they are due.

Nature and Size of Displacement

As noted above, the IR team, in its letter to the WB president, stated that 23,000 families were to be resettled. This figure was nowhere directly quoted in the text of the Report, though the claim is indirectly and ambiguously supported (174, 184). The team also accepted the MP government's estimate that about 50 per cent of the oustee families would move to Gujarat, requiring 23,000 hectares of land. This implies that 11,000 families would be completely uprooted from their villages. The remaining 50 per cent, who the MP officials claimed would not lose agricultural land, would remain in MP and require only 2,000 hectares of land (186).

There is a degree of confusion here. How is it that 23,000 families would need alternative land for resettlement, but only 50 per cent of the MP oustees (11,000 families) would require alternative agricultural land and go to Gujarat while the remaining 50 per cent would remain in MP? This confusion can be clarified by clearing up an ambiguity in the use of the phrase "to be resettled".

"To be resettled" in MP may mean that some families affected by the reservoir would lose only their houses and have to "resettle" by shifting them. Some, however, might lose their land as well and would have to be uprooted completely "to be resettled" at a distant place. One needs to be clear about this difference. For instance, in each of 70 out of 193 villages, 10 per cent or more of the land would be submerged, whereas in each of the remaining 114 villages, land submergence would be less than 10 per cent (185). The Report of the IR also notes that the most fertile villages of Nimar are not among these 79 villages; since they are the farthest from the dam site, they suffer inundation of house sites rather than land (185).

This distinction notwithstanding, the IR report makes a major issue of Nimar's landless labourers who fear relocation because their economic security, which has been based on a long-established web of human and geographical links, will be destroyed. It points out that "their sense of security comes from occupying a particular place in an integrated network of labourer-landowner relationships, which is threatened by resettlement process" (183). But a paradox, if not contradiction, is apparent in that if Nimar villages are only going to have their houses shifted to a higher plane in the same village, how is the "web of human and geographical links" being disrupted by relocation? How can there be a rupture of labourer-landowner relationships that will threaten the economic security of the landless by the resettlement process? And yet the word 'relocation' (read resettlement) indicates the total uprooting and displacement of these villages. Thus the ambiguous phrase "to be resettled" as used in the Report of the Independent Review conveys factually wrong impressions.

Submergence Pattern

This particular pattern of submergence – where a majority of the villages do not lose land but only houses (114 out of 193 villages) and a minority (79 villages) suffer loss of land and houses – arises from the peculiar shape of the Sardar Sarovar Reservoir and the underlying morphology of the river. The Narmada River runs through a deep gorge with a steep slope in part of its course. Through its course in the Sardar Sarovar Reservoir up to the dam site, it drops at least 100 metres. This results in a long and narrow (225 km by 1.75 km) reservoir shape, sparing fields lying well beyond the river banks but inundating a large number of houses along the banks. The further upstream one goes, the more clearly this submergence pattern (the same as in Nimar) is manifest, save some pockets in the upper reaches. Thus in MP, 114 villages are losing 2.77 per cent of their total land, and as much as 65 per cent of their houses.

But ambiguities persist about different types of displacement, and the terms "resettlement" and "relocation" are used in such a way that they leave the mistaken impression that, first, 23,000 families would require land, then 11,000 families would require land, and finally, that Nimar's landless would have to be forcibly relocated and would lose the economic security they enjoyed. It is a flaw of the IR report that it fails to distinguish among different types of displacement and resettlement.

The Report of the Independent Review has raised serious doubts about MP's 1992 *Action Plan* based on a sixty-four-point survey. It contends that "such information as is on hand as a result of government questionnaire appears to be of little value. Its findings are not sufficiently solid to ensure that plans based on them will address problems accurately" (194). In so far as MP's *Action Plan* assumed that all the oustees of the 79 villages that lost 10 per cent or more of their agricultural land would go to Gujarat, the IR report's criticism is accurate. But that is not the main outcome of the *Action Plan.* The plan is based on the different patterns of submergence we have discussed above. The IR was bound to find itself in a fix when it totally rejected the baseline data gathered by GoMP and then accepted, at the same time, data that only 79 villages faced a measure of uprooting while the remaining 114 would not be uprooted (GoMP, 1992: Part 1, Annexure V).

Analysis based on information provided by MP revealed that of its 23,180 oustee families, 11,931 families belonging to 114 villages would face land submergence of 2.8 per cent while for the remaining 11,249 families belonging to 79 villages it would be greater than 10 per cent. GoMP claimed that these families would require 23,000 hectares of land (since for each family, the land entitlement was a minimum of 2 hectares in Gujarat). There is an odd implication to this claim, which is that each of the landholdings in the seventy-nine villages would lose at least 25 per cent of its land. Only then would the holders, co-sharers, major sons, and landless labourers dependent on these holdings gain land entitlements. But this is implausible. We know that 9,148 families of the 58 villages in this group would collectively lose 25 per cent of their agricultural land (very heavy inundation – on 65 per cent of the agricultural land – would occur in the remaining 21 villages). Common sense dictates that each of the landholdings in these villages could not lose 25 per cent or more of its land. Therefore, each of the 9,148 families could not have land entitlement of 2 hectares whether they settled in MP or in Gujarat. The brunt of the land loss and major uprooting in the 58 villages would be borne only by those who lost 25 per cent or more of their land and the landless who worked on those lands (and of course, by the remaining 2000 families, including encroachers who lived in the remaining 21 villages). In all probability, the MP government's data would not provide accurate enough information about the identity of the individual families in these villages and their needs. The IR should have focused on these communities and asserted that they be properly identified. Concern for the landless of Nimar, however, should induce further analysis.

Further analysis would have revealed that in 103 villages of Nimar, only 2.82 per cent of the total agricultural land would go under submergence, leaving intact the long-established web of human and geographical links within these communities, as well as the network of labourer-landowner relationships. It can be clearly seen that the security of landless persons in these villages was not going to be threatened. As for the other villages of Nimar, 33 of them would lose 18.5 per cent of their agricultural land, which would mean that only a relatively small proportion of the population would experience major uprooting. This population would include landowners, their major sons, and landless labourers. Only in a small number of remaining villages, where land submergence was between 40 per cent and 63 per cent, would there be major uprooting.

In acquiescing to GoMP's assertion that 50 per cent of the oustees belonging to 79 villages would be uprooted, the IR team failed to perceive what this implied for the landless in these villages. As we argued above, many landowners in these villages would lose less than 25 per cent of their holdings and would not be uprooted.

Consequently, the landless working on these lands would have their web of relationships left intact. It would have been a great service had the Report of the Independent Review identified those groups of landless who were being threatened by the mindless policy of GoMP which indiscriminately sought to uproot all of them and send them off to Gujarat. Moreover, the IR ought to have recorded this violation of the NWDT Award, which gave oustees the right to choose the state within which they are to be relocated. We need to know more about the problems of the landless who did not need to be uprooted and of those who would be uprooted. They might also have enlightened us about those landless who exclusively depended on fishing for their livelihood to make appropriate recommendations. At this level of analysis, they could have had a telling effect and raised the issue of the quality of the basic data. It was only at this level of analysis that those individuals who would bear the brunt of resettlement could be brought into a sharp focus and we might expose the ways in which MP's basic data may not have been up to the mark.

Major Sons

While the then prevailing R&R policies in MP and Maharashtra excluded major sons from land entitlements, an existing law in MP and Maharashtra does allow major sons to be registered in the revenue records as co-sharers with entitlements equal to those of landed oustees. In some severely affected villages of Nimar, many influential and rich landowners have used this law successfully to get their sons registered as co-sharers. Seen in this light, the major son problem assumes a radically different perspective.

The stubborn persistence of Maharashtra and MP with their unfavourable interpretation of the relevant portion of the NWDTA was disturbing and called for a tentative approach to circumvent such a roadblock. It was disappointing that the Report of the Independent Review failed to explore solutions to this problem.

Conclusions of the Independent Review

This chapter started with a statement that SSP could not be treated as a paradigm case study to corroborate the critique of large dams propounded by Goldsmith and Hlildyard (1984). We have examined a few critical issues surrounding the environment and R&R, and have tried to show that the anti-SSP critique remains unsubstantiated and based on distortions. The intervention of the Report of the Independent Review in this critique remains a pivotal issue.

We have analysed enough material in the Report to bring out two of its general and troublesome characteristics. The first is that it suffers from grave methodological flaws that include conclusions which do not follow from factual premises; conclusions based on an inadequate database, errors in reasoning or failure to carry logical analysis far enough; and sources not properly or adequately cited. These could be broadly termed intellectual errors. The second type of error includes selection and omission of information, distortion of parts of documents, and omission of vital information that could lead to entirely different conclusions from those the team itself reached. Only careful and dogged reading of the text reveals these characteristics. Thus, the Report of the Independent Review has seriously undermined the trust on which major international human rights campaigns are mounted.

The IR team jealously guarded, and rightly so, its independence from all sources of power, including the World Bank. It missed no opportunity to let it be known that its mission of inquiry was independent. Indeed, the team was truly independent of the World Bank and the concerned governments. But was it impartial or objective in its inquiry? Serious questions must be asked on this score. Far from being impartial, the Report of the Independent Review has repeatedly gone out of its way to "prove" or "establish" facts that would show the projects in a negative light. We believe it has similarly covered up or suppressed several facts that could seriously undermine the anti-dam case. The team's partisanship with MoEF has been documented repeatedly.

In the process of its investigations, the Review team seriously undermined the nascent notion of international inquiry to make governments, especially those in the Third World, more accountable and more responsible on issues that transcend national boundaries. As experienced as it was, the team ought to have known that the idea of national sovereignty can readily put up barriers against even the universal concerns that are being expressed internationally. Third

World countries can now legitimately claim that such inquiries are a guise to undermine their sovereignty and to humiliate them. By joining hands with highly vocal NGOs in Europe, Japan, and North America, the review team has set back serious efforts to democratise States and to preserve, if possible, further human rights. They probably did not understand or grasp this important implication of their inquiry.

By failing to recommend deep, precise, and thoroughgoing reforms, the review team wasted an opportunity to help many tribal oustees. Encroachers in Maharashtra and MP could have gained immense help from their recommendations, as could have the marooned villages in Maharashtra. Similarly, major sons in the tribal areas of MP and Maharashtra would have secured a decent means of livelihood had they understood their legal situation. The review team would have earned the gratitude of Nimar's agricultural landless labourers in MP had it effectively called the GoMP on its plan to unnecessarily uproot them. Similarly, fishermen's families would have been thankful to the team for helping them secure fishing rights in the reservoir and free access to the market. Yet, instead of pressing for such constructive and creative reforms, the team decided that "It would be irresponsible [for it] to try to patch together a series of recommendations on implementation when the flaws in the Project are as obvious as they appear.... As a result, [we think that] the wisest course would be for the Bank to step back from the Projects and consider them afresh (IR, 1992: xxv).

Conclusion

When the World Bank, under pressure from Northern NGOs, decided to organise an independent review, we welcomed it even though we disagreed completely with the NGOs' critique. We relied on the Review team's stature and reputation and hoped that they would straighten out complicated issues and set the record straight. We also hoped that some long-standing problems in MP, Maharashtra, and Gujarat would be solved. We cooperated fully and with an open mind with the members of the Independent Review even when the anti-dam movement publicly questioned and raised doubts about their competence. Even when there were unmistakable warning signals about the team's strong biases against the project, we chose to ignore them. We trusted them and their objectivity to steer the inquiry to arrive at the truth and to solve some deep seated problems. We admit, though not without regret, that we were wrong.

The irony of the Report of the Independent Review is best captured in the last paragraph of the letter to the president of the World Bank: "Our job has been to make an assessment. We have done so. We have, in the course of our work, made many friends in India. We wish to assure them that our sole desire has been to find the truth and report it. We hope that our assessment may advance the search for constructive and creative solutions" (xxv).

We hold this to be their ultimate failure.

15

Sardar Sarovar Project: Some Contentious Issues

D.C. Sah and *Shubhra S. Tomar*

The Sardar Sarovar Project (SSP) is part of a gigantic scheme seeking to build more than 3000 dams, including 30 big dams, on the River Narmada. A mega-project costing Rs. 84,080 million in 1986-87 prices (revised to Rs. 145,985 million in 1992 prices), SSP is supposed to irrigate 1.8 million hectares of agricultural land and provide drinking water to 4,720 villages and 131 urban centres. It would generate additional agricultural output worth Rs 14,000 million, apart from generating 1450 MW of power. The project has been controversial since its inception.[1] The controversy started in the late 1960s on sharing the benefits and cost of the project between Gujarat, Madhya Pradesh (MP) and Maharashtra. This was followed by agitation for Relocation and Rehabilitation (R&R) policy for the project between 1982 and 1987. During the late 1980s and early 1990s, controversies again arose over the estimation of social and environmental costs and economic benefits. The opponents of SSP argued that its social and environmental costs were concealed and benefits overestimated. They believed that despite the heroic assumptions made, the project was an environmental disaster. They were critical of its techno-economic feasibility. Their criticism was also been directed at (i) the

1. The controversy that started with interstate sharing of benefits and costs, gradually shifted to environment and rights. See, for details, Sah (2003).

large area of submergence, (ii) the destruction of forests and wildlife, (iii) the interruption of the natural flow of river and resultant sedimentation, (iv) problems of waterlogging and salinity, (v) seismic problems due to the large reservoir, and (vi) risk of waterborne diseases to the downstream population.

Anyone following the SSP debate would substantiate the fact that second generation R&R problems due to SSP are equally important. The fate of over 40,000 project affected persons (PAPs)[2] who have been displaced does need concentrated efforts. The violation of the rights to livelihood of some of those PAPs who were allotted poor quality land has also been a subject of heated debate.

The debate on R&R of the tribals in the Narmada Valley is informed by numerous sources including print and electronic media, the State, the Greens, the Morse and Berger Independent Review, the pro- and anti-dam activists, tribals in the submerging zones and the rich landowner beneficiaries of project, the interests of Gujarat state and concerns of MP government. Each of these agencies has its own perspective and source of information. More often, the aim of these agencies is either to discredit R&R, branding it as impossibility, or proclaim that everything with respect to R&R is as planned. The facts lie between the two extremes. Of late, right to livelihood of PAPs has become a moot question wherein the insensitivity of the major agents towards grassroots needs has led the judiciary to play an active role. Human rights of the poor, for one reason or another, have always been disregarded by the Indian State using the excuse that these rights

2. It is estimated that about 40.7 thousand families from Gujarat, Maharashtra and Madhya Pradesh will be displaced from their habitats owning to construction of SSP and out of this a majority belong to tribal population. Experiences relating to relocation and rehabilitation (R&R) of SSP displaced people are far from encouraging. A review of the R&R of displaced population of development project reveals that in absence of a National R&R policy, inadequate legal provisions for land and property acquisition, and lack of institutional provisions to relocate and rehabilitate, the displaced population faces bureaucratic insensitivity and loss of livelihood. The livelihood adaptation by community either in improving management of commons (Jodha, 1981) or in creating an alternative to top-down approach of state, or in scarcity management (Desai, Singh and Sah, 1979) have earlier been attempted.

have different meanings in different societies and a Western concept cannot be imposed on us. This argument is typical of states that have little sensitivity for human misery. Unfortunate though it may seem, both the State and civil society have failed in protecting the livelihood rights of a section of PAPs. Consequently, the Supreme Court interpreted some of the rights implicit in the Directive Principles of the Constitution and made them enforceable.

SSP is also one of the development projects in India that has received attention and scrutiny to an extent that it has changed the rules of how projects should be formulated and examined for their socio-economic and environmental viability and executed so that human displacement is minimised. This paper has an aims to document some of the controversies associated with SSP and important lessons of the project, so that the mistakes committed in it are avoided in future development projects.

I. The First Generation Problems

It may be argued that the emergence of NGOs sensitive to actual needs of the PAPs has changed the political process of empowering the poor.[3] In fact the shedding of built-in inertia to act with humane

3. A few Sarvodaya activists who took the cause of the oustees of Ukai Dam in south Gujarat were concerned about the future of the PAPs of SSP. They along with other non-party activists began to work in the area and mobilised tribal population for a 'better deal'. The discriminatory provisions of the 1979 GR were highlighted. Public demonstrations of the tribal people were organised. Writ petition was filed in the Supreme Court. The Court appointed a commission to inquire into the availability of alternative land for rehabilitation. Media and researchers repeatedly raised issues for proper rehabilitation. As a result of growing pressures from NGOs world over and their concern for environment and tribal people, the World Bank evolved guidelines for rehabilitation for the projects supported by the Bank. Moreover, by the early eighties the Government of India began to pay attention to the problems of tribal oustees. In 1984, it formed a committee for evolving a policy of rehabilitation for tribal oustees displaced by development projects. The R&R package of SSP is a land for land compensation package which recognised every adult in the submerging village as a PAP and provided 5 acres of land in resettlement site,

values in case of SSP is the outcome of a combination of forces: a strong grassroots movement and initiative of civil society[4]; an in-depth understanding by academics of the socio-cultural needs of affected people to generate alternatives in policy formulation; and pressure by the international funding agency—the World Bank—to allow for various options have played an important role in developing an acceptable R&R policy. It is least likely that such interplay of forces can be accommodated in many development projects so that a just R&R becomes part of the project. Not only the existence but also a change in the nature of initiative by various participants could result in a different R&R policy frame.

In a normative sense, after accepting the R&R package, an important institutional intervention was to introduce accountability in land acquisition and relocation. Barring a few exceptions on the front of providing land, the State and the NGOs did a creditable job; PAPs did get what was due to them in terms of land and infrastructure. While a policy frame cannot solve all the problems of society, it nonetheless reflects the concerns we have and the type of society we believe in.

---

along with a house plot and temporary shed plus transportation, plinth construction, village civic amenities, one time up-keep allowance of Rs. 4500, implement grant. But this package was accepted after a prolonged struggle. This flexibility demonstrated by Government of Gujarat, in liberally defining the PAPs, was forced. To begin with the Narmada Water Dispute tribunal Award recognised only legal land owners in submerging villages as PAPs and hence only these were entitled for a compensation package. Pressures from NGOs, social activist and the Tribal protest forced the state to recognise encroaches, grown up sons of the PAPs, sharecroppers as a separate family affected by SSP.

4. ARCH-Vahini was the first NGO in Gujarat to mobilise the tribal community in order to protect their right to livelihood. Recognising that the then existing norms do not recognise the landless labourers, sharecroppers and forest encroaches as PAP, ARCH worked relentlessly for forcing a just R&R policy for four years starting 1983. In December 1987 government of Gujarat was forced by this grassroots movement to announce a new R&R policy. ARCH was joined by other two NGOs, Sarmik Vikas Sansthan and Rangpur Ashram, to help the state in relocating and rehabilitating PAPs after 1987.

Viewed against the backdrop of SSP experiences, the R&R also reveals many positive features. The land-for-land policy of compensation is one of the most important of these. By accepting land for land, subjectivity in price fixing can safely be avoided. Tribal groups, often residing in remote undulating areas, sustain themselves in low-yield equilibrium. Cultivating larger area compensates for the low yields. Through the R&R policy, the procedure of compensating on quality based pricing of land would bring not only subjectivity but also meagre compensation to the PAPs.

The R&R policy of SSP has subdivided the implementation plan into a relocation plan and a resettlement plan. In doing so, it has disassociated the relocation process from the project construction. The R&R policy states that this plan would be ready at least a year before the start of land acquisition. In the process, as experiences show, relocation takes place *pari passu* (simultaneously) with submergence. The problem with *pari passu* relocation is that project authorities turn indifferent to the needs of PAPs as time passes. But the clause cannot be avoided. Thus, a project should commence only after ensuring equal treatment and package to early as well as late resettlers.

We shall mainly focus on the R&R issue, as nearly 40.7 thousand PAP from 245 villages have to be rehabilitated. We shall confine our analysis to PAPs from 19 Gujarat villages. Before examining their present condition and possible problems faced by them, let us briefly review the R&R policy.

The R&R Policy

Even after more than six decades of Independence, the country as a whole does not as yet have an R&R. What we have today except in Maharashtra and MP is Government Resolutions or (GRs) Government Orders that vary from state to state, and in some cases from project to project within the same state, to meet exigencies of the local situation. Till recently, the common central aspect of rehabilitation was The Land Acquisition Act, 1894, enacted by the British Government of India. The underlying assumption of the Act is individual ownership of land as maintained in legal records. This Act provided compensation for the land acquired, but not for related means of sustenance or livelihood such as forest and other common resources. After receiving compensation for one's own legally owned land, one was left at the mercy of market forces for resettlement. Local and community variations, particularly the nature of and the needs of tribal communities affected by the project, were not taken care of. As a result, a tribal unaware of the intricacies of the law and the market was placed on par with a rich capitalist peasant. Such policy—a British legacy—continued till the early 1980s.

By this time, deteriorated conditions of tribals who lost land and other resources to various 'development' projects were highlighted by the media and by social science research. The State was alarmed by the Naxalbari and other militant tribal movements. To counter them, it launched several welfare programmes to improve the condition of tribals. The political parties competing in electoral politics began to woo the tribals and other deprived sections of society for votes. A few non-party groups raised issues related to rehabilitation of the tribals and their right over forest resources. All these concerns found an echo in the directives for rehabilitation of PAPs given by the Narmada Water Dispute Tribunal Award (NWDTA) in 1978.

The NWDTA defines a PAP as an 'oustee' not just one whose property would be submerged but any person who, since at least a year before the publication of the notification, has ordinarily resided or cultivated land or carried on any trade, occupation, calling or work for gain in the area likely to be submerged permanently or temporarily. This definition includes landless labourers and others who would be affected by submergence even if they do not own any land. In defining the 'family' of an oustee for the purpose of providing compensation and other benefits, the Award stipulates that every major son constitutes a separate family.

The NWDTA directed that Gujarat bear all the costs, charges, and expenses for rehabilitating the oustees. Gujarat was also directed to establish rehabilitation villages in the command area of SSP, for which it was to acquire necessary lands within two years of the decision of the Tribunal. Gujarat was also to acquire and make available one year in advance of submergence irrigable land and house sites for rehabilitation of oustees from MP and Maharashtra who were willing to resettle in Gujarat. According to the Award every displaced family from which more than 25 per cent of its landholdings had been acquired was entitled to, and was to be allotted, irrigable land to the extent of land acquired from it, subject to the prescribed ceiling in the concerned state and a minimum of 2 hectares per family. Moreover, every PAP family was to receive a house site measuring 60 feet by 90 feet, free of cost. Every resettlement site was to be linked to a main road and certain civic amenities were to be provided for every group of PAP families resettled there. These include a drinking water well and a platform for every 50 families, a primary school for every 100 families, and a *Panchayat ghar*, dispensary, seed store, children's park, and village pond each, for every 500 families,

Three underlying assumptions guided these provisions: (1) that the oustees, a majority of whom are tribals, have individual ownership of land which they cultivate with legal title; (2) the government could provide irrigable land at one place for a minimum of 50 families; and (3) the oustees would prefer to resettle in a group.

The Government of Gujarat (GoG) interpreted the directives of the Award as being applicable to the oustees of Maharashtra and MP and not to those of Gujarat. Therefore, GoG issued a resolution in June 1979 which laid down standards for the amenities to be provided at resettlement sites for such oustees of large and medium irrigation projects in the state as had not already availed of rehabilitation facilities under project-specific policies. This resolution by and large followed the directives of the Award as far as the definition of 'oustee' and civic amenities to be provided in the resettlement villages were concerned. However, there were four major differences: (1) the GR considered only adult sons as separate families; (2) joint landholders were not to be considered as separate families for the purpose of allotment of agricultural land even if they constituted separate families according to the definition of 'family' in the Award; agricultural land was to be allotted thus to unit holdings obtaining in the revenue records; (3) there was no commitment in the resolution to provide irrigable land, or land in the area benefited by a project; (4) the plot for the house site as per the GR was smaller than that provided in the Award.

Civil society in Gujarat, who took up the cause of the oustees of Ukai dam in South Gujarat, was concerned about the future of the PAPs of SSP. They, along with other non-party activists, began to work in the area and mobilised tribals for a 'better deal'. The discriminatory provisions of the 1979 GR were highlighted. Public demonstrations of the tribals were organised. A writ petition was filed in the Supreme Court. The Court appointed a Commission to inquire into the availability of alternative land for rehabilitation. The media and researchers repeatedly raised issues about proper rehabilitation. As a result of growing pressures from NGOs the world over and their concern for environment and tribals, the World Bank evolved guidelines for rehabilitation for the projects supported by it.

After initial resistance, GoG relented to the pressures, particularly from the World Bank, which had provided a loan to the project. In 1985, GoG extended the provisions of the Tribunal Award to apply to project oustees from Gujarat as well. It passed several GRs to that effect and accepted the guidelines of the World Bank as objectives for R&R of the oustees. These included clauses for PAPs (a) to improve or at least regain the standard of living they had enjoyed before displacement; (b) to be relocated as village units, village sections or families in accordance with their preference; (c) to be fully integrated in the community in which they were resettled; and (d) to be provided with appropriate compensation and adequate social and physical rehabilitation infrastructure including community services and facilities. The major provisions of various GRs, besides community and infrastructure amenities as mentioned in the Award, were as follows:

(a) Each PAP, landed or cultivating forest land or landless agricultural labourer or trader or grown up sons, is entitled to get irrigable land in the state in which he chooses to resettle, of equal size of that which he owned prior to his resettlement. In those cases where the oustee owned less than 2 hectares of land, he is entitled to get at least 2 hectares of irrigable land acceptable to him.

(b) Oustees will get *ex gratia* payment to enable them to buy 2 hectares or equal size of land from private sellers.

(c) Every PAP family is entitled to get free of cost a plot (60′ × 40′) for house site.

(d) Every PAP family will get a sum of Rs. 750 inclusive of transport charges as resettlement grant.

(e) In order to improve the economic standard of the oustees, the government would provide subsidies for agriculture and other occupations, and proposed a number of family oriented development programmes.

(f) The Project also promised to create amenities and facilities like drinking water, approach roads, village platform, primary school, *Panchayat ghar*, seed store, children's park, dispensary and pond at the new sites.

Thus, at a policy level, the provisions mentioned above are favourable to the PAPs. How far have they been implemented? What has happened to the PAPs' quality of life after rehabilitation? These questions will be examined in the next section.

Relocation

The R&R policy of SSP is one of the best in the country, but as mentioned earlier, GoG accepted it after a prolonged struggle. Relocation from the submerging villages, however, started before acceptance of the policy. Early resettlers did not immediately shift to the new sites; over 70 per cent of them took more than 5 years to make the new site their home. During this period they commuted, or members of households split, between their old and new locations. Social interactions between the hosts and new settlers were by and large cordial. Visiting each other during sickness, death, weddings and religious occasions was common. But on common property resources depleted due to the new settlers, or when social groups were dissimilar and unknown to each other, relations between them two became occasionally tense.

The commitments made by the state at the time of relocation have been fulfilled to a large extent: only 4 per cent of oustees have not received agricultural land. Since these oustees have been recognised as PAPs, they are in the process of acquiring all the benefits available to them, including allocation of agricultural land. Barring a few exceptions in Malu, Piparvati, Ambavadi, Chikda and Kali-Talavadi, they are satisfied with the quality of their land. They highly appreciate the facilities provided such as education, road and transport, health care services and institutions like input and output markets.

There is on the whole no deterioration in their living standards after relocation. Going by the condition of early resettlers, it can be argued that the tribals who were better exposed to market are gradually adjusting to their new social milieu. The support structure to facilitate their participation in the modern economy is weak. This puts them at a disadvantage compared to those who have been part of the system longer. They do feel a loss of identity and security. But that feeling is not strong and pronounced. Despite some irritants and complaints, at the subjective level they seem happy. They are, however, unhappy about fuel and fodder constraints, tin-shed transit houses and frequent disruption in drinking water facilities.

In the submerging villages, health care services were provided mainly by a *Bhuva* (tribal medicine man); alternatives to this were totally lacking. After relocation, the traditional health care services were largely replaced by modern services like primary health sub-centres, visits of health workers and private practitioners. Despite these, the PAPs relocated in 45 out of 120 new sites have to travel about 6 to 10 km to reach a health provider. Moreover, though health care preferences have been considerably influenced by allopathic services, the PAPs' perceptions about diseases are still guided by their experiences of the submerging villages. Another notably improved service is educational facility in the new sites compared to submerging villages. Not only were schools inaccessible in the latter, a sizeable population of school-going children was engaged in cattle grazing and agriculture. Consequently, literacy rates were low; only about 25 per cent males and 6 per cent females were literate in the submerging villages. However, in the new sites, after about 15 years, the literacy rates increased to 41 per cent among males and about 18 per cent among females. Nonetheless, fragmentation of families, agricultural land to major sons, lack of access to fuel and fodder, high prices of land, relocation site outside the command area, short-changing in land, fragmented and encumbered plots, non-functional hand pumps and lack of access to drinking water, hardship of living in tin sheds, institutional isolation were some of the major first-generation problems faced by the early resettlers. The state was quick to solve these problems. Serious though they are, the existing R&R policy has provisions to tackle them. More serious are the second-generation rehabilitation problems for which no such provisions exist.

As will be seen in next section, the Bhil, Nayka and Vasava of interior submerging villages are finding it difficult to adapt to new agriculture. Even during normal years about two out of every ten PAPs do not have access to minimum nutrients. Although the relocation of over 4500 PAPs had been achieved quite successfully, their rehabilitation has just started. For it to succeed, an investigation for policy review and revision is needed.

II. Second Generation Problems

Agriculture is still a way of life in tribal society where land is not only a production input but also has a cultural affinity. Despite this and a number of legislative enactments, as well as a quantum jump in financial allocation to tribal development, the heavy influx of outsiders with commercial interests, and development interventions, have eroded the resource base and socio-cultural heritage of the tribal population (Government of India, 1992a). Despite growing concerns,[5] the disruption of economic environment and lifestyle of this group remains uncompensated owing to unimaginative and insensitive relocation package (Government of India, 1990).

It has been argued (Sah, 1995b) that alienation and deprivation are inherent in the strategy of Indian economic development. Large dams in independent India have become the new icon of development and are considered to be among the main sources of irrigation on which the synergistic effect of all other agricultural inputs depends. The SSP also has similar claims.[6] Even these claims have not remained

5. It is estimated that over 16 million people are displaced from their habitat owing to construction of dams since Independence and out of this, about 40 per cent belongs to the tribal population (Mitra, 1995). Experiences relating to relocation and rehabilitation (R&R) of displaced people are far from encouraging (Mathew, 1989; Reddy, 1994; Thukral, 1992). A review of the R&R of displaced population of development project reveals that in the absence of a National R&R Policy, inadequate legal provisions for land and property acquisition, and lack of institutional provisions to relocate and rehabilitate, the displaced population faces bureaucratic insensitivity and is forced to join the growing rank of wage labour.
6. Sardar Sarovar Project is one of the largest irrigation projects, costing the state a sum of Rs. 84,080 million in 1986-87 prices. It is estimated

unquestioned,[7] especially in relation to the rehabilitation of displaced persons. However, it can be argued (Wood, 1993) that within the given constraints of the development paradigm, the compensation package to PAPs of SSP is far better than that of any other irrigation project. Although, land has been provided to all the PAPs, its optimum use may have been constrained due to adequate support.

Relocation of project affected people brings a spate of changes along with it. SSP is no exception to this. When relocation is accompanied with changed economic environment, where traditional mutual exchange economy is replaced all of a sudden by markets, the oustees may need support to (a) visualise and get adjusted to competitive markets, and (b) have resources and information to become active participants in technologically dynamic agriculture. This may become all the more necessary because agriculture in the submerging villages is unable to generate surplus, and by and large, the oustees at the time of relocation had little entrepreneurial abilities to experiment and adapt to new ideas. The oustees in the new environment may make efforts to adopt new

---

that it would (a) irrigate 1.8 million hectares of land, (b) provide drinking water to 4,720 villages and 131 urban centres, and (c) generate 1,450 MW power. All these benefits in monetary terms would be Rs. 14,000 million of additional production per year (Raj, 1990; Government of Gujarat, 1992; Patel, 1991).

7. The opponents of Sardar Sarovar Project argue that this project hides its social and environmental costs and over-estimates its benefits and believe that even under heroic assumption this project is nothing but environmental disaster. These activists have been critical of the techono-economic feasibility of the project. Their criticism has also been directed to (i) large area of submergence, (ii) destruction of forest and wildlife, (iii) interruptive approach to the natural flow of river and resultant sedimentation, (iv) problems of waterlogging and salinity, (v) seismic problems due to large reservoir, and (vi) risk of water-borne diseases to downstream population. We certainly don't want to project these issues as unimportant. Nevertheless, anyone who is observing SSP (Narmada) activities could substantiate the fact that second generation human problems created because of SSP are no less important. The fate of over 20,000 people, who have been displaced and relocated do need concentrated efforts (Alvares and Billorey, 1987; Kothari and Singh, 1988; Paranjpye, 1989; Morse and Berger, 1992).

ways in agriculture and may start developing input and output linkages with markets. But they may be unable to overcome the investment constraints and therefore to reap the full potential of their productive resources. There may be variations in technology adoption between the host villages and new sites, the farm income may he higher on the farms of experienced host villagers. Linkages with credit market were extremely weak in the submerging villages. Thus it will be worth understanding whether low demand for production credit is responsible for low technology adoption in dry relocation zones. The available agricultural technology itself may be risky in the dry relocation sites. In this situation, it may be desirable to provide long-term support for reducing yield variation across farms through direct investment in agriculture.

The process of integration with mainstream agriculture without extension, irrigation, and credit support may have been left to market forces. PAPs may be active participants in the existing markets or may avoid them because of the perception that they may become exploitative. But markets, nonetheless, play a facilitating role in rural transformation. At the micro-level they help industrialisation by resource outflows from agriculture in the form of labour and economic surplus; create demand for non-farm consumer goods and agricultural inputs; and facilitate land allocation and transfer of agricultural technology. But such commercialisation may take varying paths depending on the stages of agricultural development.

Traditionally, sectoral transformation is driven by surplus in the primary sector where irrigation works as the prime mover of such commercialisation. In the case of oustees, their acceptance, participation and integration with mainstream agriculture and its market institutions have been forced by relocation. Their socio-economic milieu may be responsible for the process of market penetration or exploitative development of interlocked markets. When commercialisation is forced by involuntary migration, the PAP farmers' response to markets may be guarded: it may be confined to the product market and be guided by village trade. In order to avoid interlocked output and input or output and credit markets, oustees may avoid participation in some of them. Therefore, PAPs' responses to forced commercialisation would vary across locations, farms and crops. The important question is how the PAPs, being unfamiliar with many of the intricacies of markets, have responded to commercialisation in agriculture.

The resettlement and rehabilitation (R&R) policy of SSP emanated over a period of time and after a long struggle by activists in Gujarat for a better deal for the displaced population (Sah, 1995b). The R&R policy frame for PAPs of the SSP seems significantly superior to similar packages available to PAPs in the past. Over 4,500 PAP families who have been displaced from 19 submerging villages have been provided with land. However, this R&R package heavily rested on the productive resource 'land' for maintaining the living standard of PAPs.[8] This paper attempts to compare productivity, farm production, input use, and the extent of commercialisation of agriculture of PAPs before and after relocation. It could be argued that such a comparison may not be correct when the soil type, topography and access to irrigation vary between the two situations. Nonetheless the comparison would still provide insights on how the PAPs' living conditions and access to food have changed.

Within the frame developed above, this paper seeks to understand the adjustment behaviour of relocated SSP oustees towards technology adoption and commercialisation, and the effect of these processes on their standard of living during normal as well as abnormal years.[9]

---

8. For a detailed discussion on first generation problems faced by PAPs, see Sah, 1994b.
9. At the end of 1984, the Narmada Planning Group of the Government of Gujarat, commissioned the Centre for Social Studies (CSS) lo carry out the task of monitoring and evaluating the process of rehabilitating the Gujarat PAPs. Since then, the CSS has been involved in this work. The CSS has, so far, submitted twenty-one six monthly reports to the Government of Gujarat. On the basis of the experience of the three monitoring agencies — Gujarat, Maharashtra and Madhya Pradesh — questionnaires were prepared. Each relocated household was visited three times a year to seek data on production, consumption and the demographic details of its members. The reports, inter alia, cover: (1) the socio-economic conditions of the PAPs and the changes therein; (2) allocation of land and problems thereupon; (3) infrastructure facilities; (4) development programmes; and (5) relationships between the PAPs and host villages. Almost each report carried recommendations related to policy and/or day-to-day problems faced by the PAPs. The government began to take note of our recommendations or observations from 1989 onwards (see Sah and Soni, 1993; Sah, 1993; 1994a; 1995a; 1996a, b).

The paper is divided into six sections, including the introduction. The second section presents the findings on allocation of agricultural land to PAPs' productivity of allotted land, consumption intake and variations in it caused by variations in farm production due to fluctuating yields. The next three sections present analyses respectively on the coping mechanism of PAPs during scarcity, the constraints in technology transfer, and commercialisation. The last section presents the concluding observations.

Productivity Gains of PAPs after Relocation

Relocation has not only increased the land assets of PAPs hut has also changed the way land is owned and operated by PAPs. From joint land operations in submerging villages, which were prevalent among about 43 per cent of the households, cultivation has moved towards individually owned enterprise (Table 1). About 93 per cent of the holdings in the new sites are operated by individuals. On the other hand, for every acre owned in the submerging villages, more than one acre was encroached forest land. In this process, though only 4.2 acres of land was owned by cultivating households in

Table 1: Land Owned by PAPs in Submerging Villages and New Sites

Particulars	*Submerging villages*	*New Sites 1993-94*
Total households	2,310*	2.981**
	Percentage	
Landless	32	6
Individual holdings	53	93
Size of holding		
Small	48	75
Medium	14	15
Large	6	4
	Acre per PAP	
Size of holding	4.2	5.7
Operated area	9.2	5.9

* In 14 out of 19 submerging villages: PAPs from remaining 5 villages had already relocated at the time of the study.

** In 120 relocated new sites by July 1994.

Source: Sah, 1993; 1996a.

submerging villages, PAPs operated 9.2 acres of land. About 32 per cent of households in the submerging villages did not own land legally. They were cultivators who had encroached forest land. In contrast to this, at the new sites all the land is owned; an average PAP owns about 5.7 acres. By September 1996 about 96 per cent of the total 4,557 PAPs had been allotted agricultural land. Only 4 per cent of the total relocated PAPs are landless. The bulk of the remaining 4 per cent landless households are in search of agricultural land in various new sites.

Table 2 gives the proportion of cropped area allocated to different crops and crop yield of PAP households in the submerging villages and in their new sites. After relocation, easy access to markets and increase in price fixation have been responsible in replacing traditional low value food crops with high value crops like cotton, paddy and wheat. In the new sites, about 60 per cent of the gross cropped area (GCA) is allocated to cereals, 23 per cent to pulses and the remaining 17 per cent to commercial crops – cotton, oilseeds, fodder and vegetables. In comparison to new sites, the PAPs in their submerging

Table 2: Cropping Pattern and Yield Changes after Relocation

Crops	*Area allocation*		*Yield*	
	Submerging villages	*New sites* 1993-94*	*Submerging villages*	*New sites* 1993-94*
(1)	(2)	(3)	(4)	(5)
	Percentage of GCA		*Kg per acre*	
Paddy	2	14		379
Maize	30	22		220
Jowar	42	22	95	212
Wheat	0	2		611
Bajri	2	@		63
Minor cereals	5	–	–	–
Tur	11	21	95	179
Other pulses	6	2	87	161
Cotton	@	12	87	187
Oilseeds	1	1	–	182
Fodder	–	4	–	NA

Source: Sah (1994a; 1995a; 1996a,b).

* Based on 610 households in submerging villages and 536 households in new sites for 1993-94.

villages had over 96 per cent of the area under cereals and pulses, and only about 2 per cent under commercial crops.

About 10 per cent of GCA in the new sites is under irrigation. Yields in such sites for paddy, cotton, maize and wheat are significantly high even in drought years. A large proportion of resettlement sites fall under the Narmada command, but 43 out of 126 of these are located where Narmada water will not reach. Agriculture in these areas is vulnerable to fluctuating weather. Special schemes and additional funds are needed for providing groundwater irrigation to PAPs relocated in these 43 sites. Nevertheless, in comparison to the submerging villages the yield obtained by PAPs in the new sites is about twice as much for the main cereals, pulses and commercial crops. At constant prices, farm production per household has increased from about Rs. 4,100 in the submerging villages to about Rs. 10,048 in the new sites in 1993-94.

Variations in Calorie Intake: Distress at the Margin

The findings of productivity gains of PAPs have been based on average production during two normal years. This optimism of superior productivity and farm production in new sites should, however, be interpreted with caution, for there are wide variations in productivity over the years as well as in a year over farms (Table 3). Such weather induced fluctuations in this drought prone region are mainly due to abnormal rainfall. Over the last 16 years, two out of every five years have been agriculturally abnormal.

Although the average farm production of households after relocation has increased two and a half times when compared with the submerging village, this production has been extremely low during 1991-92. The farm production has direct bearing on consumption pattern; the caloric, intake per adult per day has reduced during abnormal year (1991-92) and the production of households reporting lower than 2400 calories per day per adult intake has increased (Table 3). In normal years, the proportion of households reporting less than 2400 caloric intake per adult per day is about 50 per cent, and in 1991-92 it increased to 72 per cent.

Even during normal years, this proportion is sizeable. This is so because there are some locations where farm productivity remains

Table 3: Yield Calorie Intake and Perception over Time

Particulars	*Submerging villages*	*New sites*			
		1991-92	*1992-93*	*1993-94*	*1994-95*
(1)	(2)	(3)	(4)	(5)	(6)
Sample size	610	670	578	536	718
Farm production (Rs.)	4,100	3,002	7.807	10,048	3,309
CV (%)	N.A.	(120)	N.A.	(110)	(107)
Farmers' reported total crop failure (%)	@	40	@	@	30
Caloric intake/ adult/day	2.588	2.476	2.493	2.446	2.529
CV (%)	N.A.	(33)	(28)	(21)	(27)
PAPs consuming <2400 calories/day/adult (%)	53	72	48	57	33
PAPs reporting deterioration in economic condition compared to submerging villages (%)	-	2	13	25	38

@ Less than 2%, N.A. = Not available.

Source: Sah (1994a).

relatively low. Households located in new sites at Ambavadi, Amroli, Chandanagar, Dabhvan, Dhamadra, Gora, Kavitha, Lachharas. Pipervati, Sandhia and Vaghrali have reported relatively low farm income. A subset of households located in Dabhvan, Dhefa, Dormar, Jemalgadh, Kadila, Khandania, Lachharas and Rustampura have reported their calorie intake perpetually low over the years.

During the early years of resettlement, the PAPs' perceptions about their economic condition, compared to what it had been in their submerging villages, were positive. Despite the fact that 1991-92 was an abnormal year, only 2 per cent of PAPs reported a negative perception. This proportion reached 38 per cent in 1994-95. Surprisingly, though as many as 30 per cent of the farmers have reported crop toss in the case of crops like maize and jowar and the farm production declined to Rs. 3,309 and a record 38 per cent of the farmers perceived their economic condition as having deteriorated, the consumption intake has not been affected adversely. This may be due to high food stock remaining from the previous year's bumper crop, or past savings enabling consumption to remain unaffected, or to current borrowing. It reflects the increasing capability of the average farmer to meet crop failure without much loss in consumption intake. But it is not very often that a bumper crop precedes a crop failure.

Nonetheless, the effect of crop failure, low productivity and unproductive land is serious at the margins. PAPs in the new sites of Malu, Ambavadi, Pipervati, Chikda-1, Chikda-2 have protested about these conditions. In order to supplement farm income, which is not sufficient to meet family demand, a large number of family members from Ambavadi and Pipervati have sought wage employment. The situation in mid-1990s has become so bad that 22 households from Malu, 12 from Ambavadi and 10 from Pipervati have gone back to their submerging villages. An organised protest is developing in these areas with a demand to change unproductive land. Although, all the PAPs do not desire to relocate, there is a growing demand for shifting to submerging villages. Agricultural productivity is low and avenues of employment are poor in this area. As a result many PAPs cannot depend on agriculture alone. These PAPs prefer to lease out their land and (a) go back to submerging villages to do cultivation; (b) migrate to Surat, Anklehwar and dam site to work as wage labourers; and (c) move to other new sites where their close relatives have relocated. These tendencies indicate the inadequacy of some households to sustain only on their own agriculture.

This is the situation where households face perpetually low yield due to unproductive agricultural land. The adjustment mechanism, however, is not significantly different for households who face scarcity.

Coping with Scarcity

Crop failure has a number of simultaneous and sequential repercussions, which disrupt the economy of the affected area. In submerging villages, access to forest and river on the one hand, and social institutions and mutual exchange relations of the closely knit society on the other, insulated the low level economy from shocks, helping it to sustain; even during the worst of droughts, the household consumption was only marginally affected. The fall in crop production at new sites, however, directly reduces household income and availability of food for consumption. It also reduces avenues of farm employment. While current income declines, the need for cash to purchase food increases. Households in such situations make various adjustments to minimise the impact of crop failure. These include migration of some family members, reduction in the use of purchased inputs, sale of milch animals, reducing expenditure, and changing the consumption pattern to less expensive items. It is hypothesised that households with (a) higher agricultural income, (b) additional purchasing power through agricultural labour and casual work, (c) milk production, and (d) larger out-migration cope with drought more easily without suffering a severe impact on their consumption level.

Table 4 reveals that calorie intake is independent of avenues generated through agricultural labour, higher milk production, service, availability of loan, and cropped area. But these findings clearly establish that (i) households engaged in casual labour are worst affected by crop failure and their calorie intake is significantly low; (ii) distress sale of cattle leads to significantly lower calorie intake in households making such sales; (iii) households whose agricultural income is higher have significantly higher calorie intake; (iv) households with fewer mouths to feed have significantly larger calorie intake; and (v) households with larger out-migration have significantly higher calorie intake.

Table 4: Correlation and χ^2 values between Explanatory Variables and Calorie Intake Per Capita

Explanatory variable	*χ2 value*	*Correlation*
(1)	*(2)*	*(3)*
With per capita calorie intake		
Casual labour work	22.78***	-0.19***
Milk production	1.48	0.04
Cattle sale	4.75*	-0.08*
Agricultural income	7.48*	+0.10**
Total members	65.16***	-0.31***
Migration	20.43***	+0.13***
Agricultural labour	2.58	-0.02
Service	1.70	0.05
Loan taken	–	–
Cross cropped area	3.38	0.01

+ Based on round III data with N = 565.

*, ** and *** Significant at 10, 5 and 1% level respectively.

Source: Sah (1994a).

The findings on â coefficient (Table 5) reveal that among the variables explaining variations in calorie intake, the three most important are the size of the household, out-migration, and cattle sale. These explained the bulk of the variance (16.2%) in calorie intake across households. A low R^2 also indicates that some of the important variables like past savings, food stocks and current borrowings are missing from the explanation. Although these marginal products should not be considered as precise estimates of quantum of effect of explanatory factor on calorie intake, the findings of Table 5 indicate that (1) in 1991-92 when calorie intake reduced by about 18.50 per cent or roughly 480 calories/adult/day, some members of the household decided to work as casual labour; (2) if household size increased by one member, it reduced the calorie intake by about 3.7713 per cent or by about 98 calories/adult/day); (3) if the household's agricultural income per head increased from low to medium level, its calorie intake increased by 2.395 per cent or by about 65 calories/adult/day; and (4) as soon as the intake reduced by 3.096 per cent or by about 100 calories/adult/day, the household took a decision to migrate some (around 10%) of its family members.

But the farmers' coping mechanism operates without any external support and relies mainly on migration, cattle sale, and casual labour. With traditional sustaining mechanisms distorted due to randomness of resettlement,[10] all these options in the new site may create further

---

10. A total disregard for social implications of relocating PAPs from 19 submerging villages to 125 plus new sites is paradoxical. The relocation process could be seen from two different dimensions: first, the traditional social links and economic institutions at the submerging villages had an important place in the life of PAPs. These arrangements had helped in sustaining themselves in the submerging villages. The randomness of relocation in absence of these arrangements at the new sites has made the PAPs vulnerable to exploitation. The other dimension to the process is the appropriateness of the existing policy. The people associated with resettling the PAPs defend the policy and the process by reiterating that the decision of shifting is an individual decision. The strength to the argument is drawn from the fact that the village is not a homogeneous unit; even in a phalia the people do not always want lo shift together. And since the PAPs were shown land in different locations, their moving to the new sites is voluntary, for PAPs

Table 5: Effect of Changes in Some Important Factors on Calorie Intake/Adult in 74 New Sites

Factors	*Coefficient* [a]	*t statistics*	*β rank* [b]
Cattle Sale	– 0.1851*	– 2.332	III
Agricultural income	0.0240*	1.900	V
Migration	0.0031***	3.972	II
Casual labour	– 0.0040*	– 2.160	IV
Family size	– 0.0377***	– 7.684	I
	R^2 = 0.162	F = 22.82**	N = 565

a Log (calorie = log a + b_1 cattle sale + b_2 income class + b_3 log (migration) + b_4 casual labour + b_5 (family size) + U_1
where b_1 = elasticity/coefficient; a = constant; and U_1= error term.
b Showing the importance of factors in explaining the variations in the calorie intake.

Source: Sah (1994a).

social disruptions. Moreover, these options together explained less than 20 per cent of the variations in calorie intake, indicating that many of the distress options have not been revealed by the respondents. This situation may become a cause for concern and may result in depletion of productive assets. The farmers' coping mechanism, therefore, has to he strengthened by external support in the form of (i) short term immediate income generating avenues giving respite from depleting food grain stocks during drought, and (ii) long-run efforts to insulate the farm economy from fluctuating rainfall. Moreover, the state has also to develop mechanisms that enable short-term relief programmes to be taken up immediately.

have preferred the land they are cultivating in the new sites. The argument could have been accepted had there been some method in the randomness of the relocation. It is often argued that the policy provides avenues and instrument of choosing the new site. The fact is PAPs certainly had a choice; to choose the best amongst the worst. Thus the migration in the absence of social and economic arrangements is not only involuntary but also an outcome of a policy incapable of safeguarding the risk of social disruption. The misery could have been avoided with adequate policy frame and planning.

Technology Adoption

This section, using a case study,[11] tries to bring out the contrast, if any, between PAPs and the host villagers with respect to technology adoption, yield and income in the developed cotton zone, semi-developed paddy zone and in the undeveloped maize zone of the resettlement sites.

Farm Income

The sample farms in new sites are predominantly medium size whereas the farms located in host villages are small and medium. This is because agricultural land belonging to PAPs has not yet been fragmented by divisions in households. On the other hand, population pressure has resulted in the division of large farms into smaller holdings in the host villages.

Cotton, an important commercial crop, accounts for about 20 per cent of the GCA in the host villages. The importance of cotton in the new site is relatively low and this differential area is allocated to food crops like jowar and paddy. More importantly, allocation of GCA to hybrids (HBs) and high-yielding varieties (HYVs) of seed in the case of cotton, jowar and paddy is significantly higher in host villages compared to new sites.

Gross income in rupees per acre aggregated for all crops is significantly higher on farms located in the host villages. It is estimated at Rs. 1,953 per acre in the new site and Rs. 2,464 per acre in the host village (Table 6). The difference grows larger as one moves from maize- and paddy-dominant zones to developed cotton zone. This

11. The evidences for this analysis are from primary data collected for a case study from (a) an irrigated cotton growing area where market and technology are well developed; (b) a semi-developed paddy growing area where water markets have become defunct recently; and (c) an unirrigated under-developed maize growing area. From these three areas, Tentalav and Gadkoi villages were purposely selected from the developed irrigated cotton zone, Ambavadi village was selected from the undeveloped maize zone, and Suka village group was selected from the semi-developed paddy zone. In all, 200 farmers (100 host villagers and 100 PAPs) were randomly selected for the case study (Sah, 1994a).

difference is mainly because of a mix of superior cropping pattern, higher crop yield and better prices received by the farmers located in the host villages, specifically in developed cotton zone. Evidently the new settlers have much less negotiating capacity when it comes to new technology and institutions; this is true whether it relates to choosing a cropping pattern, yield of new crops or bargaining for remunerative prices for the crops.

Table 6: Gross Returns of the Sample Farms

(*Rs. per acre*)

Crop Zones	*New site*	*Host village*
Developed cotton zone	3,264	4,482
Undeveloped maize zone	1,222	1,466
Semi-developed paddy zone	1,579	1,750
Gross income per acre	1,953	2,464

Source: Sah (1994a).

Input Use

The rate of fertiliser use is significantly high on farms managed by the host farmers compared to farms managed by the PAPs. The difference is sizeable in the case of vegetables, paddy, fodder and cotton. This difference is extremely pronounced when the comparison is made keeping crop variety into consideration. All the HBs and HYVs of crops grown by the host farmers receive fertilisers at considerably higher rates. The case of HYVs of paddy, HB jowar, and HB cotton is specifically worth noting where either the crop variety is not grown by the PAP farmers, or if it is, this is with relatively low fertiliser inputs. All these crop varieties are new to PAPs who in their submerging villages were mainly growing maize, coarse cereals and pulse under dry conditions using local seeds.

The level of input use also widely deviated from the recommended rates; the seed rate and plant population are extremely high in the case of HB cotton compared to recommended levels; and in the case of paddy both the seed rate and plant population are significantly lower than recommended. Similarly, the rate of fertiliser use on cotton by the sample farmers is relatively higher than the recommended rate, whereas on paddy and maize it is much lower.

Fertiliser Use Technology

The findings on fertiliser use technology reveal that the farmers do not at all practise basal application of fertilisers. This also means that phosphate use is almost non-existent among the sample farmers. This pattern reveals a severe nutrient disequilibrium among the sample farmers located in both new sites and host villages. Those who reported basal application do so without the use of seed-fertiliser drill. These farmers reported spreading phosphate by hand just after sowing. This, of course, is not an efficient way of applying phosphatic fertilisers. Top dressing of fertiliser has been widely prevalent in the case of cotton and paddy but not on maize. This is true for the sample farmers located in both new sites and host villages. Urea is the main product used for top dressing, and is applied in two or more splits by over 60 per cent of cotton and paddy growers.

Despite low fertiliser use, about half of the paddy growers felt that the top dressing doses were adequate. However, about 80 per cent of maize growers felt that the rate of application is inadequate. On the other hand, despite using more than recommended doses of fertiliser on cotton, three-fourths of the cotton growers in the developed cotton zone believed that these doses are adequate. A significant number of sample farmers who felt that the doses applied were adequate on cotton, maize and paddy felt so because they thought that for proper crop growth this much of nutrient use was required. Those farmers who believed their top dressing doses to be inadequate hesitated to increase its use because of financial constraints, or believed that rainfed crops would become susceptible to crop failure and pest attack with higher doses of urea. Rainfall has a high impact on top dressing in all the three crop zones. With normal rainfall, over 80 per cent of the fertiliser users reported an increase in fertiliser use.

Extension Support

The foregoing analysis bears evidence of low diffusion of HYVs and HB seeds, no seed treatment, total reliance on nitrogenous fertilisers and almost no use of phosphatic fertilisers, little reliance on plant protection chemicals, very high or very low seed rates, and a disregard for maintaining recommended plant population. It is not surprising that the number of farmers who claimed to know the recommendations regarding these inputs was extremely small. The major source of information among the sample farmers is non-institutional – fellow farmers and the input dealers in the taluka. A minority of farmers (less than 2 per cent) reported that the recommendations came from village level workers (VLWs). Access lo VLWs, nevertheless, is significantly higher in the host village compared to the new site. The availability of VLWs is also high in the developed cotton and semi-developed paddy zones.

Efficiency of input use is an important component of yield-input cycle. Each farmer develops his own optimal input use pattern from either his experiences or from extension contact. Sub-optimal input use and poor quality of extension support *ex ante* reinforce the hypothesis that yield-input relationship is insignificant. Breaking the vicious circle of low yield ? low income ? low surplus ? low input use needs concentrated effort for increasing the demand for yield increasing inputs.

Commercialisation in Agriculture

Commercialisation is a social process which brings into focus the relationship between producers on the one hand and agents engaged in input-output marketing, labour, credit agencies, etc. on the other. This process is governed either by a *normal mechanism* driven by market incentives or by involuntary forces like indebtedness, state coercion or involuntary migration. Forced commercialisation is linked with interlocked product, credit and labour markets. Moneylenders, landlords, and traders exploit this situation in two ways. Normal commercialisation is associated with larger production gains and their just distribution. However, both these forms of commercialisation coexist. Thus it proceeds unevenly across regions and over time. This may also happen in the case of PAPs, because of their varying response to various incentives and constrained access to markets.

In a normal commercialisation process, the amount of marketed surplus depends on prices, level of output, size of holding, domestic requirement and cash need of households. Thus, if we divide agricultural households into (a) various levels of consumption,

(b) different size classes of holdings, and (c) varying levels of irrigation, it can be hypothesised that commercialisation in normal circumstances may not have negative implications for disadvantaged groups: the subsistence sector with small holdings, the group where normal agriculture does not offer an adequate living standard, and the less endowed. On the other hand, if commercialisation is involuntary, the surplus generated in the product market is dictated by a dominant group. In this case, there may exist (a) a negative relationship between standard of living and marketed surplus, (b) a positive association between access to free market and price received by different size classes of holdings, and (c) an interlocked market, for example, credit and output markets.

Input-Output Linkages

In this sub-section, an attempt has been made to examine the demand for market purchased inputs and marketing of production surplus by PAPs located in new sites. This analysis will help us understand the extent and nature of market integration the relocated households have achieved. These findings should, however, be interpreted against the background that in the submerging villages such market links were only marginal and confined to Zone I.

Table 7 gives the extent of market purchased input use by PAPs and the proportion of total purchase made during the *Kharif* season in 1993-94. The results show that purchased inputs like hired labour, seeds, fertilisers and pesticides have now found place in the production process of PAP farmers. These inputs were hardly purchased in the submerging villages. However, the bulk of the purchases are made during the *Kharif* season. In terms of nutrients, total fertiliser consumption is about 20 kg of NPK per acre. Keeping in mind the experience of PAP farmers with chemical fertilisers and the level of irrigation – only 11 per cent area is irrigated – this is not a mean achievement. However, per acre expenditure on seeds and plant protection chemicals is relatively low.

The findings of Table 7 also indicate the importance of socio-economic factors in influencing the input use. The PAPs have been uprooted from three submerging zones (Joshi, 1987; Sah, 1993). Five submerging villages of Zone I were relatively developed with

Table 7: Input Market Linkages in New Sites, 1993-94

(1)	*Labour (day)* (2)	*Seed (Rs.)* (3)	*Urea (kg.)* (4)	*(DAP) (kg.)* (5)	*Pesticide (Rs.)* (6)
			Per acre		
Average					
Average use	5	16	29	9	41
Percentage use in Kharif	58	60	85	79	75
Factors governing use@			χ^2 *value*		
Submerging zone	–31.2***	+27.4***	+47.9***	+51.7+++	NS
Tribe	+43.3***	–41.5***	–53.7***	–75.2***	NS
Farm size	+27.9***	+25.5***	–25.3***	+42.2***	+18.1***
Farm income	+9.8**	+42.1***	+21.2***	+31.6***	+36.7***
Calorie intake	NS	-9.1*	NS	–11.5**	–12.5**

For estimating χ^2 and correlation coefficient (– or +), the level of factors and inputs are as follows:

Submerging zone: rock fill dyke villages=l; Semi-developed villages=2; interior undeveloped villages=3.

Tribe: Non-farming *Nayaka* = 2; *Bhil* gatherer farmer = 3; Settled *Vasava* = 4; Risk taker *Rathwa* = 5; Market oriented *Tadavi* = 6.

Farm size: <5 acres = 1; 5-10 acres = 2; >10 acres = 3.

Irrigation: Nil = 0; < 25% = 1; 25-50% = 2; > 50% = 3.

Migration: No = 1; Some family members migrated = 2.

Calorie intake: <2000 cal per day per adult = 1; 2000-2400 cal/day/adult = 2; > 2400 cal/day/adult = 3.

Purchased inputs: Nil = 0; Low = 1; High = 2.

NS = Not significant. *, ** and *** Significant at 10, 5 and 1% level respectively. + Positive correlation; – Negative correlation.

Source: Sah (1996 a).

established land, labour, credit, input and output markets. Services like school, transport, health care were easily accessible, and institutions like Panchayats were present in the villages of this zone. The major tribe in the villages of Zone I was *Tadavi*. Zone II comprising six submerging villages had *Bhil*, *Vasava* and *Tadavi* as main tribes. This zone was relatively underdeveloped with respect to various services and amenities. A Panchayat existed in only one village of the zone. Markets as institutions supporting agriculture were also present only in one village. Zone III, comprising eight interior submerging villages, was the most backward of the three, with *Bhil, Nayaka* and *Rathwa* as its major tribes.

The economy of Zone III was non-monetised exchange-based and the PAPs were unfamiliar with modern institutions and agricultural markets. Nonetheless, relocation brought substantial changes in input use pattern. The use of hired labour was relatively high in the case of PAPs from Zone I and also amongst the *Tadavi* and *Rathwa* (Table 7). As against this, the PAPs relocated from Zones II and III and tribes like *Nayaka, Bhil* and *Vasava* have resorted to use of only family labour in agriculture. On the other hand, the findings of Table 7 reveal that inputs like purchased seed, and chemical fertilisers are more intensely used by PAPs relocated from Zones II and III and by the tribes who have had no exposure to input markets (*Nayaka, Bhil* and *Vasava*). Factors like cultivated area and proportion of area under irrigation are positively and significantly associated with input use. Consequently, those farmers who used purchased inputs at a higher level are also farmers whose farm income is relatively high.

This pattern of input use reveals that once the new techniques are introduced, their adoption spreads across farms irrespective of impeding social factors like lack of previous exposure to markets and techniques. Tribe and submerging zone barriers are broken. The process is, however, governed by economic factors like endowment of the user; irrigation, farm size and technology status. It is, therefore, feared that despite relatively high diffusion of input use across deprived social groups, its efficiency and productivity may still be constrained by social factors, at least during the initial stages of rehabilitation. This calls for designing extension support based on the social reality of the displaced groups.

On the other hand, market linkages in the form of sale of produce are extremely weak. Except for cotton, vegetable, *tur*, and to some extent paddy, the proportion of farmers marketing their produce is extremely low (Table 8). Among these crops, the proportion of farmers marketing their produce ranged from 5 per cent in the case of paddy to 90 per cent in the case of cotton, an overwhelmingly market-oriented crop. A large proportion of sales take place in the village itself. The price received by the farmers for their produce range from Rs. 58 per 20 kg for maize to Rs. 108 per 20 kg for wheat and Rs. 275 per 20 kg for cotton. Although a large part of the surplus is sold to the village trader, it is not a distress sale; the produce is sold over months after the harvest. For example, in the case of cotton some produce was stored for over four months before it was sold.

Table 8: Output Market Linkages in New Sites, 1993-94

Crops	*Proportion of farmers selling*	*Crop sold*	*Sale to trader*	*Average price*
(1)	*(2)*	*(3)*	*(4)*	*(5)*
	Percentage			*Rs./ maund*
Paddy	5	50	92	74
Maize	4	5	90	58
Jowar	2	2	75	68
Wheat	5	4	50	108
Tur	36	38	97	202
Other pulses	8	18	100	109
Oilseeds	10	9	100	190
Cotton	90	100	52	275
Vegetables	25	65	–	75

Source: Sah (1996 a).

Commercialisation

Commercialisation not only integrates the village traders and wholesalers into a commercial structure, the agricultural economy also develops strong backward and forward linkages. To support the process large flows take place between and within the agricultural and non-agricultural sectors. This also creates avenues for input absorption and increased demand for credit. In forced commercialisation, the product and input markets could become exploitative. Some of these issues are examined in this sub-section.

Labour: In response to technology and price incentives in the new sites, as seen in earlier sections, a substantial area has shifted to paddy, cotton, vegetable, fodder and oilseeds, and the yields of traditionally grown crops have nearly doubled. Thus from subsistence farming the PAPs have turned into market-oriented farmers. More equitable distribution of agricultural land and higher farm yields have resulted in improved allocation of the workforce to agriculture in terms of both intensity of employment and employment during the *rabi* season. Using such findings, it can be estimated that labour absorption per acre in agriculture is about 53 man-days per annum.

About 91 per cent of total labour requirement is met by family labour. Only 9 per cent of total labour absorbed in agriculture at present is hired. A low hired labour component indicates relatively equitable land distribution and gainful employment on their own farms among the PAP households in the new sites. It also shows that in only a few operations at any time does demand for labour exceed its supply from own sources.

Credit: Although institutional and private informal credit sources co-exist in this drought-prone zone, what is unique in this area is that the proportion of PAPs availing of any working credit from formal or informal sector is significantly low. Despite their larger land assets, less than 3 per cent of the PAP farmers were members of a co-operative society, and none had access to co-operative short-term credit. Only a few PAP farmers had access to bank finance, and fewer have shown any interest in private production credit sources (Sah, 1995a).

This low reliance on formal and informal production credit in the new sites is not surprising when one introduces behavioural factors in the analysis. The dominant phenomenon of exploitative interlocked credit, land, and output markets is widely feared by South Gujarat farmers. That the small borrowers of South Gujarat turned into bonded debtors for generations, is well recorded (Sah, 1995a). This interlocking of markets is also well understood by tribal resettlers in new sites. Hence, they adopt a cautious approach of 'self-financing' current production to avoid the risk of losing land. It should also be noted that the PAP farmers are latecomers to the credit nexus, where besides assets, social links also play an important role. Credit co-operatives are faction-ridden and other things remaining equal, the supporter of the dominant faction gets easier access to credit and other benefits than those who belong to the dissident faction and/or the newcomers.

The role played by petty village traders in forming perceptions about the risk involved in approaching institutional credit sources is also worth understanding (Sah, 1995a). During the period of depleted food stocks, current consumption is generally supported through loans in kind from petty traders. To enhance their own business, they try to gain confidence of PAPs by eroding the credibility of institutional lenders. They spread propaganda that it is much better to go to a trader for a loan, without any fear.

To sum up, the oustees in the dry zone regard production credit as an input with a very high cost – the cost of their land. They also believe that production credit does more harm than what it contributes to productivity. This belief is fairly well founded. Their refusal of all formal and informal sources of production credit has not been detrimental to the transfer of production technology in the dry region. Rather, it is the lack of public capital investment in agriculture that is the deterrent.

Water Markets: Sale and purchase of water for irrigation in the new sites takes place in two different ways. Water markets are prevalent in some of the new sites located in Dabhoi laluka of Vadodara district. In these areas, the water sellers are not tribal resettlers but are Patel farmers. On the other hand, in some of the new sites like Kasundar, Khokhar and Karmdi a few PAPs have common property water resources managed on co-operative basis. About 10 per cent of the area in the new site is irrigated, of which over 90 per cent is through the private market. The members of the co-operative groundwater societies are generally those PAPs who technically can receive water and they alone manage the water distribution. A scrutiny of water sale in these co-operatives reveals that the water source cannot serve all the relocated PAPs. Sometimes defunct water sources were revived by interested PAPs coming together – as the case was in Kasunder;[12] and sometimes the land

12. Kasunder PAPs came to know after receiving land titles that the land has encumbrance of an order of about Rs. 18,000. Eight of them decided to clear this encumbrance and utilise bank resources lo restart the defunct water source. Eight out of twenty PAPs are at present utilising this source. There are other than encumbrance-related causes which also contributed in non-participation of other 12 PAPs in the venture. For example, untimely availability of water to all the PAPs, infights among the oustees relating lo water distribution, problems of leakage in administration of the venture, problems relating to transit of water to relatively far away locations (an household in a host village had to be given priority in giving water just because he allowed the pipes to pass through his land). These 12 oustees are demanding the Sardar Sarovar Narmada Nigam to invest in groundwater exploration so that the problem related to dry farming they are facing now can be tackled.

allotted to a PAP had come along with an energised well – as the case was in Karmdi.[13]

In many new sites of the Dabhoi area agriculture is totally irrigated. However, crop failures are frequent if irrigation is not availed of. Host villagers, generally Patel farmers in these areas, prefer to service first their neighbours and friends, and then sell water to PAPs. The open market rate of water in the area is fixed, depending on the HP of the motor used. For diesel energised wells, the rate depends on total discharge of the well per hour. This is also constant across farms in a village. It can be argued that if the market is free and operating efficiently, this system would enable the large majority to get water at crucial stages of moisture stress and sustain the income of individual buyers. Price advantages, if any, to the seller in such situation level out due to competition and become uniform and close to its marginal cost. When the purchaser has to decide from whom to purchase water, if there is more than one supply source, the location of the source, relative cost-effectiveness of various sources, timely availability from various sources, the tendencies of the seller, problems in receiving water from various sources – especially those related to transport of water – are the major considerations.

Although the seller is obligated to provide timely and adequate water, this actually happens when the relations between the host villager-sellers and PAPs are cordial. About 90 per cent of PAPs who

---

13. In Karmdi two motor driven borewells were part of the land transfer to PAPs, one 7.5 HP and another 10 HP motor. The cost of these energised wells was part of the deal. With these two sources about 80 acres of land belonging to 16 PAPs is irrigated. A progressive farmer, Devsingh Rathaw looks after the administration. On an average, the electricity bill for utilising these resources comes to about Rs. 440 per PAP. Water distribution is according to reach of water across the water sources. By any chance, if a need arises not in accordance with the distribution pattern, the project affected person generally purchases water from other sources, if available. Due to circumstances if a project affected person has to lease-in or lease-out his land, he is allowed to use irrigation up to 2.5 acres of such transacted pieces of land. The farmers generally have access to irrigation in about 3 to 4 acres. Unless land transaction has not taken place, the host village farmers are not encouraged to purchase water. The PAPs who do not receive water from such venture, purchase water from nearby host village.

purchase water are dissatisfied with their relations with Patel sellers. Moreover, the sellers also try to sell water to more farmers than they can service adequately. In this process the losers are the PAPs, who are asked to wait for a day or two. Though this results in some crop losses, total crop failures are rare in these irrigated farms. The lands where water does not reach are generally sharecropped between the water seller and PAP landowner. All the cost is borne by the landowner and the water seller provides water and receives one-third of the output. This is not a very economical proposition for the water purchasers. But to maintain soil productivity, avoid crop failures and earn some net profit, they have to accept it. Nevertheless, in this sharecropping contract, the seller has a higher stake and ensures timely delivery of water.

A small minority of PAPs have developed good social relations with Patel farmers and willingly pay charges before using water. In such cases the seller defers the payments for a crop season in case of genuine default. This is an indicator of strong relationship within the two agencies. The sellers in such cases prefer PAPs over the host villagers. But developing such social relations is not easy for all the PAPs. Unlike privately owned land, groundwater is a common property resource, privatised as soon as it is captured and ends up being appropriated. In fact, the right to use groundwater thus is provided by the right to own an energised groundwater source. The emergence of water markets, however, does provide opportunities for equitable access to water (Shah, 1991). The PAP farmers tend to gain more with the water market than without it. Such a venture (a) brings technological changes through HYVs, thus increasing income and risk bearing capability; (b) increases employment opportunity with higher wages; (c) leads to appreciation in land value; and (d) reduces the cost of operation to the entrepreneur.

In the dry zone, by and large, the resources have not been adequately developed. Risk associated with capital investment, constraints in investment (Sah, 1993), and slow spread of HYVs and HB technology restrict the scope of groundwater development. Economic efficiency is a major issue in water market development. Where the water seller enjoys monopoly power, the benefit of irrigated farming would be appropriated by the seller (Sah, 1993). In contrast, in an efficient market, more water will be sold to more buyers and the price of water will be close to the cost of pumping. Higher dependence on protective irrigation, high cost of installation and government restrictions, non-availability of other sources, insufficient delivery system by other sources are the major determinants of monopoly power enjoyed by a seller in this area. But the field realities operate in such a way that the few available water sellers work together to further their own economic interests. Thus water charges levied by these sellers reflect the degree of monopoly power enjoyed by them.

Output: Although the analysis on output market links was attempted in three dimensions – size-class specific, irrigation specific and calorie intake specific – only the last of the three has been presented here. When the phenomenon is analysed across different levels of calorie intake groups (Table 9), cropping intensity is found to be about 112 per cent in the case of low calorie intake households and 110 per cent in the case of high calorie intake group. Farm production among low calorie intake group at Rs. 1,433 per acre is also higher than the Rs. 1,263 per acre of households with high calorie intake. But the proportion of paddy marketed is as high as 55 per cent of the total production among the low-calorie intake group compared to the negligible proportion among the high-calorie intake group. In the case of maize this proportion is 14 per cent in the low-intake and only 2 per cent in the high-intake group. Similar is the situation with the proportion of produce marketed in the case of jowar. But there is no significant difference in the proportion of *tur* production marketed among the three calorie intake groups.

Thus, among different calorie intake groups, there is no significant difference in either cropping intensity or yield per acre. But surprisingly, the proportion of total cereal production marketed is significantly high among those who have relatively low intake.

Perceptions about Economic Condition

During a normal year about 90 per cent of the PAPs have reported that their economic condition has either improved or remained same

Table 9: Variations in Marketed Surplus across Three Different Calorie Intake Groups

Particulars	*Calorie intake/adult/day*		
(1)	*< 2000* *(2)*	*2000-2400* *(3)*	*> 2400* *(4)*
GCA per farm (acre)	*7.5*	*7.5*	*6.9*
Cropping intensity (%)	112	112	110
Production			
Rs./farm	10,758	11,386	8,755
Rs./acre	1,433	1,502	1,263
Marketed surplus (%)		*Percentage**	
Paddy	55 (75)**	80 (74)	@ (70)
Maize	14 (58)	3 (58)	2 (56)
Jowar	5 (71)	1 (63)	1 (65)
Wheat	1 (125)	7 (–)	@ (90)
Tur	34 (203)	39 (208)	39 (200)
Oilseeds	@ (–)	16 (190)	8 (190)
Cotton	100 (273)	100 (278)	89 (274)

* Percentage of total output.

** Price received in Rs. per 20 kg.

@ Less than 0.5%.

Source: Sah (1996 a).

compared to what it was in the submerging villages. Not surprisingly, during abnormal years a large proportion of PAPs reported significant deterioration in their economic condition. These findings should be interpreted with caution. They do not necessarily reflect the pre- and post-relocation comparison of economic condition of PAPs. In fact they indicate the importance of contemporary experiences in shaping these perceptions. Farm production of households who are dissatisfied is significantly lower, and PAPs who were unable to feed their family from self-production also responded deterioration in their economic condition. Those PAPs who have to resort relatively more to casual and agricultural labour are also the PAPs who have negative perceptions about their economic condition. On the other hand, the PAPs who perceive an improvement in their economic condition have significantly higher income and calorie intake, depend much less on market purchase for food, and do not have to fall back on casual and agricultural labour compared to those PAPs who believe that their economic condition has deteriorated. These tendencies will further strengthen if farm income can be insulated from the vagaries of the monsoon through appropriate policy interventions.

Performance of Different Tribes

Evidence reveals that after relocation, the Nayaka, Bhil and Vasava have less diversified cropping pattern and lower yields than the Tadvi and Rathwa. The former tribes have less than 24 per cent of gross cropped areas allocated to non-traditional commercial crops and superior cereals, where as the latter group has between 33 per cent and 44 per cent area allocated to these. Production is also higher on farms operated by the Tadvi and Rathwa. Among the relocated households there is a small group of PAPs who were landless in their submerging villages. These groups were mainly in interior villages of Zones II and III. In their submerging villages they depended on agriculture as a subsidiary occupation; their main occupation was animal husbandry and labour. After relocation about a third of they have not adjusted to agricultural practices at the new sites. The input used by the tribal groups like Nayaka, Bhil and Vasava reveals that they have yet to establish strong links with the input market. The bulk of their farming is with family labour and home grown seeds. The use of purchased input, seed, fertilisers and chemicals on their farms is about a third to a half of the inputs used by the Tadvi and Rathwa. Fertiliser use is about a third on the farms of Nayka, Bhil and Vasava compared to Rathwa managed farms. The same is the situation with other market purchased inputs, plant protection chemicals and hired labour. A low input use is reflected in the agricultural production of the PAPs from submerging Zones II and III. The farm production of the Nayaka, Bhil and Vasava is about half that of the Rathwa, who not only produce relatively efficiently but also do not have to rely on large market purchases for their food needs. The calorie intake amongst the tribes from interior zones also reveals constraints. This variation in calorie intake is mainly governed by the variation in pulse and cereal consumption. The low farm production is also reflected in the capability of different groups to spend on things like consumer items, health and education and assets.

The anxiety expressed in the last section, that the condition of the oustees of Zones II and III after the relocation is unlikely to be as satisfactory as observed on the farms of Tadvi and Bhil and that of the PAPs of Zone I has come true. The Bhil and Nayaka are worst hit due to their inability to assimilate with the mainstream.

The findings of the present analysis can be summarised as follows: With a liberal relocation package, agriculture for PAPs after relocation has become the major economic activity. During normal years gross income from agriculture aggregated for all crops is more than double the income of farmers in the submerging villages. But the income is significantly lower at the new sites than the income of farmers in host villages. The difference increases as one moves towards the developed irrigated zone. This is mainly due to superior cropping pattern, higher crop yield, and better prices received by the farmers in the host villages. With respect to technology adoption, rates of fertiliser use are significantly low on PAP farms compared to farms located in the host villages. Moreover, they do not at all practise basal application of fertilisers. This pattern of fertiliser use reveals a severe nutrient disequilibrium among the PAP farmers. With relatively stagnating technology, the market links also reveal a negative pattern: the proportion of total cereal production marketed is significantly high among those groups who have relatively low calorie intake.

Total dependence on agriculture, nonetheless, has its own problems especially in the dry villages. Erratic weather induces wide yield fluctuations. During abnormal years up to 40 per cent of the PAPs face total crop failure; yields of major crops register a fall of up to 35 per cent; access of households to foodgrains reduces by about 20 per cent and the level of consumption and calorie intake falls by about 17 per cent. In scarcity years the PAPs' own coping mechanism hinges on (a) decision of household members to work as casual labour; (b) migration of some family members and (c) cattle sale. But these adjustments are not enough to maintain the calorie intake of the households. Moreover, agricultural growth in this dry zone is riddled with moderate increase in production and recurring crop failures. Even during normal years erratic yields constrain caloric intake. Households in some new sites have year after year reported relatively low farm income. A subset from these new sites also reports that more than a fourth of the households are perpetually living on per adult calorie intake less than 2400 per day.

In the context of agrarian transformation,[14] these findings are disturbing.[15] Agricultural growth is constrained not only by lack of technology transfer but is also accompanied by unfavourable marketing linkages. With little surplus left due to fluctuating yields, technology adoption even during normal years remains at a low level because of associated risks. The single most important reason behind the persisting low level of agricultural development in these scarcity prone new sites seems to be technological stagnation. It is neither the backwardness of the farmers nor lack of socio-economic institutions but lack of suitable technologies for the given environment which is responsible for the stagnating agriculture. It is not that cultivators in these regions have not heard about fertiliser responding HYVs, but the fact is that under prevailing moisture stress situations these technologies are not superior economic alternatives to varieties grown

---

14. In a developing economy like ours, one of the two sectors (agricultural sector) is traditional. It produces wage goods (foodgrains) and has excess supply of labour. The non-agricultural sector is modern and draws capital and labour from the traditional sector. The productive accumulation in the economy can be faster if the agricultural sector performs. This is so because only then (a) strong input-output linkages between the two sectors would establish; (b) terms of trade between sectors would support surplus transfer, and (c) demand for non-agricultural goods in the traditional sector would increase. The overall growth of the economy, in the long run, would be constrained if (i) technology in the traditional sector remains stagnant; (ii) market links between sectors remain weak or the terms of trade remain unfavourable for resource transfer to the non-agricultural sector; and (iii) contribution of various parts within the agricultural sector is unequal. The above constraints, in the short run, would also increase social and economic inequality.
15. The findings of this analysis may be seen from other dimensions too: the traditional social links and institutions at the submerging villages had important place in the life of PAPs. These arrangements had helped sustaining the PAPs in the submerging villages under harsh environment. Even in worst of the situations, the system was enabling a minimum level of consumption. The randomness of relocation in the absence of these traditional arrangements at the new sites has made the PAPs vulnerable to exploitation and subsequently risk aversion.

there. This underlines the need for allocating resources for adapting technologies to the agro-climatic environment.

On the other hand, despite the wide variation in calorie intake, the inequality issue has remained neglected.[16] In bureaucratic circles, only political compulsions keep these issues alive. In India, the importance of agricultural growth in influencing rural poverty has been firmly established.[17] The negative relationship between growth and poverty, however, be accepted with caution,[18] for there may be a time lag between growth in farm yield and reduction in rural poverty; but there is no such lag between deceleration in farm productivity and reduction in availability of food grains and hence consumption levels. It means that no sooner than the current year's crop failure is

---

16. One can argue that all necessary support already exists in the systems, and it is the PAPs who are not taking advantage of the facilities. This is a faulty argument, for the PAPs arc not facing a normal process; their unfamiliarity with markets, institutions, and technology makes them vulnerable. Thus it becomes pertinent on the part of the policy makers to facilitate the rehabilitation process by understanding the, farmers' own adjustment behaviour first and then strengthening it by external support.
17. The Indian concept of poverty has moved from identifying a minimum normative absolute living standard in terms of per capita consumer expenditure — may it be the normative food basket or the minimum calorie intake towards explaining the poverty. It should, however, be noted that cross-sectional analysis of calorie intake by individuals of PAP households contain noises despite the fact of a large sample. Moreover, this approach of poverty measurement does not give adequate emphasis to quality of life: the PAPs' access to services, their environment relating to law and order, access to justice, the environment where they live, the air they breathe and the water they drink. In the absence of these concerns, the measurement of poverty is partial, reflecting only to food insecurity of PAPs.
18. Although it was observed that high agricultural growth reduces rural poverty and higher consumer price aggravates it, the process may be affected if (a) early growth takes place under extreme inequality and its benefits may not 'trickle down' to large rural masses; and (b) oiigopsonic influence of landowners distorts the rural labour market, dampening the on-farm employment opportunities and wage rates.

established beyond doubt, the current consumption and recurring expenditure on inputs are adversely affected. But the reverse may not always be true. Efforts to accelerate agricultural growth, therefore, must be supplemented by short-term measures of employment generation and augmenting availability of food grains.[19] In a period when purchasing power is low, revamping the Public Distribution System, though necessary, may not alone help except in neutralising the trade affected distortion in foodgrains prices.

The solution to the problem is difficult because of its complexity. Three processes have to start simultaneously: (1) creating avenues of irrigation and efforts for adapting modern seed-fertiliser technology for the agro-climatic conditions of the new sites; (2) organising PAPs around the institution so that they respect the rules of open markets; and (3) strengthening the farmers' adjustment mechanisms of coping with scarcity by creating more non-farm employment avenues.

The need to understand the operating constraints in adapting technology for local environment, technology transfer, irrigation management, and scarcity management cannot be over-emphasised. After a successful relocation of over 4,500 PAPs, these processes are constraining their rehabilitation. If mid-course corrections through appropriate policy changes are not in-built and mandatory on R&R Package,[20] impoverishment risks due to second generation problems of involuntary migration cannot be ruled out till Narmada water reaches in command area.[21]

III. Missing Rights in SSP

When the R&R policy of SSP proclaimed that the tribal community

19. Households located in Dhefa, Kukarda, Ambavadi and Thapavi are totally dependent on agriculture. Because of little off-farm employment opportunities and low yields, the calorie intake of households located there is the lowest.
20. Even the National Policy on R&R has not raised these second generation issues (Sah, 1995b).
21. By the time this paper was being developed, end of 2010, new sites have started receiving Narmada water for irrigation. Hopefully, the second generation problems of yield and income risks may have reduced significantly now.

would, strictly, be settled in the area of their choice, it empowered the state to break the social fabric associated with the village, peer group and family.[22] The relocation process could be seen from two different angles; first, the traditional social linkages and economic institutions in submerging villages had an important place in the life of oustees. These arrangements had helped in the sustenance of the society in submerging villages. The randomness of relocation and absence of the traditional arrangements at the new sites could make the oustees vulnerable to exploitation. Second, the existing policy is inappropriate. Those associated with resettling the oustees may defend the policy and the process by reiterating that the decision to shift is an individual decision. The strength of the argument is drawn from the fact that the tribal village is not a homogeneous unit; even in a *falia* people do not always want to shift together. And since the oustees may be shown land in different locations, their moving to the new

22. A total disregard for social implications of relocating PAPs from 19 submerging villages to 120 plus new sites is paradoxical. The relocation process could be seen from two different dimensions; first, the traditional social links and economic institutions at the submerging villages had important place in the life of PAPs. These arrangements had helped sustaining themselves in the submerging villages. The randomness of relocation in absence of these arrangements at the new sites has made the PAPs vulnerable to exploitation. The other dimension to the process is the appropriateness of the existing policy. The people associated with resettling the PAPs defend the policy and the process by reiterating that the decision of shifting is a individual decision. The strength to the argument is drawn from the fact that the village is not a homogeneous unit; even in a falia the people do not always want to shift together. And since the PAPs were shown land in different locations, their moving to the new sites is voluntary, for PAPs have preferred the land they are cultivating in the new sites. The argument could have been accepted had there been some method in the randomness of the relocation. It is often argued that the policy provides avenues and instrument of choosing the new site. The fact is PAPs certainly had a choice; to choose best amongst the worst. Thus, the migration in absence of social and economic arrangements is not only involuntary but also an outcome of a policy incapable of safeguarding the risk of social disruption. The misery could have been avoided with adequate policy frame and planning.

sites is voluntary, as they have preferred the land they want to cultivate in the new sites. The argument could have been accepted if there were methods in it of avoiding the randomness of the relocation.

It may be argued that the R&R policy of SSP provides avenues and instruments of choosing the new site. The fact is that oustees may only have the choice to choose the best amongst the worst. Thus, migration in absence of proper social and economic arrangements may not only be involuntary but also an outcome of a policy frame incapable of safeguarding the risks of social disruption. This can be termed as lack of interaction within a rigid policy frame.

Lack of interaction among agents was evident not only in site selection but also in gender issues. While interpreting 'member dependent on land', unmarried women, widows and divorcees need to be identified as adult members. This is not just a question of justice and equality. It also reflects the attitude of policy makers towards the needs and participation of women. This attitude is further manifested in the form of ignoring the gender perspective in choosing relocation sites, selecting sites for house plots, locating drinking water sources (pumps and wells) at the new sites etc. In tribal communities, female participation in day-to-day decision making and in economic activity is equal to that of the male. The patriarchal norms of non-tribal societies restrict mutual decision making. Revisions recognising women's needs and concerns, and ensuring their participation in decision making, have to be incorporated in the R&R policy. In fact, R&R must be recognised as participatory processes and this must be realised by policy makers through appropriate and effective interventions. The term 'participation' has been used here as the right of the displaced to participate in a process which affects their livelihood. It also presupposes that there exists mutual trust between institutions involved in R&R and the affected community. This may result in better living conditions with respect to choices of migration, identification of new sites, agricultural land, house plots and location of social amenities. The proposition was well founded, but the reality shows that though the interaction between state and the civil society existed, it was missing at the community level. Exclusion and marginalisation was the obvious outcome.

Experience of SSP shows that the well intended policy statements had remained unimplemented owing to a failure in identifying instruments for their execution. The R&R policy of SSP also lacks necessary linkages that make such a policy implementable. For example, with respect to land rights of the PAPs and their cultural losses, the policy does not ensure how the right for entitlement of land free from encumbrance would be enforced; nor does it spell out how cultural losses owing to relocation and social and economic integration with the mainstream would be taken care of. Almost always these statements show well intended concerns, but they emerge as rhetoric that clearly lacks the necessary institutional arrangements for their implementation. And thus, when land rights are being discussed in the R&R policy, many issues have remained outside the purview of the discussions, like checking (a) the size of allotted agricultural land; (b) if a canal, road, or any other project is on the land allotted to a PAP; (c) whether the same land which was shown to PAP has been allotted to him; (d) for encumbrance on it; and (e) if the right of a minor in the father's property clashes with the right of a major son.

These omissions raise doubts in the minds of PAPs about their status as landholders. Therefore, the necessity to identify institutions which will carry out such verifications cannot be overemphasised.

The PAPs, despite being tribal are not a homogeneous entity.[23] They vary significantly in terms of their languages, customs, beliefs, perceptions, needs and capabilities. The SSP has failed to address the needs of groups which had remained unexposed to the market economy and modern agriculture – the Bhil, Nayaka and Vasava of Zones II and III. The extension wing neither differentiates between hosts and PAPs, nor among different groups in tribal population.

Relocation allows only movable heritage to shift along with the displaced. Relocation of PAPs as a community demands reconstructing

---

23. Displacement from Gujarat is from 19 villages located in three Zones. Zone I villages were most developed among all the submerging villages. The main tribe living in this zone, Tadvi, being near to the nontribal area, had experience of modern markets and modern agricultural production technology. The villages in Zone II where Tadvi and Dungaree Bhil were the main tribes was underdeveloped. On the other hand, Zone III comprises of 8 interior villages (with Rathwa, Dungaree Bhil and Nayaka as the main tribes) , was most backward among all the three zones.

immobile cultural icons at the new site. This is possible when (a) it is recognised that it is a community that has to be relocated rather than individual PAPs or families, and (b) freedom of cultural expression (the relevant cultural necessities) becomes part of the development process. Both the above prerequisites are difficult to achieve. For one thing, we keep on forgetting the concept of heterogeneity amongst the PAPs, and for another, there may not be many social scientists who would like to be associated with the process of identification of 'relevant necessities' – social, cultural and economic – to be included in the development process. Empowering the PAPs is a process which may not fascinate many social scientists. The research agenda in this case is guided by tangential pressures from funding agencies and the *publish* or *perish* syndrome from peers. Having cautioned about these dangers, the help of trained social scientists – anthropologists, sociologists and economists – in the process of planning itself is not an undesirable option. In fact, such inclusion may improve the chances of identifying meaningful alternatives.

For the relocation sites, the R&R policy of SSP states that amenities such as water supply for drinking and cattle, fodder and grazing land, schools, approach roads, electricity connection, health centres, religious places, cemeteries and graveyards, *Panchayat* house should be made available. Often these facilities are linked to a minimum number of relocated families. The relocation process, as experiences have proven, does not always contain such a minimum number of specified families at a new site. This easily leads to non-compliance in providing these amenities. Another major problem in the new sites has been the lack of maintenance of created services. This not only marginalised the inhabitants but resulted in the exit of ARCH-Vahini from the R&R scene.[24]

India is a signatory to the ILO Convention 107 adopted in 1957, concerning the protection and integration of tribal population. The

24. A large number of problems were cropping at the field; there were widespread complaints of shortages in drinking water, frequent break downs in hand pumps, poor quality of roads and drainage in resettlement sites, water-logging and out-break of disease during monsoon *et cetera*. These were communicated by concerned NGO to state for necessary actions. But inaction from state resulted in deteriorating well being of PAPs (Patel: 2001).

Convention affirms the right of tribal people to their land and proclaims that if they are displaced due to national interests, the displaced will be provided land suitable for their present need and future development. We in India also believe that people are free and equal in dignity and that all people contribute to the diversity and richness of Indian culture. Any doctrine of cultural superiority, therefore, is legally invalid and socially unjust.

Despite the fact that relocation package of SSP reflects the idea that relocation will not adversely affect the standard of living, and also reflects the development of human rights, there have been violations of human rights reported in the displacement due to SSP. The first serious protest against such violations originated in Manibeli (PUCL, 1993). The PUCL report viewed the process of relocation infringement on right to life, right to information, right to struggle, as well as violation of NWDT recommendations. This was followed by a Gujarat PUCL (1993) report which found the findings of PUCL report factually incorrect and concluded that allegation of coercion were baseless. The Gujarat PUCL cited the findings of Fact Finding Team's report on Manibeli incident (Acharya *et al.*, 1992). The Fact finding team observed that there was a rupture of dialogue between the two groups who were socially and culturally the same, which was aggravated by the intervention of outside agencies. Police intervention became inevitable to help the group who wanted to shift to a new site. Bulldozers were used more for operational purposes than for intimidation. The committee found no visible signs of violence nor was violence reported by any party during the operation. The police were not used for any dismantling or shifting operations. Other human rights violations were cited, ranging from rights of non-PAPs (Soni and Appa, 1993); right to struggle; violation of Court Orders (HRCN, 1993); right not to be forcibly evicted; right not to be forced to stay; right to information and right not to be physically harassed (Bhatia, 1997). These reports from pro- and anti-dam groups are equally persuasive. But when Bhatia reports forced eviction from the Antras submerging village of Gujarat where 30 families were refusing to shift; violence against Budiben – a PAP of Antras and NBA activist – and her subsequent rape to psychologically terrorise her to shift; forcible retention of PAPs from Malu who wanted to shift to their submerging villages Gadher and Vadgam; the drama was being staged for evidences to stop the dam. Neither was the text mediocre nor were the underlying sentiments trivial. But the way whole set of evidences served the ideology, the sense of patronage towards PAPs was more dominant than the possibility of rehabilitation.

But more disturbingly, evidences confirm that abnormal weather in the region results in severe yield fluctuations and *Rights to Livelihood.* The food security implications are serious if the PAPs have also experienced yield stagnation in two consecutive years. Even during normal years, the risks involved in technology adoption are so high that PAPs, by and large, have preferred to avoid adopting the new agricultural technology in dry areas. As a result, the cushion available from good years is not enough to neutralise the risk of food insecurity during abnormal years. Erratic weather induces wide yield fluctuations. During abnormal years, nearly half of the PAPs face total crop failure. The yield of major crops registers a sizeable drop, access of households to food grains reduces by about 20 per cent, and the levels of consumption and calorie intake fall by about 17 per cent (Sah, 1997). In scarcity years, the PAPs' coping mechanism revolves around (a) reducing farm expenditure; (b) decision of household members to work as casual and agricultural labour; (c) seasonal migration of some family members; and (d) borrowings and distress sale of cattle. But these adjustments are not enough to maintain the households' calorie intake. The role of the state in supporting the PAPs' coping mechanism has been marginal. Moreover, agricultural growth in this dry zone is riddled with moderate increase in production and recurring crop failure. The pressure on land is immense and agriculture alone cannot sustain the households' consumption in a weather cycle.[25]

The PAPs' consumption response to income changes is similar to the one observed in case of the mainstream society: a decline in the share of food in total family budget. Although cereals and pulses still account for over three-fourths of the total calorie intake, superior

25. Apart from poor quality land in the case of about 15 per cent PAPs, 43 new sites are outside command area of the project. In both these regions the livelihood adaptations are only community induced. The state and the civil society has no role in this adaptation.

cereals are replacing coarse cereals among all the income groups, more so during abnormal years. This substitution is more an upshot of the availability of superior cereals through PDS rather than a matter of choice. What is most disturbing is the fact that even during normal years about a fifth of PAPs have reported per capita intake of less than 2,000 calories per day. During abnormal years the calorie intake of half of the PAPs remains below this norm. Not only is the incidence of poverty high during such years, the intake gap within the vulnerable group of the poor also further deteriorates and inequality indices show a sharp increase. In order to bridge the poverty gap, a mixed policy focussing on rapid agricultural growth and expansion of non-farm employment is desirable.

These second generation problems of involuntary migration were known to the policy makers, and yet inaction on their part for strengthening the coping mechanism of PAPs only indicates their insensitivity. The state, it seems, instead of stepping in at the time of the short-term shocks, believed that its interaction with the community at such times crisis lay outside the purview of R&R. Nevertheless, in order to utilise the findings of information support for policy reorientation, it would be mandatory on the part of the project authority to participate in the discussion and dissemination of field realities. This dissemination would lead to (a) identifying the deviations from expected norms; (b) understanding implication of the observed deviations; (c) identifying the causes of such deviations; and (d) evolving measures to be adopted for minimising the associated risks. Devoid of such participatory process, the PAPs, as has been the case in SSP, would find themselves in a perpetual state of poverty.[26]

26. Although pastoral nomads of north Gujarat, narvies from Saurashtra, road workers from Panchmahals, query workers from tribal Bharuch, and tribal workers from south Gujarat are seen in major industrial towns of Gujarat, the seasonal migrants seldom cross district boundary and work as farm labour in irrigated area and casual semi-skilled casual labour in industrial towns. It has been argued that migration may be motivated by desire to seek skill and leads to development through a process of migration lead to urbanisation leading to socio-economic development which ultimately result in overall development of the area. The other proportion argues that inclination to migrate is

IV. Shades of Civil Society

Individual researchers (Cullet, 2007) may have taken an emancipated view of civil society involvement in SSP, and may have given due credit to the contribution of NBA. But in doing this, he has dwarfed the achievements of Gujarat civil society by not highlighting the role others played.

"The World Bank loan brought to the SSP the whole set of policies that provide the framework for the implementation of bank-funded projects...The involvement and subsequent withdrawal of the World Bank from the project is a landmark event... The involvement in SSP was, by all accounts, one of the biggest public disasters for the World Bank. As a result, a number of initiatives were taken to ensure that similar problems would not resurface... After the World Bank withdrawal from the project, the development of SSP took a different turn... From the point of view of *Narmada Bachao Andolan*, the withdrawal of the World Bank was a boon and a challenge. From March 1993 onwards, much less international pressure could be applied on the Central and State governments. Eventually, in view of the difficulties faced in making itself heard, NBA decided to resort to public interest litigations, which had come to be seen by the mid-1990s as one of the best avenues for ensuring justice and securing the realisation of human rights" (Cullet, Chapter 1).

The SSP is product of varied and mixed experiences; good at the level of policy, acceptable in the implementation of relocating PAPs, and poor in case of their rehabilitation. The success of the R&R package at various levels is a product of coming together of many

---

determined by difference in the income between source and destination of migration, and may result in equating expected income. In Ambavadi and Piparvati, two villages where seasonal migration is relatively higher; about 53 per cent of relocated households have reported out migration compared to 14 per cent in the sample as a whole. The out migration in these two locations is owing to low productivity of land and inability of land to support all the family members. During the summer of 1996, about 1 out of 5 members of household in these two locations had migrated for 1 to 4 month in search of jobs as masonry helper, nursery hand and sugarcane cutter in Surat, Ankleshwar, Dam site and Vadodara.

minds. The R&R policy of SSP has evolved over a period of time and after a prolonged struggle by activists in Gujarat. The involvement of NGOs – especially ARCH-Vahini, Anand Niketan Ashram and Rajpipla Social Service Society – in R&R in Gujarat can be seen in two phases. The first phase ends by 1987 when these NGOs were able to organise tribals to force the policy makers to come out with a just R&R policy. The second phase is of post December 1987 when the R&R package was accepted by the government of Gujarat and these NGOs decided to become active partners in the relocation process. Notwithstanding the fact that left to the state, the implementation of R&R package would have been as indifferent as has been in other projects in the country. A fair amount of success in the relocation of PAPs certainly would not have been possible but for the support provided by these NGOs, who had earlier fought for a just R&R policy for SSP.

It is quite possible that the success of the SSP relocation process will emphasise the inevitability of NGO involvement to ensure a fair deal to PAPs elsewhere. This may be a premature conclusion, for it is possible that what led to the success of a process may end up in its demise when the process is replicated. The working of NGOs reveals a heterogeneous pattern: Those working in rural reconstruction are a mixed lot; some, like ARCH-Vahini had delivered services keeping with their ideology of social reform, but many more are simply participants in what may be termed as a 'market of development programmes'. The differences in the actions and ideology of NGOs are shaped by the character of the people running them. Such variations result in indifference in delivering what the organisation is supposed to deliver with sensitivity. Importantly, the concept of sensitivity is not static; it varies not only across organisations but also within an organisation over time. Thus it would be futile to presume a consistent approach. A policy decision to involve NGOs, arrived at in good faith that their grassroots level understanding would help to successfully implement programmes, may be self-defeating and may not ensure sensitivity, as demonstrated by some NGOs in the case of SSP.

The post-1988 period had seen immense pressures on NGOs. On the one hand, they had to sustain the political will and see that this will was translated into actions. On the other hand, the ability of pro-state NGOs to influence and work with the oustees depended solely on fulfilment of promises they had made to them before relocation. These two roles were not congruous. The bureaucracy at the lower level in its turn exploited this contradiction. Allocation of unproductive land and short-changing in land and subsistence allowance were some of the problems which did not surface for over a decade. Some PAPs who were relocated between 1983 and 1987, when the R&R package was not accepted by the state, also faced a problem of discrepancy in the compensation package. They not only got a meagre monetary compensation for their land but their grown-up sons, who were recognised as separate PAPs after 1987, had to wait for another five to seven years to receive their package for want of suitable land in the nearby new sites. Fragmentation of families was the tragic outcome in many cases. No matter how hard the affected oustees and the involved NGOs tried to settle these cases, non-availability of land and bureaucratic insensitivity delayed the outcome[27]. The blame for this was invariably placed on the NGOs.

The political will was activated and the bureaucracy did strive for a fast and smooth relocation. The bureaucracy at the grassroots level either did not get the real feedback from the leaders who were the contact points for dealing with PAPs in over 120 locations, or it preferred to ignore the representation made by NGOs regarding their grievances. This created a momentary calm, as the real grievances of oustees did not surface immediately. Many of the leaders were also co-opted to voice what the bureaucracy wanted to, that all was well on the R&R front. The incentive for such co-optation was the benefit to the leaders in the form of contracts for construction of facilities at the new sites, and access to the inner circles of decision making. As long as the dissatisfaction did not surface, especially that concerning the quality of land and amenities at the new sites, the NGOs, especially ARCH-Vahini, did co-operate with the bureaucracy.

Nonetheless, a group of activists, who in 1989 formed the Narmada Bachao Andolan (NBA), took the stand of 'No Dam' on grounds of ecological and human costs, soon after the acceptance of the R&R package in mid-1988. Their struggle, though started around

27. This is one of the major problems faced by PAPs in the new sites. The origin of this problem was lack of participation of PAPs in the process of land selection.

displacement, reopened the question of the project's sustainability. This paradigm shift opened a new dimension in the debate on Development. Sustainability did not remain confined to tribal rights, loss of livelihood and resettlement, it also included ecological and environmental issues associated with the project. Their belief regarding the unsustainability of Sardar Sarovar, strong views about the impossibility of rehabilitation, and lack of faith in capabilities of the administration to handle a large involuntary migration – originating from not only Gujarat but also Maharashtra and Madhya Pradesh – caused a vertical split between the involved NGOs. A group spearheaded by Medha Patkar seriously questioned this development approach.

The other group, headed by ARCH-Vahini and consisting of a majority of NGOs from Gujarat, remained pro-rehabilitation. They believed that rehabilitation was possible provided there were adequate checks and balances in the resettlement process. "Tribals," says Vahini, "have shown clear preference for good quality of land, security of tenure, and much greater access to market and other infrastructural facilities like roads, transport, electricity, schools, even rudiment of health service. Their access to the market, their integration with the market economy is limited" (Patel, 1997). Making all these things happen in the relocation process was, thus, desirable.[28] These views,

---

28. As for the poor quality land, Sardar Sarovar Punarvasavat Agency (SSPA) reports that to mitigate the problem necessary measures are under consideration. A technical committee under the chairmanship of Director (Canal) of Sardar Sarovar Narmada Nigam Ltd. (SSNNL) was appointed Executive Committee of SSPA. Subsequently, a task force with Director (R&E) and CEO (SSPA) has already been constituted by GOG to take corrective measures. To begin with, 21 sites were jointly inspected by field officers of SSNNL and SSPA and remedial measures were outlined for each new site. The remedial measures have been taken in 43 sites and the work except for major structure in one or two cases has been completed. Impact of these measures is yet not clear. Although, the present analysis reveals wide spread dissatisfaction amongst the PAPs with respect to land quality, official documents do not show that PAPs have registered their complaints with SSPA. The SSPA received complaints only from seven new sites. As a result of tractorisation partial or entire relocation in these new sites has been in progress. Still this is an issue that need to

though pragmatic, are explicitly at a variance with the Andolan's ideology. The defence of tribal rights from Vahini took various forms of colloquy, from rhetoric and polemic (Patel, 1997), to defending the state against the Independent Review (ARCH-Vahini, 1992). But nowhere has ARCH-Vahini's commitment to serve the interests of the oustees has ever been in doubt, neither when it was working closely with the state in the project nor when it withdrew.

From this pro-rehabilitation position, when ARCH-Vahini withdrew itself from the Project in 1995, the question may not have been so much of a failed paradigm. This drastic step more likely would have been taken due to failed processes, which were unable to insulate the PAPs from various problems in their resettlement. ARCH-Vahini disassociating itself from R&R after almost a decade of involvement may be also because its pro-rehabilitation self was in contradiction with its pro-state self. A large number of NGOs, many of them with questionable backgrounds as discussed above, were roped in for R&R process. As a result the quality of the R&R work suffered and ARCH was unhappy with the increasing problems faced by the PAPs. As the state started relying on the Judiciary to take briefs for R&R problems, the Grievance Redressal Authority was getting more importance than Civil Society. This made ARCH seriously think about their role in R&R. This debate notwithstanding, what should be remembered is that in normal circumstances the second generation problems faced by the PAPs could have been sensed and raised by ARCH-Vahini much earlier. But their preoccupation with relocation did not allow them to play the facilitating role for such identification; relocation pressures were too intense and too prolonged for them to have either time or energy to take up rehabilitation issues.

Though development could have varying meanings, the Indian State used it to achieve a single goal, the goal of capital accumulation for rapid growth. This process demanded sacrifices on the part of the citizens, invariably the poor who have to make the sacrifices for the sake of larger community interests. Development projects like SSP are the products of this conventional paradigm. At the operational

---

be resolved as early as possible. As the progress in identification of poor quality land and corrective measures was too slow, the Supreme Court came in to action.

level, more often than not, there is conflict of interest between the poor and the rich, traditions and development, civil society and state, upstream degradation in Madhya Pradesh and downstream irrigation benefits in Gujarat and between natural resources and its sustainable distribution. This was a prolonged process where people with varying capabilities interacted for narrow advantages for a particular section of the society. In this debate, the decreasing availability of water has been an important argument for large dams by their proponents. They argue that irrigation is an important input to growth in agricultural production and rural prosperity. Moreover, the increased demand for purchased inputs strengthens the linkages between agriculture and industry. Many do not find this process acceptable. Since this is a conflict between unequals, they resist the process itself. They believe in people-oriented development, which calls for a process where access to natural resources is not adversely affected, and where planning and its implementation improves access to food and services in such a way that social, human and environmental losses are minimised. The resistance to Indian development strategy, thus, has come from many quarters.

An important dissent emanated from those who reject the existing paradigm of development itself[29]. The approach of NBA falls in this

29. In a developing economy like ours, one of the two sectors (agricultural sector) is traditional. It produces wage goods (foodgrains) and has excess supply of labour. The non-agricultural sector is modern and draws capital and labour from traditional sector. The productive accumulation in the economy can be faster if agricultural sector performs. This is so because only then (a) strong input-output linkages between the two sector would® establish; (b) terms of trade between sectors would support surplus transfer, and (c) demand for non-agricultural goods in the traditional sector would increase. The overall growth of the economy, in the long run, would be constrained if (i) technology in the traditional sector remains stagnant; (ii) Market links between sectors remains weak or the terms of trade remain unfavourable for resource transfer to non-agricultural sector; (iii) Contribution of various parts with in the agricultural sector is unequal. The above constraints, in the short run, would also increase social and economic inequality. Although, it was observed that high agricultural growth reduces rural poverty and higher consumer price aggravates it, the process may be affected if (a) early growth takes

category. This paradigm shift opened a new dimension in the debate on development. Sustainability did not remain confined to tribal rights, loss of livelihood and resettlement; it also included ecological and environmental issues associated with the project. Their belief regarding the unsustainability of Sardar Sarovar, strong views about the impossibility of rehabilitation, and lack of faith in capabilities of the administration to handle a large involuntary migration – originating not only from Gujarat but also from Maharashtra and Madhya Pradesh – caused a vertical split between the involved NGOs of Gujarat. A group spearheaded by Medha Patkar seriously questioned this development approach. The group resisted development on the ground that it had adverse implications. The Andolan's approach during the early 1990s changed the very concept of project formulation. No project in the future can go without scrutiny for its assumptions. This is no mean achievement. In the heart of this resistance lies the callous manner in which environmental issues were dealt with in the SSP. They argued that restricting the natural flow of an irrigation system like Narmada, other than immediate economic pay off, has far-reaching consequences. It not only results in submergence and involuntary migration of a large population and downstream salinity and waterlogging, but also results in adverse upstream environmental degradation like deforestation, soil erosion, sedimentation, catchment area treatment and biological and zoological degradation. Against

---

place under extreme inequality and its benefits may not 'trickle down' to large rural masses; and (b) oligopsonic influence of landowners distorts the rural labour market, dampening the on-farm employment opportunities and wage rates. Indian concept of poverty has moved from identifying a minimum normative absolute living standard in terms of per capita consumer expenditure — may it be the normative food basket or the minimum calorie intake — towards explaining the poverty. It should, however, be noted that cross-sectional analysis of calorie intake by individuals of PAP households contain noises despite the fact of a large sample. Moreover, this approach of poverty measurement does not give adequate emphasis to quality of life: PAPs' access to services, their environment relating to law and order, access to justice, the environment where they live, the air they breathe and water they drink. In absence of these concerns, the measurement of poverty is partial, reflecting only to food insecurity of PAPs.

this backdrop, the rest of the chapter traces the environmental compliance debate over SSP.

V. Environmental Compliance

Singh (2003) argues that environment clearance for large dams was introduced only in 1978. The requirement became statutory in 1994 when Environment (Protection) Act, 1986 was modified. Between 1978 and 1980, Department of Science and Technology was the nodal agency for environmental clearance for development projects. Only after this clearance, the Planning Commission accepted the proposal. The National Committee on Environment Planning and Coordination was the assessing the project's statements. With the formation of Department of Environment (DOEn) in 1980, the task of clearance was transferred to the new department. But the Department of Forests and Wildlife had the responsibility to accord forest clearance for the forestland coming under submergence. With setting up of Ministry of Environment and Forests (MoEF) in 1985, it was this ministry that had responsibility of conducting an environmental impact assessment and granting environment clearance as well as forest clearance. The DOEn and Department of Forests and Wildlife became part of the new Ministry. It was only from 1994 the environment clearance became mandatory though forest clearance was made mandatory in 1980.

The Environmental Appraisal Committee set up by MoEF performed the function that was originally performed by National Committee on Environment Planning and Coordination. Based on recommendations of the Environmental Appraisal Committee, MoEF decides whether a project should be cleared unconditionally, with some conditions or should be rejected. Another committee on forest recommends on clearance of forestland under submergence. The Environment (Protection) Act was amended in 1994 to make public hearing mandatory in the assessment process.[30] Within the Central Water Commission, an Environment Monitoring Committee was constituted in 1990 to oversee the implementation of environmental standards.

30. For a short duration, the public hearing was made optional but from 1998, public hearing was once again made mandatory (Singh, 2003)

Environmental Issues

The SSP came under critical environmental appraisal when Narmada Water Dispute Tribunal made its Award in 1979. For, the location of the project site, flow of water from catchment area including Narmada Sagar Project, reservoir level, dam height, submergence of forest area, sharing of benefits between Madhya Pradesh, Gujarat and Maharashtra etc. had for the first time become clear. In 1983 both Sardar Sarovar and Narmada Sagar Projects were before DOEn for environmental clearance. The application was supported by benchmark studies conducted by MS University of Baroda. However, the environmental clearance was not accorded to the project. The SSP authority was informed about the difficulty in the study as well as additional studies required.

Despite this, in 1985 World Bank approved the credit and loan agreement for SSP. In June 1987 MoEF issued a joint clearance for SSP and Narmada Sagar Project. This clearance was accorded with condition that more detail would be furnished for: (i) catchment area treatment, (ii) command area development, (iii) compensatory afforestation, (iv) resettlement and rehabilitation, (v) flora and fauna study, (vi) carrying capacity of the area, (vii) sismicity, and (viii) health problems. While consideration on Narmada Sagar Project was deferred, the World Bank in 1990 made it clear that the loan for SSP would be cleared with condition that,

(a) Progress of R&R and environment would be assessed as the construction of the dam progressed; and
(b) Satisfactory environmental assessment was conducted.

The condition (a) above is important, for it means that the Narmada Control Authority would ensure that the environmental measures were implemented *pari passu* with the progress of construction of dam. The interpretation of *pari passu* by the Project Authority, it seems, was not 'as the construction progresses' but that at the end when the dam was ready the environmental bindings on the project would also be cleared. This made the Project Authority less vigilant of the emerging environmental issues. Consequently, for four years between 1983 and 1987, when MoEF did not clear the project, the Project Authority showed little apprehension and went ahead with construction without environment clearance. After four years, when the clearance was given in June 1987, MoEF still had questions about the aspects of major environmental issues. This delay and conditional clearance became the issue of debate within as well as outside. The anti-dam movement used these contradictory evidences to prove that SSP is a planned environmental disaster (Alvares and Billorey, 1988; Ram, 1993). The May 1987 note of MoEF to Prime Minister of India[31] did detailed analyses for SSP and the Narmada Sagar Project separately. Major environmental concerns (from the listed eight concerns above) were unexpectedly found satisfactory except for the Resettlement and Rehabilitation (R&R) of oustees. As the environmental clearance for the SSP was accorded, in October 1988 the Planning Commission also granted approval for the Project.

An analysis of the stand taken by MoEF in June 1987, whereby it accorded a conditional clearance and its letter to Prime Minister of India in May 1987 whereby the environment compliance was found satisfactory (except for R&R) shows inconsistency between the two. Both the anti-dam activists and the Morse-Beger Independent Review that the World Bank instituted in 1991-92 cashed in on this inconsistency. The Independent Review of SSP conducted a detailed analysis of issues such as hydrology, upstream environment, downstream environment, command area environment and health.

Within the context of environmental implications, it may be relevant to place on record the hydrology debate that is so very vital for the design of SSP. As already discussed, this design is a balance of inflow and rate of discharge from the reservoir. The NWDTA (1979) estimated that on average 75 per cent dependability, 28 MAF of water would be available from the project. The Tribunal also allocated this water between the four states of Madhya Pradesh (65.2%); Gujarat

31. The note to Prime Minister (May 1987) by the Department of Environment and Forest strongly complained about lack of data and callus compliance of various human and environmental norms. The letter concluded that 'In the overall interest of natural resource optimisation, reducing to the minimum the impact on human beings and for minimising the ecological damage, an objective review of the design parameters seems desirable (for a full text of the letter, see Appendix 3, Alvares and Billorey, 1988).

(32.1%); Maharashtra (0.9%) and Rajasthan (1.8%). There are two major issues arising from the Award. First, what if the yield was less; and second, its effects on the downstream canal design and irrigation. Both these issues are contentious, for they influence the water availability and hence the design of the project. Critics of SSP have argued that the actual yield of 28 MAF is an overestimation and from other sources of GoI, they estimated that the yield could be just 22 to 23 MAF.

Using the data provided by the Central Water Commission, Ram (1993) estimated the variations in actual yields between 22.2 MAF and 22.8 MAF. He based his findings on the actual flow of the Narmada for 15, 22, 33 and 42 years. This reduction in yield makes the allocation to Gujarat actually 7.3 MAF against the 9 MAF designed for the SSP. At this stage, the controversy of building Narmada Sagar Project along with SSP is irrelevant,[32] for even with the Narmada Sagar Project the assumed yield is suspect. The controversy of yield of SSP was also fanned by the May 1987 note to Prime Minister sent by MoEF. It estimated that at 75 per cent dependability, runoff in Narmada should be only 23 MAF instead of 27.2 MAF as assumed by NWDT (see Appendix I). This naturally means less water for irrigation and drinking to Gujarat. Ram (1993) argues, "As the entire planning for the Sardar Sarovar Project is based on an assumption of using 9 MAF of water, any decrease in this amount will result in decreased benefits. Indeed, the benefit-cost ratio of project, which at 1.13 is already at marginal acceptability according to Government of India and World Bank norms, will decrease below acceptable limits if the smaller quantum of assured water is taken into account."

Independent Review and after

Following their detailed analysis, the Morse and Berger's Independent Review (1992) of Sardar Sarovar Project recommended to the President of World Bank that,

32. it is argued in the Tribunal Award that this yield is possible only if Narmada Sagar Project — which at that point was to be constructed simultaneously with Sardar Sarovar Project — is actually constructed simultaneously.

> "We found discrepancies in basic hydrological information related to these works. We therefore examined the stream flow data and did our own analysis. We found that there is good reason to believe that the Projects will not perform as planned. The problems relate to the sequence and timing of stream flows and the capacity of the dam and canal to store and divert water. The effects of Sardar Sarovar upstream, downstream, and in the command area, therefore, will be different from what has been assumed to date whether or not the upstream Narmada Sagar Projects are built as planned. A realistic operational analysis upon which to base an environmental assessment is lacking. This alarmed us and it should alarm others, especially for a mega project with such far-reaching implications as Sardar Sarovar.

For the area upstream of the dam there are piecemeal studies that suggest that the impact on biodiversity will be minimal. But there has been no attempt properly to assess the cumulative effects of the impacts arising from the Narmada Sagar Projects.

Although the Narmada Sagar Projects are not within our Terms of Reference, the resulting cumulative impacts will almost certainly be serious. The Bank has placed itself in a. difficult position by agreeing to proceed with Sardar Sarovar Projects before the environmental implications of directly related projects upstream are understood.

Programmes in the upstream region for compensatory afforestation and catchment area treatment are under way. We believe that these programmes, however successful in the short term, are likely to fail because of the lack of participation by local people. It is our view that achieving the necessary cooperation is not likely to be possible within the construction schedule imposed by Sardar Sarovar."

"In spite of non-compliance with Bank resettlement and environmental requirements, the Sardar Sarovar Projects are proceeding—in the words of Chief Minister Patel of Gujarat—as "an article of faith." It seems clear that engineering and economic imperatives have driven the Projects to the exclusion of human and environmental concerns. Social and environmental tradeoffs have been made that seem insupportable today."

"Important assumptions upon which the Projects are based are now questionable or are known to be unfounded. Environmental and social trade-offs have been made, and continue to be made, without a full understanding of the consequences. As a result, benefits tend to be overstated, while social and environmental costs are frequently understated. Assertions have been substituted for analysis."

"Every decision as to the Sardar Sarovar Projects has always been, and will continue to be, a decision for India and the states involved. Together, they have spent a great deal of money. The foundations of the dam are in, the dam wall is going up, the turbines have been ordered and the canal is completed to the *Mahi* River. No one wants to see this money wasted. But we caution that it may be more wasteful to proceed without full knowledge of the human and environmental costs.

We have decided that it would be irresponsible for us to try to patch together a series of recommendations on implementation when the flaws in the Projects are as obvious as they appear to us. As a result, we think that the wisest course would be for the Bank to step back from the Projects and consider them afresh. The failure of the Bank's incremental strategy should be acknowledged."[33]

The World Bank's withdrawal from SSP, following the recommendations of the Independent Review, created a debate that aroused contradictory interests of various groups. The battle was not only between pro- and anti-Sardar Sarovar agencies but also between (a) submergence in MP and irrigation beneficiaries of Gujarat,

---

33. These conclusions are part of Independent Review (Morse and Berger, 1992, xxii, xxiv-xxv). Gujarat based Non-government organisations especially Arch-Vahini termed these conclusions as dramatic and sweeping. "Morse-Berger" writes Arch-Vahini in their letter to Mr. Lewis Preston, President, World Bank in September 8, 1992, "have obviously been able to convince majority of the readers about the truth of the conclusions they make in the letter. This is one of the most worrying aspects of the Morse-Berger style of reporting because majority of the momentous conclusions in their letter do not follow from the facts they have assembled or not assembled in the text of the Chapters. At least some highly significant conclusions they reach in the Chapters curiously do not find place in the letter to you. And it is out of the contents of such letters that protest letters are formulated and sent off."

(b) powerful state and grassroots activists who understood the R&R problems of the displaced, (c) environmental implications of SSP and the past performance of other big projects in India, and (d) sociologist vs. economist perspective of development.

How valid is the claim of Independent Review on Hydrology that "there is good reason to believe that the project will not perform as planned"? As already mentioned, not everyone involved in SSP has accepted the analysis of the Independent Review. For example ARCH-Vahini (1993) in its response to Independent Review has raised questions on the reliability of data used by the Review: Arch argues: "...and then very blandly, without giving any evidence and arguments, they (Independent Review) assert that NWDT's assessment appeared to be unfounded or questionable. A very vague reference is made to an independent hydrologist they had hired. Even his/her name is not given, nor his/her findings and assessment are described or discussed *vis-à-vis* the findings reported by the country's apex body – the Central Water Commission". Notwithstanding this and also the fact that hydrology was not in the Terms of Reference of the Independent Review, there were inconsistent stands taken by different agencies like the Central Water Commission, MoEF and the SSP Authority. Whether or not the Project Authority was overstepping its role as one that would decide on environmental compliance, the sad part of this incident is the erosion of independence of the environmental appraisal agency.

VI. Conclusions

The pre-1987 struggle of civil society in Gujarat is a testimony of how displacement can become an important element to account for in development. And once R&R policy was accepted, civil society participated in the relocation process. The institutional arrangements made in the process ensured that the relocation was humane. Despite early problems, relocation in SSP was a success. But there were many second-generation problems that only got tackled once the Narmada water started flowing in the resettlement sites.

On the other hand, the environmental debate on SSP was polemic in nature. The politics of SSP, like any political discourse, has been full of circumstantial divergence. In so far as environmental standards are concerned, the critique developed by MoEF in 1987 was no different than what the Independent Review (1992) aserted. The question is why the Project Authorities were adamant to ignore the standard set by the indigenous but independent agency. Was the Project Authority in a hurry to complete the construction of dam while MoEF was forcing a debate on patchy environmental compliance? This is a moot question. What we nevertheless learnt from SSP is that, whether it is R&R of oustees or Environment Assessment, if the independence of the monitoring agency is not respected, the human and environmental cost in a development project will be high enough to render it socially unsustainable and environmentally unfriendly. This debate has also opened windows to argue about alternatives; it throws open the priorities of planners to either optimise economic profit or minimise the social and environmental cost of development.

These controversies, and many others, have not only been part of learning processes but have also been documented. But in such documentation (see, for example, Sah, 2003; Cullet, 2007), in the zeal to be pro-poor and highlighting such evidence that provides their view-point a stand, injustice has been done to many evidences, like the stand taken by ARCH-Vahini. Consequently, such documentation at times sounds – to put it mildly – biased.

Appendix I

Source of Design Controversy of Hydrology of Sardar Sarovar Project[34]

A NOTE TO PRIME MINISTER FROM DEPARTMENT OF ENVIRONMENT AND FOREST, MAY 1987

"The Narmada Sagar and Sardar Sarovar projects are being designed as multi-purpose projects to maximise irrigation and power benefits. The Narmada Sagar FRL has been fixed at 860 ft. to provide a live storage of 8.8 MAF. The following points are worth noting regarding the design criteria:

- Irrigation in Narmada Command will be provided only to 1.4096 lakh ha by utilising just 1.4 MAF of the live capacity. There is reason to believe that even with variations in dam height down to FRL of 814 ft. the irrigation potential in Narmada Command will not be affected at all. The adverse impact would be on the generation of firm power which would become 22 MW instead of 118 MW eventually.
- The positive impacts of fixing the FRL at 814 ft are that the forest submergence gets dramatically reduced to 8,075 ha, total land submergence is reduced to 35,628 ha and the number of villages affected becomes 77 with a population of 20,200 only.

Reconsideration of fixing the dam height for the above reasons alone is justified. This becomes all the more important considering the fact that the 75 per cent dependable run-off has been estimated to be only 23 MAF instead of the 27.2 MAF assumed by the Narmada Tribunal. The Mean Draw Down Level (MDDL) of the Sardar Sarovar Project can be modified to provide an additional live capacity. A rough estimation is that the MDDL can be lowered by about 50 ft.

There is thus a strong case for an objective review of the storage planning of Narmada Sagar and Sardar Sarovar so that the water resources are optimally utilised while minimising environmental damage."

34. *Source:* Appendix 3, Alvares and Billorey (1988).

16

Status of Agriculture and Animal Husbandry among Narmada Oustees: Narratives from Some Re-settled Sites

Biswaroop Das

In spite of ongoing changes in the market, institutional structures and a variety of intermediation, agriculture and animal husbandry continue to play an important role in the Indian rural economy. Though the relationship between these two sectors varies with changing socio-economic context, it remains an important basis of rural livelihood systems across many regions. This gains further significance when perceptions of villagers are examined in terms of their dealings with such sectors, especially in post-rehabilitation situations being negotiated with planned interventions. This chapter, aims at comprehending the agricultural practices and livestock rearing scenario of 'resettled' *oustees* from parts of Maharashtra, Madhya Pradesh (MP) and Gujarat states especially from within the villages located in the Narmada basin and evacuated partly or fully in order to create space for the large-scale, multipurpose Sardar Sarovar Project (SSP). Before doing so, it traces the broad contours of practices related to agriculture and livestock rearing as it prevailed at some of the earlier sites, in order to provide terms of reference to view the continuities and discontinuities between the earlier and present sites.

I

Even some years prior to the first half of the twentieth century, agricultural practices in these settlements were largely an expression of 'geographic determinism' together with 'social cohesiveness' of much of the population inhabiting the Narmada valley and its floodplain area. With slopes, topsoil and periodic variations in floodplain fertility, yields and types of crops grown across various zones in this region depended on rain-fed irrigation and topography. Livestock formed an integral part of such agriculture, with bullocks used for ploughing and cows reared for their offspring and for milk. In many settlements across the region, especially among the tribals, rearing of goats was common. And with plenty of vegetation and crop residue, fodder was not much of a problem for the herds of goats and other cattle. This was, however, not a uniform pattern. Some of the tribal groups residing in the hilly interior regions did not even practise routine settled cultivation or cattle rearing.

Despite such variations, production in agriculture and associated institutions was not very complex. The agricultural kit of the farmer consisted of simple implements such as plough, hoe, *karab*, *karabdi* and sickle. Cultivation was in accordance with rainfall and the crops raised were of low commercial importance. In fact, across large tracts of settlements along the slopes and hilly terrain, subsistence agriculture was the norm. However, the higher fertility of the alluvial soil across some stretches within the floodplain of Narmada had a better yield of specific crops including some vegetables. The crop-cycle was one-crop and double crop. The former was on land where topsoil was thin, loosely bound and slopes or the immediate landings of major undulations. The latter was in pockets across plains characterised with relatively thick and more mature topsoil.

Since land under double crop was limited, a section of the population, dependent on one-crop land or only cultivating the forest land, was also engaged in other occupations. In some settlements people worked as contract labourers cutting wood and bamboo from the surrounding forests. Contractors from Nandod paid a sum of one rupee per 100 cut bamboos. Loading was done by other labourers and the bamboos carried in rafts along the Narmada to different landing points from where they were sent as far as to some southern states. The timber item mostly sent out was *sag*, which has high commercial value.

Though agriculture was the mainstay of survival, by the first half of the 1970s, petty trade had emerged in some surplus food grains, sale of minor forest produce like *timru* leaves and gums extracted from specific trees, and seasonal sale of vegetables, melons etc. This formed subsidiary activity, directed at nearby shops or small townships. Also, with increased activities associated with Kevadia colony and other works related to Sardar Sarovar, some occupational diversification had begun to emerge – especially at the five 'rock-fill dykes' villages within Gujarat – Navagam, Panchmuli, Khalvani, Vadgam and Mokhdi. However, despite these location-specific changes, activities other than agriculture continued to remain at a low in the valley. Within this context, it would be pertinent to understand the broad characteristics of agriculture, allied activities and people's dependence on other jobs and sectors that existed in the villages that are now partly or fully submerged. Following such description, the narration will shift to the status of agriculture and animal husbandry at a few select locations where the project affected people have been settled for the last 15 to 20 years (as of 2003).

II

Agriculture, Animal Husbandry and Other Activities at the Earlier Sites

Agriculture

At most of the old sites, agriculture was mainly rain-fed. The monsoon decided the yield of food crops and to a large extent their fate of production. With a major part of the cultivated land having a thin and gravelly top-soil, the one-crop cycle was predominant. Such land best responded to the coarse crops of *kodra*, *banti*, as well as *jowar*, maize and pulses like *arad* and *tuver*. The second or winter crops were taken on the more fertile plots along the river bank, on small valley plains and slopes on natural watersheds. These crops generally included *jowar*, *val*, gram and castor. By the early 1970s small amounts of groundnut and cotton too were being raised in some pockets of specific settlements. There were locations like Katnera in the Kukshi tehsil of Dhar district in MP, which had other sources of irrigation available besides rain. This enabled some farmers to raise even three crops a year. Similarly, farmers in the Ter *faliu* of Gadher on the river bank could raise two crops a year on patches of alluvial soil.

The land cultivated was under three broad types of tenure. First was owned by an individual both through purchase or inheritance, the second was government barren land, and the third was forest land under the jurisdiction of the State Forest Department. Although not in full legal ownership, households cultivated the government's barren land and forest land with varying intensity. Significantly, the overall incidence of such cultivation was higher in the interior tribal villages – especially within Gujarat – than those near the dam site (Das and Charan, 1983).

While the cropping pattern, determined especially by variations in land quality and rainfall, varied across the region, the area under food crops constituted the largest share. In fact, across the submerging villages in Gujarat its share was as high as 62 per cent of the gross cropped area. This was followed by 20 per cent of area under pulses and the rest under some non-food crops. Of the area under food crops, as much as 57 per cent was under *jowar* followed by 20 per cent under maize and another 7 per cent each under *bajri* and paddy. Other food crops grown in a limited area included a wide range of coarse grains like *kodra*, *bhedi*, *banti*, *kang*, *bavta*, *samel* etc. *Arad* was grown in around 47 per cent of the area under pulses followed by limited acreage of *tuver* and gram. Some groundnut and cotton were grown as non-food crops with a sprinkling of castor and *tal* (Das and Charan, 1983; Sah, 2003).

Since at these sites a large number of households cultivated forest and other government barren land, the share of land owned per household was much lower than the actual area operated. In most cases in the Gujarat villages, more than 50 per cent of land operated by a household was forest land. Significantly, more than half the land actually tilled was cultivated jointly with a sizeable section of the landless population. According to Sah (2003), the share of landless households in the submerging villages was as high as 32 per cent. This however did not mean that such households were not working any land. A large section among them was cultivating forest land. However, with increasing size class of land actually owned, a rise in the incidence of cultivating forest land was recorded at Vadgam in 1982 (Das, 1983).

Evidently, with no substantial difference in cropping pattern, lack of irrigation and hardly any use of chemical fertilisers, land productivity in these settlements remained poor. Moreover, barring few pockets with the advantage of being able to raise multiple crops, especially on the alluvial patches of Narmada's flood-plain, the rest of the region was more or less engaged in subsistence agriculture- linked poorly with the market. Stagnant yield and lack of commercialisation continued to keep much of the agricultural activities at a constant low, and hybrid seeds and fertilisers did not enter much of the region for a long time. Hence, dependence on subsistence agriculture giving limited yield on select crops cultivated on 'encroached' areas could not have sustained farming as well as non-farming households in the region all through the year. Need for ready money was negotiated and responded through (a) sale of some non-food crops like cotton and a limited amount of food crops; (b) sale of minor forest produce including that of *timru* and *ashitra* leaves used in rolling *beedies* and a variety of gums etc. to petty agents and shopkeepers[1] ; (c) occasional sale of logs extracted from the forest which was possible in spite of it being an illegal activity; (d) working as labourers for contractors harvesting bamboo in nearby forests; (e) making charcoal and selling it to petty traders; and (f) augmenting money needed to respond to life cycle events like marriages, deaths, births as well as crises emerging from crop failure by selling cattle especially goats.

Occupations Other than Agriculture

In spite of variations across the submerging settlements in the states of Gujarat, Madhya Pradesh and Maharashtra in terms of intensity of cultivation, those residing in the vicinity of forests and forest lands had a life that revolved around a range of activities. While none of the activities pursued by them was dominant in terms of being the primary source of earning, together these helped, albeit to a limited extent, in responding to a set of minimal requirements of the

1. Sale of *jowar* was more frequent among food crops while among pulses sale of *arad* and *tal* was pronounced. These crops were purchased generally by traders running petty shops in the villages and also doubling as money lenders. People like Sattarbhai at Vadgam and Sevantibhai at Gadher were such traders who also bought minor forest produce from people and sold them to traders in Rajpipla, Bharuch, Surat etc. at a much higher price.

households. Forests and the river were important sources of food, fuel and fodder with a large number of cattle grazing on forest. Some cattle that could be sold acted as insurance during crisis and abnormal agricultural cycles. From December to April, when activities associated with agriculture were lean or nearly absent, there were households from where members went to work at the dam-site as casual workers for daily wages. The extent of migration for jobs to farther areas from many of these villages appeared low and at places non-existent.

Animal Husbandry and Livestock

Rearing of cattle in these settlements has been an important activity for a long time as a needed support for agriculture and also selling in times of crisis. Economic viability of livestock rearing was ascertained mainly by the reasonably vast expanse of forest land supporting fodder requirements. Additionally, the stock was also linked with production of dung for turning into manure used for enriching the soil. All cultivating households had bullocks, buffaloes or cows. Ownership of buffaloes was however much less than that of cows. On an average, a cultivating household had around seven heads of cattle not counting goats which were many. Taking cattle for grazing was the responsibility of *govals* of which there was at least one household in every *faliu* in most of these villages. In return for their services, they were given a fixed amount of food and money proportionate to the number of cattle taken for grazing per household.[2]

While cow milk was generally consumed and seldom sold, *ghee* made from it was often sold to the village shopkeeper. Goats and sheep were sold at times of monetary crisis and also amongst some groups consumed for meat on specific social occasions. In some villages, traders visited off and on to buy goats and sheep from households interested in selling them.[3] Bull calves were nurtured, castrated five or six years after birth and used for ploughing and levelling of land for cultivation. They were also used for pulling carts

---

2. The rate varied between rupees five and six per cow, bullock or a buffalo for a month of grazing.
3. Many of these traders coming to the submerging villages in Gujarat were from Rajpipla, Tanakhala and from as far as Bharuch.

for tasks like transporting manure, grass and crops like cotton sold to petty traders in nearby shops and settlements.

This narrative indicates the heterogeneity as well as homogeneity, especially in terms of economic behaviour, prevalent among groups in the region prior to their rehabilitation. The economic homogeneity was reflected by a broadly uniform cropping pattern determined by rains and varying characteristics of topsoil on slopes and patches on the plains; the near absence of 'higher' technology and lack of adoption as well as access to extension services; pronounced lack of commercialisation in agriculture; dependence on a wider range of activities, albeit with a low intensity, that were linked and integrated so as to sustain a livelihood that was poor but yet endowed with survival strategies reflecting negotiations through a multi-pronged dependence on forests, cattle and cultivation – including that of raising some coarse crops on the forest land.

In spite of this, it would be naive to assume that the economic behaviour across these settlements was dominantly homogeneous. Some villages were linked with markets through petty sales of limited surplus grains, crops like cotton, poultry, livestock and on a rather feeble scale, some minor forest produce. Even in some interior villages inhabited by specific tribal groups, the sole shopkeeper acted as a link to the outside market. In addition, farmers who could cultivate vegetables and crops on the floodplains of Narmada were more exposed to the market. On many such plots, especially in MP and Maharashtra, cultivation alone determined the contours of their livelihood systems. It can be concluded that economic environment was more or less homogeneous in specific pockets especially along, as well as in the vicinity of, high slopes and hilly terrain, while in the basin and the plains they were different across the region. This indicates that all those being rehabilitated did not exactly belong to a similar situation, for they were located differently in economic terms. This helps in understanding and analysing the manner through which oustees were responding to and negotiating their lives and livelihood systems especially within the context of agriculture and animal husbandry at three of the randomly selected resettlement sites.

III

Site-1: *Kundi Unchakalam*

Located in the immediate vicinity of the host village Unchakalam, Kundi Unchakalam is one of the sites where 46 Project Affected Families (PAFs) from MP and 16 families from within Gujarat were resettled. While all those from Gujarat are *Rathwa* Bhils from Hanfeshwar, the PAFs from MP belong mainly to the caste groups of *Balai*, *Bhilala* and *Mankar*. The Gujarat oustees began their journey to this new site in May 1991. Till 1993, they were still cultivating land and keeping cattle in the submerging village and also cultivating parts of land received at the new site. By March 1993, all households had fully shifted to Kundi Unchakalam. Of the families from MP, 90 per cent came from village Katnera in Dhar district and the rest came from the settlements of Akadia, Jaser and Jahangira in Badwani. By the time the first batch of these oustees came in 1993, PAFs from Gujarat had already started cultivating land.

Of the PAFs from MP many continued to feel unsettled, for parts of their families still lived at the old sites. This also included families who were earlier landless but having received land here, still wished to hold on to their house-sites at the submerging villages and periodically take up higher wage works available in agriculture and other petty sectors there. Those PAFs who received plots at different locations found it difficult to cultivate their land and some, with their plots at villages other than where they resided, had opted to lease them out. In some cases, they cultivated them jointly with farmers belonging to settlements where their land was located.

Not all, however, were unhappy with their land. For many it was a gain as having shaped the new plots and worked on improving their quality they began to raise crops and at the same time continued as part of their families still living at the earlier sites. For some it was an added amount of land over and above part of the land that their fathers and uncles still cultivated at Katnera or Akadia. And thus evolved a new arrangement where participation in agricultural operations and cultivation on both sides could draw mutual support of labour and resources whenever possible. Within this context, it is also important to note that of the PAFs who had received land at the new site as many as 36 per cent were landless, 16 per cent had land ranging between 1 and 3 acres, 23 per cent had between 4 and 6 acres and the rest above 7 acres. Thus for some the land gain at the new site was an additional asset.

Significantly, a little over 60 per cent of the PAFs had to stay in temporary tin sheds for long – something that was shorter for the later batches. Much of the land for the second and the third batches, as per their perceptions, was not very good, and that they had to work towards improving it to the extent possible by treating it with manure year after year. During this initial period they faced much hardship. In the absence of adequate returns from the plots, ongoing attempts of improving them and the absence of irrigation, many had to seek wage labour in agriculture and other sectors. Some periodically visited their original sites and worked as agricultural wage labourers, while others combined such jobs with work related to digging and clearing ground within around 50 km of Kundi Unchakalam especially at other new resettlement sites.

In comparative terms, the crop cycle and pattern of cultivation at the two sites have been different ever since. Substantial amount of agricultural land at Katnera and around had irrigation. Through water drawn from bore wells and canal lines, many plots could raise as many as three crops a year. Nurtured on a regular basis, the soil quality better facilitated ploughing. The major crops raised were groundnut, maize, *jowar* and *bajri*. Of the commercial crops, cotton was regularly cultivated, easily yielding two harvests a year with maize ripening within a period of four months. Wheat was raised on a limited scale while some tobacco was also grown for consumption as well as sale. Some oilseeds and pulses too were raised regularly. In addition, those having motors with 3-5 HP capacity could easily avoid paying electricity charges making harvest of water nearly free. With cropping intensity in and around the region being high, the extent of availability of wage labour in agriculture was better. The average availability of agricultural work at villages in these pockets was about 120 days a year, in contrast to around 80 days at the new site and surroundings.

Given the land quality, and after prolonged efforts for improvement and successive raising of crops, per unit return from land was still considered much lower than that on the submerging sites. Significantly, as many as 32 per cent households had access to irrigation at the old site. After having lived at the new site for a little more than a decade, as many as 92 per cent of them complained of lack of adequate irrigation facilities. This, they felt, was one of the main reasons why they could not raise adequate food crops from most of these allotted plots.

A major subsidiary canal of the Narmada that passes through the immediate vicinity of the village had, however, brought about some changes in the pattern of cultivation among the PAFs. Some of them lifted water from the canal to irrigate their plots. Several farmers from the host village were also using this facility. Together with the resettled oustees, they bought motors for Rs. 17,500 for drawing water from this source. This was a major outlay for many of the PAFs for an input that had come nearly free at their earlier locations. Cost of irrigation for a five acre plot came to around Rs. 12,000 for a one-crop cycle. Chemical fertilisers cost another Rs. 2,500, seeds Rs. 1,800, ploughing with tractors around Rs. 7,000 and Rs. 1,500 for labourers' wages. Thus the total investment in five acres of land during one annual cultivation cycle added up to around Rs. 24,000 or a per acre cost of Rs. 4,800. Returns against that came to around Rs. 40,000 giving a household a net annual income of around Rs. 16,000 a year excluding food grains for home consumption. Besides working in their own fields, some of them also went to the host and other villages to work as agricultural wage labourers.[4] Women joined men in seeking work. They were generally involved in weeding, plucking cotton, harvesting maize, transplanting paddy etc. Wage works in agriculture fetched an adult male around Rs. 35 a day and an adult female around Rs. 30 a day for nine hours of work.

The pattern of cultivation and agricultural practices, particularly among the PAFs from MP in this village, were also determined to a large extent by how they had negotiated their economic and social life, especially in the last eight to ten years. Significantly in this

4. Not only resettled PAFs but also farmers of the host village of Unchakalam were seeking agricultural wage work, especially in large farms of Patels at villages such as Totarmata and Suryaghoda. During 1994-95, when cultivation was still difficult, and some were still working on making the allotted plots fit for cultivation, a group of PAFs, particularly from MP also went to Karjan to seek work in the secondary and tertiary sectors.

context, 70 per cent of these households had actually faced only partial land submergence at their original villages, with amount of land lost varying across groups. Only around 30 per cent had lost both land and dwellings. Thus, not only did they continue to raise crops at both sites but those with no land and dwelling at the submerging village remained part of its cultivation cycle as wage workers in agriculture while at the resettled site they cultivated their own land and also worked as wage labourers As many as 62 per cent of the resettled plot owners had their parents, wives and children still living in the original village, indicating that their social lives were still organised and rooted in networks evolved and sustained at their original sites. For most of them, their plots in Gujarat seemed mainly to serve as complementary farm sites to works and sectors still surviving in their old villages. To this extent, agriculture and associated tasks in production and organisation of cultivation at the new site remained subordinate to similar tasks at the old site, despite their having lived at Kundi Unchakalam for around a decade.

As mentioned above, all 16 PAFs from Gujarat at Kundi Unchakalam had moved to the site before than the PAFs from MP and settled down by 1993.[5] Initially cultivating at both sites for a year and a half, they quickly settled in the new village, albeit owing to the fast approaching submergence. In contrast to PAFs from MP, all these households were landless and cultivated forest land by paying fines to the department.[6]

At Hanfeshwar, there was no irrigated land and crops were entirely rain-fed. Maize, *jowar* and *bajri* were the major food crops and *tuver* and *arad* the main pulses. Some dry cotton was also raised and sold to traders at Kadipani and beyond. Returns from land as well as sale of crops were low. None knew about chemical fertilisers and pesticides. If the rains were good and timely, depending on the crops sown, they could raise 8-10 quintals of maize, 12-15 quintals

5. Belonging to the group of *Rathwa Bhils* they were from Hanfeshwar – one of the 19 submerging villages in Gujarat. Entirely under submergence, all families from its 12 *falius* had been resettled at 22 new sites within a stretch of approximately 14 km between Dabhoi and Naswadi.
6. Cultivating an acre of such land was possible by paying a fine of Rs.100 to the forest department.

of *jowar*, 5-6 quintals of *bajri* or about 10 maunds (20 kg) of cotton from one acre of land. *Jowar* and maize were always stored in *kothis* for consumption during lean periods. The limited cotton grown did not fetch more than Rs. 300 per quintal.

Importantly, and in contrast to the PAFs from MP, all but one oustee household from Hanfeshwar got their plots in one piece in the same village, which were of good quality. None was interested in changing the plots and a few of them, whose land at the old village was only partly submerged, continued to cultivate the un-submerged part of it. While the overall intensity of cultivation at the new site had increased, maize and *jowar* were still the major food crops. There was however a fall in the cultivation of *bajri* while cultivation of pulses and oilseeds had increased. Most of the farmers also grew some vegetables. However, there were notable changes in the acreage and intensity of growing cotton. It was raised as the most prominent commercial crop by all. Returns from cotton were high, with its selling price having reached up to Rs. 2,000 per quintal.[7]

From the accounts of oustees from Hanfeshwar, it appears that they were able to organise cultivation of their plots at the new site reasonably well. In spite of having lost some periodic income from forest produce such as *timru* leaves, gums, some timber, *mahuda* flowers etc., they were able to turn into 'land-based' households reasonably well. Also significant is how they could quickly adopt new technologies and inputs in agriculture which was evident from all of them having chosen to irrigate their land from five bore wells dug by some of the families at a cost of Rs. 1,15,000 each.[8]

Along with agriculture cattle rearing was an important activity for the PAFs from Gujarat as well as MP. Bullocks, cows and buffaloes were the principal cattle, though goats too were reared by many. It is noteworthy that the government gave the PAFs grants for buying

---

7. The cost of cultivating an acre of cotton was about Rs. 2,500 (land preparation Rs. 400; DAP, Urea and sulphate Rs. 850; harvesting Rs. 800; and other labour and transportation costs Rs. 250). On an average 25 *maunds* or 5 quintals of cotton could be raised with these inputs and a farmer could net an earning of Rs. 7,500 from one crop of cotton sown in an acre.
8. Owners of these tube wells were selling water at the rate of Rs. 30 and Rs. 50 an hour on motors of 5 HP and 7½ HP capacity respectively.

bullocks. Cows and buffaloes were reared at the submerging sites too. As many as 90 per cent of the PAFs from MP and 62 per cent of those from Gujarat had cows; owning buffaloes, however, was more prevalent among PAFs from Gujarat. There was a similarly distinct pattern in respect of poultry. Nearly all households from Hanfeshwar in Gujarat reared hens. Of the PAFs from MP only 30 per cent did. On an average, a PAF from MP had about five heads of cattle. The cattle collectively owned by the Gujarat PAFs added up to 170, of which around 65 per cent were goats. With around 26 buffaloes, 16 bullocks, 22 cows and 104 goats, a PAF from Gujarat owned on an average around 10 heads of cattle.

A reduction in the ownership of milch cattle and a significant fall in the number of goats per household appeared to characterise cattle ownership. However, the pattern varied across groups with PAFs from MP having milch and draught animals somewhat lesser than the submerging sites as against the Gujarat PAFs who reported substantial reduction in the number of cattle compared to their earlier sites. The rate of such reduction also varied across cattle types, for the fall had been drastic in case of goats and sheep. It must be noted here that while animal rearing for the MP oustees had been equally important at both sites, it was more important among the oustees from Hanfeshwar at the submerging site. Easy availability of fodder from the forest and river water in the vicinity was able to support a higher number of cattle per household. At the new site, however, cultivation had taken precedence, with animal husbandry placed in a subordinate position.

Access to fodder, initially curtailed and later meagre, had been responsible for the lower number of livestock heads – especially goats at the new site. Evidently, the area under *gauchar* has been shrinking all over the state and the new sites were no exception. The presence of a larger number of cattle grazing on *gauchar* land in the host village was initially absorbed by the original residents. However, with the passage of time and intermittent periods of strain, the new settlers' right to graze cattle on the original village *gauchar* emerged as a contentious issue and for some time cattle belonging especially to the MP oustees were not allowed to graze on the Unchakalam *gauchar*. Emphasising the effects of lack of grazing land, some oustees from MP said that there had been a rise in cattle deaths and given the high cost of fodder, maintaining large herds was economically unviable. Even meeting part of the fodder need from fields, the annual cost of feeding an adult cow or buffalo came to around Rs. 4,000. This meant that the cost for a household to maintain an average of five heads of cattle (excluding goats) was around Rs. 20,000 a year – a prohibitive amount for many, especially during periods of low returns from agriculture and associated wage labour.

The PAFs from Hanfeshwar in Gujarat stated that in spite of increasing pressure on the *gauchar* land, they could access the host village's common grazing area, even though to a limited extent. But since their plots were irrigated and more or less without undulations, they could harvest more fodder. However, they felt that having lost their access to forest and thereby the variety of free fodder at the submerging sites, they were unable to raise more goats at the new site.

What emerges from the above narrative is that across as well as amidst different groups of oustees at Kundi Unchakalam, perceptions varied regarding the 'new' status of agriculture and livestock rearing. Such variations resulted from the manner in which the groups in general and individual PAFs in particular have taken to cultivation and animal husbandry. However, it appeared certain that agriculture was on its way to becoming the core activity at the resettled village, and placed within the context of sustained exposure to market pressures, it had already led to changes in the ways that the PAFs had opted to cope with their social lives at the new site of Kundi Unchakalam.

Site-2: *Parvetha*

At Parvetha, 199 PAFs from Maharashtra and 48 from Gujarat were resettled at the new site. All those from Gujarat were *Tadvis*. Most of them were from the Ter *faliu* of Gadher. Three families were from Limdi. From Maharashtra too, 88 per cent of the PAFs were *Tadvis* and the rest *Vasavas* from the submerging village of Manibeli. A few of the families were from Akadia village adjacent to Manibeli.

Around 47 per cent of the Gujarat PAFs at Parvetha came during 1985-86. The last group of five families had shifted by 1992. PAFs from Maharashtra, however, had begun to arrive from 1982, when six households from the Mandir *faliu* of Manibeli came in. Initially for about two years after shifting, PAFs from both regions cultivated land at both places. From 1984 there was a steady stream of families coming from Manibeli and by the end of 1986 around 62 per cent of the families had come in. Another batch of 21 households came during 1992. The last batch arrived by the beginning of 1995.

Data on land given to oustees from Maharashtra indicate that a little less than 66 per cent of the families had received their plots as compact parcels while the rest had got fragmented plots. In the case of Gujarat PAFs, 51 per cent of the households had said about their getting fragmented holdings. Of the PAFs, 38 per cent and 35 per cent respectively from Maharashtra and Gujarat were still dissatisfied with the land they got and desired change of their plots. As per their perceptions, the reasons for dissatisfaction ranged from allotted land being of poor quality to their locations being far from their residences.

Around 70 per cent of the PAFs from both states cultivated forest and other government land at their earlier locations.[9] Agriculture in such areas was totally rain-fed. However, barring a few PAFs from Maharashtra, the rest had stopped cultivating any land at their original villages after having shifted to the new site. Those still cultivating land at the submerging sites had faced only partial submergence.

Characterised by more or less similar geomorphologic and soil characteristics, the annual crop cycle and cropping pattern at Manibeli and Gadher were similar. Farmers mostly adopted a two-crop cycle – (i) monsoon or *chaumasu* and (ii) winters or the *siaru*. The latter was not as extensive as the former. The crops mainly grown during this season were castor, maize, gram and some select vegetables. Most of these winter crops responded best in the floodplains of the Narmada.

---

9. Proportions of households cultivating forest and other *kharaba* land were similar at the original sites in Maharashtra and Gujarat, though the extent of their cultivation varied. Evidently, more than 31 per cent of PAFs from Gadher in Gujarat were cultivating such land that measured beyond 10 acres while the corresponding figure in case of the Maharashtra PAFs was slightly less than 13 per cent. For a large proportion among them, the average landholding size of the forest and other government land cultivated was well within three acres – a figure which had been nearly double in the case of families at the site within Gujarat.

Such patches of land existed in the Ter *faliu* of Gadher as well as at the confluence of Dev and Bhanumati – the two rivulets meeting the Narmada near Manibeli. Prominent among the *chaumasu* food crops were *jowar*, *bajri*, maize and paddy while *arad* was prominent among pulses. Other coarse food crops included *kodra* and *banti*. Significantly, the part of floodplain available for cultivation never exceeded 10 to 12 per cent of the total land under cultivation. Of the crops, castor was sold at 50-60 per *maund* to petty traders and shopkeepers based at Tankhala, Jetpur and Haripura. Some *arad* too was sold when harvests were in excess of consumption needs and of seeds for the next crop. Crop yields were moderate and fluctuated with variation in intensity of rain. No chemical fertiliser was used and all crops were aided by manure.[10]

Agriculture was a major economic activity. But livelihoods at these sites were sustained through periodic earnings from the collection and sale of minor forest produce – a variety of gums as well as *timru* and *ashitra* leaves used in rolling *beedis*; and intermittent jobs with the contractors in forests. In addition, a variety of fruits like *Sitaphal* and *Rayan* and flowers were plentiful and good quality wood was easy to access. Not only ploughs and hoes, but also a variety of other implements that required wood were easy to make. Some money could also be made through stealthily selling pieces of good quality timber and from the cattle – especially goats which were cheap to maintain and could be sold frequently in order to meet any sudden need for money. Life was thus organised around multiple activities with a range of coping strategies available.

At the new site, nearly all PAFs had become part of Parvetha, albeit still continuing their social relations with households settled elsewhere from their earlier village sites as well as those who continued to live there in houses outside the submergence limits. Agriculture as an enterprise now was not only a core support sector

10. Though people had heard of chemical fertilisers, many had not seen such products even at that time. Some PAFs from Gujarat fondly remembered one Bapu Kadar who had brought some fertiliser for hybrid *jowar* that he was planning to experiment with during 1981 and a large number of farmers had gathered to see the product that he had brought from the primary agricultural co-operative society at Jharia.

for them but had also emerged as being qualitatively different at the new site. Patterns of cultivation practised earlier at the old sites had changed over the time, and so had the inputs, crops as well as patterns of investment in agriculture. The concept of intensive cultivation through technology adoption had been absorbed by nearly all and irrigation, hybrid seeds, fertilisers and insecticides were considered as major inputs for higher returns.

Irrespective of a clear spatial division between localities inhabited by the PAFs from Maharashtra and Gujarat, social interactions among them reveal a prevailing cordiality in relationships.[11] Also nearly all were found to share similar concerns and to adopt common strategies while dealing with cultivation and livestock rearing. What the majority lacked was access to irrigation. Less than 30 per cent of the PAFs could draw water from the nearby Narmada canal.[12] However, as

---

11. It should be noted here that in spite of no apparent conflict, there was a 'we' and 'they' feeling between the two groups reinforced mainly by the fact that of a majority among Gujarat oustees identifying themselves as *Bhagats* and abstaining from consumption of alcohol and meat as against a good number of oustees from Maharashtra eating meat and consuming liquor. Stereotyped as generally aggressive by their Gujarat counterparts, the PAFs from Maharashtra were often considered to be temperamental and less organised while investing in any productive enterprise. Significantly, the number of people from among the Maharashtrians working as peons at different locations such as Tilwadi, Bodeli, Baroda, Kevadia, Dhandhuka etc. were higher and a larger proportion among these oustees were more vocal and assertive. This may have been due to their having a higher exposure and acquired contacts outside of Parvetha.
12. Water was drawn through pumps hired from farmers in surrounding settlements like Vasna, Undiya, Kaledia and the host village of Parvetha. Generally a group of cultivators shared the costs of hiring such machines, which included Rs. 100 for its carting to and fro, Rs. 30 as rent per hour of its use and an additional Rs. 22 as the cost of diesel per litre that helped run a pump approximately for an hour. Once borrowed, the pump was used continuously in turns and the rent charged sometimes went up to Rs. 40 an hour especially during periods of high demand.

against tilling the land only with ploughs at the earlier villages, nearly all plots here were ploughed by tractors.[13]

After ploughing with tractors, bullocks were used for levelling the plots. Of the commercial crops, cotton was raised by many and its seeds were sown with DAP which needed to be diluted with water for best results. Hence, sowing was done mainly in June with the onset of monsoon.[14]

---

13. Rent for a tractor was fixed at Rs.180 per hour during which around an acre of land could be ploughed. The area ploughed, however, was less if the soil was too hard and full of gravel. Only one household in the village (Raghubhai's) owned a tractor and rented it out, though people also hired tractors from places such as Kothia, Vasna, Kosindra and Naswadi.
14. Different varieties of hybrid cotton were being cultivated. Till the year 2001, the farmers could procure hybrid cotton seeds from the Nigam, but after it staggered its distribution, people procured cotton seeds from Bodeli. The price of a kilogram of hybrid cotton seeds ranged from Rs. 750 for 'JK-1', Rs. 520 for cotton '888' to Rs. 400 for cotton '8' and Rs. 350 for cotton '10'. The higher the grade, the higher was the investment in pesticides and fertilisers. It is important to note here that the intensity of sowing cotton was also associated with the types of seed. For example, while a kilogram of cotton '888' was enough for an area of 1.75 *bighas*, a kilogram of 'JK-1' was sown over a plot measuring an acre. Cost of all inputs in an acre of 'JK-1' cotton seeds sown added up to around Rs. 3,500 (Rs. 450 for DAP; Rs. 750 for seeds; Rs. 700 for pesticides and some urea; Rs. 600 towards wages for the hired hands and Rs. 800 for irrigation). With these inputs one could get around 35 *maunds* of cotton which could be sold at Rs. 410 per *maund*. On an average, thus, one could net a profit of around Rs. 9,000 from an acre of hybrid cotton during a year, provided there was no crop failure.
    Beginning with its sowing along with DAP, the cotton had to pass through a strict regime of specific operations. Three weeks after its being sown, the field needed weeding every week for about three weeks. Subsequently, urea was applied and three weeks after that, pesticides sprayed intermittently. In between, urea was applied again and land softened by moving wooden *patas*. The first crop of cotton was harvested by the beginning of the tenth month from its sowing. Harvesting was done with intervals of 8-9 days between 'pluckings', which continued for 90 days. After the cotton was fully harvested, the land was left idle for a period of 15 days and thereafter wheat was sown, which responds better when sown immediately after cotton.

A variety of supports were available to PAFs for cultivating commercial crops such as cotton, food crops like maize and pulses like *tuver*, *arad* etc. Crop loans could be raised from Bank of Baroda branches at Kaledia and Naswadi and some hybrid seeds could be procured from the Nigam at subsidised rates. PAFs with access to irrigation facilities could raise as much as five times more than those without such facilities. Generally, the dry land farmers could raise a loan of up to Rs. 5,000 and the 'wet' land farmers up to Rs. 25,000 at an interest rate of 12 per cent p.a. Around 76 per cent of all plots were cultivated and worked on by family labour except for the initial ploughing by hired tractors. Respectively, around 10 and 16 per cent of PAFs from Maharashtra and Gujarat were hiring wage labour for jobs such as weeding, ploughing and harvesting, while the rest put in family labour in their own fields and also worked as wage labourers in others' fields. Some among the last of the groups mentioned also mortgaged parts of their plots when in dire need of money, and had to depend entirely on wage works in agriculture.

In spite of the intensity of cultivating cotton as well as increase in area under cotton at the new site, there had not been much change in the types of crops raised. The evident changes were (i) a large proportion – especially among the PAFs from Maharashtra – had now opted for cultivating paddy; (ii) a larger number across PAFs from both states now grew *bajri*; (iii) a handful had started experimenting with growing wheat at the new site, (not much had been grown at the earlier site); (iv) tobacco, that was grown at the submerging villages by some families was hardly grown; and (v) vegetables continued to be grown by the PAFs from Gujarat more than their Maharashtra counterparts.

With changing intensity of cultivation of some of the select food crops, pulses as well as crops like cotton, knowledge and use of hybrid seeds, pesticides and fertilisers had increased among the households. While this made agriculture an enterprise demanding higher investments, it also contributed towards a growth in risk factors in the event of a crop failure. Limited accessibility to other sources of income had led a section of households to adopt practice of raising select mixed crops. It may not be out of place here to point out that as compared to the submerging sites, the number of households borrowing money for different purposes including consumption was found to have increased at the new site. This was more pronounced among the PAFs from within Gujarat. And such loans were raised mainly from shopkeepers, petty traders, money lenders as well as from relatives and friends.

Having come at different points of time over two decades, the oustees from Maharashtra and Gujarat had resettled at Parvetha and gradually negotiated with the land and other supports received from the SSPA and the Nigam. While many of them could respond fast and engage themselves with agricultural practices prevalent in and around the new site, others took time to adopt cultivation as a core activity and agriculture as the prime enterprise. In addition, the plots received by a section had, according to them, not been up to the mark and with irrigation not having reached all households yet, the status and potentials of cultivation varied across groups. Manifestation of such variations appeared in some households who made efforts to earn from additional sources such as wage labour in agriculture and/ or petty sales or such other jobs in nearby establishments.[15] However, the availability of such opportunities and possibilities of absorption in them being few and far between, the primary dependence of all PAFs remained on cultivation. And hence, problems associated with

---

15. Summer being the lean season for cultivation, a section went to distant cities like Surat and Baroda looking for odd jobs, especially digging and headload carrying at construction sites. For wage labour in agriculture, not available in Parvetha, many went to work in the fields owned by *Patels* at Vasna, Kosindra etc. Tempo jeeps running on roads connecting these sites charged Rs. 2 per person for one way of around 5 km distance covered.

    Both men and women and in some cases children too went for such jobs in the fields. While during summer it was difficult to get these jobs, those supplementing their farm income through such works during the winter as well as rainy seasons were able to get around sixty days of work a year and were paid Rs. 25 per individual a day as wages that was equal for men as well as women. An individual could at best make an additional income of Rs. 1500 a year from such wage works. It should be noted that the subsistence economy with a near-uniform low yield, which was prevalent across fields at the submerging sites could not support long periods of wage work in agriculture.

this sector alone emerged as the major axis around which mediations were sought and coping strategies worked out, the failure of which could make life difficult for some households.

As at other sites where PAFs from different locations had been resettled, rearing livestock was an important activity at Parvetha too. As a supporting activity to agriculture, ownership and rearing of cattle had long been practised. However, there were a few significant differences in terms of cattle ownership among the PAFs across various groups, especially when compared to the submerging sites. While in overall terms there had been a reduction in cattle ownership per PAF, the proportion of households owning bullocks had increased – a trend witnessed by and large across many of the resettled sites and caused primarily by the Nigam subsidising their purchase to oustee households as part of its overall rehabilitation package. The proportion of households owning cows, especially among the Gujarat PAFs, was much higher, though the rate remained nearly uniform in case of ownership of bullocks. In fact, as against 53 per cent of PAFs from Maharashtra, as many as 83 per cent from within the Gujarat village owned cows. The proportion of households owning buffaloes however had fallen across both groups and in overall terms their number too was much lower in the village. Only 9 per cent and 11 per cent respectively of the Maharashtra and Gujarat PAFs owned buffaloes at the new site as against 32 per cent among both groups at the earlier sites. A substantial difference across both the oustee groups however appeared in case of ownership of goats and sheep. Practice of rearing goats had been a pronounced activity in the submerging region and this was clearly reflected from the data suggesting that more than 52 per cent of households from among both the submerging sites were rearing goats. The proportion of households with goats had fallen and only 6 per cent and 13 per cent respectively of PAFs from Maharashtra and Gujarat owned goats at the new site. It is worth noting that during the few preceding years there had been nearly no sale of any cattle, including goats by oustees settled in Parvetha.

Evidently the modalities of keeping cattle had changed at the new site. As opposed to its being grazed by the *govals* over large tracts of forest land with a good deal of fodder, most of them were now kept tied and stall fed.[16] Inaccessibility to *gauchar* and increased cost of maintaining livestock had led to a reduction in cattle owned per household. The fodder was purchased generally from the surrounding villages. In fact, a number of Patel households living in an around Naswadi town grew only select varieties of fodder to sell to those in need.

Having come from a somewhat similar socio-ecological background, the PAFs at Parvetha, whether from Gadher or Manibeli, seemed to have negotiated practices related to agriculture and livestock rearing in a more or less similar manner. While a tiny group had also entered some select non-farm lower level jobs, another group, much larger, supplemented its earnings from wage labour in agriculture and activities associated with construction at nearby and distant urban centres. Since irrigation had not reached a large group of PAFs, agriculture was yet to become sufficiently robust to yield a higher marketable surplus. And yet, this was the core activity around which lives of the PAFs remained linked and organised. Though not lacking in potential, animal husbandry seemed to have become less intense in the changed context. However, both the activities, being complementary to each other, were expected to change when irrigation came to the fields in a major way.

Site-3: *Suka*

Located at a distance of around 7 km from Kevadia colony, the resettlement site at Suka housed 178 households who came from Panchmuli – one of the five 'rock-fill dykes' villages within Gujarat. During the early 1940s, Panchmuli was a tiny settlement housing around 55 families in its Pitha, Nishal and Nichlu *falias*. Revenue from farmers here was raised by the *thakur* of Rampara who was responsible for its collection from five villages in the area – Gadkoi, Vadgam, Vanji, Garod and Panchmuli.

16. *Govals*, part of an important institution of maintaining livestock, whose livelihood was supported by the community of cattle owners, no more existed at the new site. Having received land as oustee families, many of them had settled at various new sites like the other PAFs. For example, the two *govals* grazing cattle for households in the Ter *faliu* of Gadher had received five acres of land each and resettled at the new sites of Sarsinda and Kachchota.

A large share of land on which the resettlement site at Suka stood was a vast expanse of around 750 acres of cropland owned by a *patidar* named Nandlal of Lachchras. The crops grown on large parts of this land included *bajri*, *jowar*, groundnut and cotton, and many labourers from nearby settlements like Kareli, Timbarva, Nagadhpur and Sungam were engaged as wage workers on this land. Parts of it were irrigated by water from the river and some hybrid cotton and groundnut too was grown. The entire land was purchased by the state government for resettling a section of the PAFs from Panchmuli.

The shifting process towards Suka coincided with the beginning of land acquisition at Panchmuli for the construction of the rock-fill dykes in 1980, and by the following year, 24 per cent of the oustee families had already reached Suka.[17] This first batch was led by Rama Lallu – a prominent farmer at Panchmuli then. For a year they continued to cultivate land at both sites, and by 1983 a little more than 87 per cent of all PAFs had reached Suka while the rest followed intermittently over the next few years. Barring a few families, all had come out of their temporary stay in tin-sheds within a period of three years.

Subsequent to their resettlement, it did not take long for the PAFs from Panchmuli to begin cultivation and augment their returns from agriculture. This was mainly because much of the land they were given was cropland earlier. Hence, the first batch had already started raising crops in 1982. Irrigation was available to 66 per cent of the households, a majority of whom had access to canal irrigation, followed by borewell. Maize, paddy and *jowar* were the major food crops raised. The frequency of growing wheat had declined, though pulses and oilseeds were now more intensely cropped than at the submerging site. Of commercial crops, cotton was grown by around 89 per cent of the PAFs and its intensity of cropping had increased at the new site, as reported by a large number of farmers.

Evidently, there was no significant variation in cropping pattern at the new site *vis-à-vis* the submerging site, though with the use of chemical fertilisers and increased access to irrigation there had been

17. Most of the PAFs from Panchmuli who headed for Suka were from its Nishal *faliu*. From the other four *falius*, different groups went to villages such as Akotadar and Karmaliyapura near Dabhoi and Lunadra and Raisingpura near Naswadi.

an overall change in the cropping intensity of some of the major crops. The yield of paddy was aided by the use of urea, which was also used for maize and cotton. A majority of the farmers had their plots ploughed by tractors hired from owners within and outside Suka. Himmat Tadvi, Jayanti Tadvi and Sanabhai Tadvi were three of the big farmers with tractors at the new site and as many as nine of them had dug borewells for irrigating their fields and for selling water to others.[18]

However, a section of the PAFs – especially among those without access to irrigation, found it difficult to organise cropping. Even partial crop failure due to lack of rains made them vulnerable. In order to cope with such situations at times they mortgaged parts of their land especially to those with higher risk-taking capacity by virtue of larger holdings and access to irrigation. In the few preceding years the incidence of mortgaging parts of their land by a number of farmers seemed higher. Consequently, many of them were cultivating the remaining land and simultaneously working as agricultural wage labourers in and around the village. This is substantiated by the fact that as many as 25 per cent of the PAFs and members in their families were also working as agricultural wage labourers elsewhere while working in their own fields.

Many respondents reported that maintaining cattle had become difficult especially due to lack of affordable fodder. They compared this with the plentiful fodder at the submerging site where, tended to by the *govals*, cattle grazed freely on forest land. All households were dependent on crop residue from their fields which had to be supplemented by purchase of fodder. On an average, they had to spend a minimum of around Rs. 5,000 per annum towards maintaining a bullock – an expenditure that went up further in times of shortage of crop residuals. Many were growing, on almost 15 per cent of their five-acres of land, specific kinds of grass for use as fodder. Other than the number of bullocks, there seemed to be relatively low average ownership of cows and buffaloes per household. There had been a

18. Though groundwater was available at a depth of 50 feet, it was found unfit for irrigating fields. Reasonably good quality water was available at depths ranging from 250 to 300 feet and required 71/2 HP motors to lift the water.

major decline in the number of goats owned.[19] However, what also seems to have happened was that some of the farm households were gradually adjusting with the changed situation and negotiating livestock rearing in a different way. This became evident from the fact that during the preceding few years, while only 4 per cent of the households sold some of their cattle, as many as 12 per cent bought additional cattle; and this in spite of adopting a changed feeding practice (stall feeding) for the cattle – most of which were kept in separate sheds.

What was striking about Suka was that although, over the time, there had not been a major change in the types of crops taken, the changeover to a relatively pronounced intensive cultivation from the earlier not-so-intensive mode had indeed been smoother and faster for a large proportion of households. This was due to much of the resettled site having been cropland earlier and to a larger proportion of households having access to water for irrigation. Though the modalities of working out a balance between the extent of livestock ownership and returns from the land were still underway, an obvious correspondence between land productivity and intensity of owning cattle had begun to emerge. While this should appear in a similar fashion at other sites too, in the case of sites such as Suka such arrangements were likely to get stabilised faster than others.

IV

Agriculture across the submerging villages, especially within Gujarat, made no substantial difference to the cropping pattern and investment in agriculture. Neither could much of the land be double cropped nor could much of 'extension technology' be absorbed, leading to a net result that primarily reflected a poor agricultural base. Limited yield and absence of surplus had led to a cyclical effect of the one reinforcing the other. A 'cumulative causation' of such a process kept

19. The milk produced was generally used for home consumption and for making *ghee* – the latter being sold to petty traders and shopkeepers. During the time of the fieldwork only six households were supplying milk at Bhavyara – a village with a milk collection centre for a dairy and located at a distance of 3½ km.

agriculture distanced from the input as well as output markets and it remained vulnerable in the absence of irrigation. In spite of such a fragile organisation of production in agriculture, life in these pockets was still sustained by cultivating extra land in the forests, collecting and selling minor forest produce, supplementing food in form of fish from the river and roots from the forest, rearing and selling more cattle, especially goats, and exchange of labour in the fields. All these together formed a combination of avenues that was negotiated as a coping strategy.

But there were sites, like some in MP, where agriculture had reasonable linkages to the market, and the cultivation pattern yielded some surplus through sale of specific crops. Some sites had the advantage of allowing crops in the flood plains of Narmada. Others had irrigation. Across many of them, farmers could raise more than one crop annually per plot. Across such sites coarse food crops were not grown as much as they were in the villages further up the slopes in what was mainly part of Maharashtra and MP. Yields here were higher, aided by fertilisers, pesticides and hybrid seeds, and production organisation somewhat different from villages which could sustain themselves because a number of avenues were available.

Social differentiation in terms of tribals and non-tribals too continued to shape varying sets of milieu. In case of the tribals, a process of seclusion kept many of them isolated compared to the non-tribals whose interactions with market and associated institutions were more pronounced. Lack of exposure and limited interactions with 'mainstream' institutions excluded many of the tribal villages from a robust integration with the rest.

Belonging to such a mosaic of varying socio-economic environments, the oustees were expected to respond to an intermediation aimed at their rehabilitation essentially through provision of land and allied supports. The very essence of such a programme seemed to aim at creating a dynamic land based resource by placing and simultaneously creating a viable agricultural system over a period of time. This however did not seem to happen fully, especially across select groups. Observations related to some such distortions can be listed as follows:

(a) Corresponding restitution of income gained from alternative employment avenues had failed to emerge at some places despite an overall increase in yield per acre. This was mainly due to dependence on a single sector, agriculture, the returns from which had not got stabilised owing to lack of irrigation as well as varying quality of land received by a group of oustee households.

(b) Yield fluctuations at sites and across groups without access to irrigation also led to migration of a section to sectors characterised by uneven wage structure and uncertainties of income. This also compelled some to seek wage labour in agriculture at sites where irrigation and other inputs resulted in higher yields and income. Such migrations were particularly pronounced during distress years that witnessed yield reduction and crop failure.

(c) Those who had received land that was not as productive, those with fragmented plots and those having some disputes over parcels of land that they believed to have entitlements to, were not able to adopt agriculture as the core activity. In addition, groups still cultivating land at the submerging as well as new sites did not seem to have linked up and appropriately integrated with the context of the new sites. Together, these factors had, at select sites, contributed to a growing inconsistency in production, investment and consumption.

(d) While around 20 per cent of the oustees responded to have received poor quality land, for a small group inability to cope with waterlogged plots continued to result in a poor yield. Exposure to risks associated with such poor returns led to enhancing the vulnerability of such groups specially in the absence or lack of employment in the non-farm sector.

(e) In addition to lack of access to irrigation, differential rate and intensity of exposure to new technology in agriculture as well as its availability and affordability had led to the formation of different sets of groups among the oustees at several sites. Some of the tribal groups especially took much longer to adapt to the modalities of a system that viewed agriculture as an enterprise rather than a sub-sector providing sustenance only at a nominal level. Some of them had not yet been able to come to terms with it, as constraints that only ensured lower returns from agriculture had often jeopardised such possibilities.

(f) The limited yet rising rate of leasing out of land, and continuing dependence on cultivation of land at the submerging villages among some PAFs, also appeared as an important phenomenon indicating that restoration of the social basis of practising agriculture as 'family-centred collective work' had not received enough attention while rehabilitating some such groups. Also, at certain sites, adult sons had received plots farther away, making it difficult to maintain immediate social transactions and exchange.

(g) Lack of grazing land and compulsions of buying fodder had reduced the average rate of ownership of cattle among a sizeable number of households across groups. This had affected the dry land farmers more. The gross cropped area being low, the crop residue produced could feed only a limited number of cattle. In addition, a notable reduction in the number of goats per PAF also made a section of the poorer PAFs more vulnerable, especially in times of crop failure or reduction in produce.

(h) With the receding role of animal husbandry especially among some of the tribal groups and also those who continued to depend on rains for irrigation, the role of livestock rearing in supplementing family income was reduced. Due to the prohibitive cost of stall feeding the cattle, returns from them appeared uneconomical to many.

In spite of some deviations and distortions as identified above, it should be noted that though located differently in terms of exposure to the market, a large majority of the oustees have done better at the new sites through agriculture. A sizeable section has responded positively to the available financial and allied institutions, market and technology. Indeed they were able to prioritize cropping strategy that had helped them to respond better to market fluctuations in input costs. At places farmers in the host settlements too were able to gain further knowledge and profited through selective adoption of newer experiments and innovations opted by the PAFs. After initial inhibitions and inertia, more and more of the oustees seemed to have responded better towards their changed situation. The most important point however was related to agriculture which even while had moved at the 'centre-stage' of their lives and livelihood systems, was still short of creating a smoothening environment of growth and better income. And, access to irrigation and extension services as factors aiding towards creation of a vibrant agricultural system was yet to reach all. Such essential inputs if injected systematically carry potentials of creating a momentum of countering investment constraints arising out of real as well as perceived risk factors. This becomes all the more important in situations where livelihood systems have to be shaped, sustained and enhanced through a sector as important as agriculture on newer sites of land given to the oustees. Once the inertia could be released with support of necessary inputs "especially, water, credit and output marketing on the one hand and institutions responsible for dissemination of production on the other" (Sah 1998: 80), it could be possible to support a large number of milch and draught animals. Combined together, agriculture and animal husbandry would then create a more effective and sustainable base towards forging stronger links with market and help further rise in income of the oustees.

(*This is an abridged version of the report on the Status of Agriculture and Animal Husbandry at the 're-settled' sites prepared for the SSPA. Field work for the study was undertaken during the second half of 2003.*

Part 5
Implementation

17

Resettlement and Rehabilitation in Sardar Sarovar Project: Policy Evolution and Implementation

Sudarshan Iyengar

Introduction

The Sardar Sarovar Project (SSP), a multi-purpose water project on the river Narmada in western India, has been undoubtedly one of the most controversial development projects in the world. Like many other development projects in India and in other developing countries, one of the weak planks of the project was the resettlement and rehabilitation (R&R) of involuntarily displaced population, a large section of which was tribal. It is by now widely accepted that development-induced displacement is disruptive. Although planners and administrators have been slow in learning this, social scientists have been well aware of how development projects can make certain sections of the population worse off. In fact, according to Cernea (1991), social scientists have been historically much better at recording development's tragedies than preventing them. An interesting account has been provided recently by Lobo and Kumar (2009). Covering the period between 1947 and 2004, they present a comprehensive account of land acquired for water resources, industries, mines, HRD, transportation/communication, and urban development projects and focus on the people displaced and affected by them. But during the times when SSP was being planned, the planners and administrators, even when aware of the tragedies it might cause, felt that the family getting displaced was making a sacrifice for the sake of the community (Varma, 1985). Nevertheless, sensitivity towards the displacement problem has increased all over the world and governments have been trying to internalise its social cost by making provisions for adequate compensation. Implementation of good Resettlement & Rehabilitation (R&R) policies continues to be a serious problem even to this day.

The case of SSP with respect to R&R had become more controversial because there was an opinion that the project was economically non-viable (Paranjpye, 1990). Similarly, a number of issues had been raised with regard to SSP's environmental impacts and sustainability (Alvares & Billorey, 1988; Kalpavriksh, 1988). Morse and Berger, 1992; Paranjpye, 1990). A strong international lobby had evolved that agreed and supported Alvares and Billorey's view that SSP was a planned environmental disaster. At another level there have been arguments about the relevance and correctness of imposing a non-tribal (mainstream society) concept of development on tribals who may have their own philosophy of development. According to this school of thought, mainstream society should not interfere with traditional tribal societies and their modes of production. Even the Morse and Berger Report endorsed what they call the tribals' cultural economy. This thesis has also been propounded by the Narmada Bachao Andolan (NBA), an organisation of activists and some tribals of the Narmada Valley. In fact NBA had raised, or rather challenged the conventional concept of development and said that it spelt destruction (Patkar, 1988). Since the SSP affected a large number of tribals, it was argued that tribals were being forced to accept a tragic future on account of a project that was inherently hazardous in terms of its social, economic and environmental impacts. The case for R&R in SSP had been thus extremely vulnerable. Probably it was this type lobbying at the national and international levels that prompted academics to believe, even in 1996, that the SSP amounted to planned destruction and that it had elements of a great and horrible act of deceit towards the tribals in the Valley (Krishna Kumar, 1996).

However, the entire controversy over the SSP in general and the R&R efforts in particular deserves a more close and objective assessment than has been attempted by most. Many important things have happened for the first time in the history of R&R in India and perhaps in developing countries. Compared to an uninspiring history of R&R in the country until recent times, significant policy reforms have been achieved in SSP. Similarly, implementation in the initial phase of the post policy reform period (1987-1993) was moving in right direction. It would be relevant and important to understand the processes and events that took place in R&R in SSP since 1980, when the implementation of the dam project began in earnest. It would also be useful to know what plagued the implementation subsequently. This paper is an attempt to trace the policy reforms in R&R in SSP, implementation of the new policy, status of R&R and the issues that are relevant for the future. The rest of the paper is divided into three sections. In Section I the evolution of the R&R policy in SSP is traced. In Section II, implementation of the new policy is reviewed for two time periods, 1988-1993 and 1993 onwards. In Section III, the problems and prospects in R&R in SSP are discussed.

I

While describing the process followed by the World Bank (WB) in arriving at the normative content and the operational guidelines on R&R, Cernea (1991) has observed that the first step in any development project was sociologically informed formulation of an R&R policy. The process of arriving at the guidelines for R&R in SSP is reviewed in the above context.

The Basic Provisions for R&R

The Narmada is an inter-state river which originates in Madhya Pradesh (MP) and traverses a total distance of 1312 km before merging into the sea in the Gulf of Khambhat on the west coast. Through most of its course, it flows between the mountain ranges of the Vindhyachal in the north and the Satpura in the south. It is one of the oldest and most venerated rivers (Paranjpye, 1991; Varghese, 1994; Varma, 1985). Because of its peculiarly narrow valley occupying or fringing on the territories of Gujarat, MP and Maharashtra, the 214 km long SSP reservoir submerges lands in all those three states.

The two major riparian states, Gujarat and MP, disputed their respective shares of water. A number of meetings and committees could not resolve the issue. Hence, as per the provision in Article 356 of the Constitution of India, a Tribunal headed by a Supreme Court Judge was set up in 1969 (for a brief and interesting historical account see Varghese, 1994). Obviously, policy decisions regarding rehabilitation of the project affected population also rested with the Tribunal. The Narmada Water Dispute Tribunal (NWDT), as it came to be known, gave its final report in 1978 (the major decisions and salient features of SSP are given elsewhere in the present book).

The NWDT made a basic provision for the allotment of a minimum of 2 ha of irrigable agricultural land to each land holding family losing 25 per cent or more of agricultural land. The term "family" included husband, wife, minor children and others dependent on the head of the family. Further, every major son of 18 years and above was treated as a separate family (Report of NWDT Vol. II). This basic provision itself was path-breaking in the history of R&R in India. In most projects, the persons losing land were given cash compensation which hardly enabled them to buy alternative land or any other equally productive asset. A large number of studies have documented this and the consequent impoverishment of the displaced population (Alexander *et al.* 1991; Dalua, 1993; Fernandes and Ganguly Thukral, 1989; Goldsmith and Hildyard, 1984; Jayaratne, 1990; Mahapatra, 1990; Muthayya and Mathur, 1974; Reddy, 1992; Sudarsen and Kalam, 1990; and Ganguly Thukral, 1992).

Although the provision of irrigable agricultural land was made to compensate for the loss of land, the NWDT provisions had serious lacunae. Firstly, it failed to take into account the characteristics of the tribal settlements and the economy where a large number of families did not own but only operated land for production. The land belonged to the forest department or it was government waste land. Describing this feature in the SSP submergence area, Patel notes, "the fact of future displacement of these people by development projects had nearly brought to the surface a serious and an embarrassing problem of unacknowledged tribal right to the resources of livelihood in their ancestral land" (Patel, 1994: 1 and ARCH-Vahini, 1988a: 3). Secondly, the NWDT provisions also ignored female headed households, widows, single woman families and the landless.

In this context Ganguly Thukral notes, "The Award (NWDT) not only ignores women as individuals and heads of families, but is also unjust to the landless man and his family. Even the major sons of the landowning peasant are recognised [...] as being entitled to compensation, preferably in land. The landless head of a family is, however, not recognised, let alone his son". (Ganguly Thukral, 1989: 59).

Obviously the NWDT Award's major provision was sociologically ill informed. However, this was not all. The three state governments were not in a mood even to abide by the NWDT provisions. These were signs of retrogression.

Activist Group Enters the Fray

With the declaration of the NWDT Award in 1978, GoG began the project work as well as the process of rehabilitation. It issued a Government Resolution (GR) in June 1979, making in it a crucial change which almost completely killed the spirit of the NWDT provision. The entitlement of a minimum of 2 ha of irrigated land to 'each land owning family' was changed to 'each land holding'. In tribal areas, both for cultural and historical reasons and laxity in revenue administration, most land is jointly held. A joint holding may, therefore, be operated by two, three or more independent families. Implementing the GR, the government offered government waste land, which was more than 100 km from the native villages to every holding (and not family). The tribals living in first five affected villages refused the government offer. The government responded by offering cash compensation and leaving the land purchase to affected families (ARCH-Vahini 1988a).

Armed with its 1979 GR, the government was ready to ride roughshod over the poor, helpless and unorganised tribals when, in 1980, members of ARCH-Vahini started frequenting the submergence villages.[1] The panic-stricken and desperate tribals were reluctant to

1. A few young doctors and individuals were trying to set up a primary health care system that could be run mainly by the community health workers about 25 km downstream of the dam site. This group had come in contact with some families living in submergence villages. The group, known as Action Research in Community Health and Development (ARCH), thus first came to know of the plight of the tribal families living in the immediate vicinity of the SSP dam site and facing imminent displacement. In the meanwhile, another set of

hear and do what the fresh-faced students advised them. Even the ARCH-Vahini activists were new to this game. They knew nothing or very little about land laws, the NWDT Award, World Bank guidelines of 1980 etc. Had they known the complications they would have probably given up (Patel, 1994). The tribals would have continued to tolerate them as sympathetic observers but for a case in which the government retracted on a written promise it had made in implementing the 1979 GR. This was towards the end of 1983. Until then ARCH-Vahini was content with monitoring the implementation of a totally watered-down policy of GoG. A silly slip by the government led to the filing in court of a case of 'promissory estoppel'. The tribals saw a ray of hope. They quickly organised and the government stood exposed.

The test case in high court made the tribals realise that the government was cheating and that it was vulnerable. On March 8, 1984 a huge rally marched to the Kevadia colony – the project headquarters at the dam site village Vadgam. The major demand put forth was the implementation of the NWDT provision not only for the land owning families and co-parceners but also for those who cultivated forest and government waste lands (ARCH-Vahini, 1988a and Patel, 1994). The might of the affected tribals increased and, perhaps for the first time in the history of execution of a development project involving displacement, the government was under pressure from below. This was followed by stoppage of project work on a dyke appurtenant to the main dam. Public interest litigation was filed in the High Court and subsequently in the Supreme Court. The Supreme Court granted interim injunction. The Government violated it. The Court appointed an independent Commission which indicted the government of Gujarat.

---

students about to finish college education, who were influenced by the late Jayaprakash Narayan's (JP) nationwide movement for 'Total Revolution' in the mid-1970s, also got interested in getting involved in the rehabilitation issue of the SSP. This group belonged to the 'Chhatra-Yuva Sangharsh Vahini', the student youth struggle force that was organised by the JP in 1975. Members of ARCH and Vahini came together in 1980 to form an alliance, ARCH-Vahini which for the next seven years was to lead a fierce struggle for reforms in the R&R policy of the SSP.

The World Bank and International Non-Government Organisations

By 1983 ARCH-Vahini had learnt about the World Bank's interest in the project as well as its rehabilitation programme. In August 1983, ARCH-Vahini wrote a detailed letter describing and quantifying the dependence of tribals on forest land for agricultural production and for their survival. The objective was to lobby the World Bank in favour of tribals. The Bank, till then, knew nothing about the problems because the government did not present the ground reality to the Bank team. Simultaneously, ARCH-Vahini also kept informed the development policy advisor of Oxfam, a British NGO which supported ARCH's health programme. The Oxfam adviser was quick to grasp the problem and launched an international campaign for the rights of tribals in SSP. The international NGOs, upon learning of the bank's involvement in the project, lobbied and created pressure on it through their country Executive Directors.

The World Bank too was learning quickly about the lapses it had made in the past on the R&R front. It evolved comprehensive guidelines for R&R in the projects it financed in 1980. Further, in 1982, another operations manual was prepared for the tribal people. It was in this backdrop that an international expert on R&R, Thayer Scudder from the California Institute of Technology, was sent to India on a Special Mission in November 1983 to assess the R&R situation in the SSP.[2] Scudder's visit was instrumental in bringing to the Bank an understanding that the NWDT provision for R&R would fail to cover a large number of tribal families (treated as 'illegal encroachers') who survived on government and forest land. The Bank also realised that with the implementation of the then existing policy, the overarching principle of the Bank regarding R&R, which was to regain the standard of living after resettlement, might not be in place.

Scudder came again in September 1984, when he came to know

2. Although, the Independent Review Mission, appointed by the then President of the World Bank B. Conable and headed by Bradford Morse, had a view that sending Scudder was an afterthought and the Bank had already completed the appraisal of the project and had totally missed out on the R&R.

the field reality. He met the members of ARCH-Vahini who apprised him of the ground reality and provided him all the facts and details he needed (Patel, 1984: 6). In his November 1983 report, Scudder had made two important recommendations. One was the entitlement of the encroachers and landless families, and the other was the release of denuded forest land for cultivation if the tribals so desired. In the September 1984 meeting, he sought and got details on both these aspects. Armed with these data on ground reality, the WB team started negotiating the Loan Agreement (LA) with GoI.

The May 1985 Loan Agreement

In May 1985, WB entered a Loan Agreement with GoI and the three states of Gujarat, Maharashtra and MP. Besides techno-economic aspects of the project, R&R conditions were negotiated. Scudder was the chief negotiator. He built upon the foundation of the NWDT Award and secured land entitlement for the encroachers and the landless. GoI also agreed to release the forest land if demanded by the displaced tribal population. The concession was a significant achievement because only in 1980 GoI had passed the Forest Conservation Act which forbade any other use of forest land without prior permission of the Ministry of Environment and Forests (MoEF). Hence, WB's achievement in this respect in the Loan Agreement was significant. The third most important provision in the Agreement was that the displaced families were to be given the right to choose the land they approved.

Evolution of the R&R Policy in Gujarat through GRs

With the signing of the Loan Agreement with the Bank, GoI and the three state governments were bound by certain commitments to the tribals in the Valley. GoG responded by issuing a new GR on May 30, 1985.[3] It was compelled to recognise the right of the encroachers.

3. It should be mentioned here that when the government of Gujarat was negotiating the Loan Agreement with the bank in 1985, it was also arguing a case in the Supreme Court and its affidavits it continued to name the tribals dependent on forest and government land as thieves, low-breakers and encroachers (ARCH, 1988a: 6).

The GR, however, did not specify which land would be allotted to the encroachers. The government wasteland offered was far from the submergence villages and it was not acceptable to the affected tribal families. The conflict continued.

A further concession was made for the land owning families. In its November 1, 1988 GR, the government made provision for a minimum of 2 ha of land to each land owning family and not to each holding. The government also offered choice. The affected tribal families could choose from government land from three locations. However, if they failed to select any they would have to buy land on their own. The family could buy the equivalent of its original land holding. A concession was granted here. The difference between the market price of the land purchased and the cash compensation for the land was to be given as an *ex-gratia* payment. This was a qualitative improvement over the NWDT Award, which stipulated that the excess cost of land had to be borne by the displaced family. However, the GR still violated the spirit of the LA, which gave an entitlement of a minimum of 2 ha per family of its own choice. In fact, by introducing the condition of selecting between three options, the government sought to bring back the old 1979 policy. People refused to accept this new GR and the struggle continued. GoG continued on the one hand to express its commitment to the LA and on the other hand offered something less than the provisions in the LA to the displaced families.

In April 1987 the visit of the World Bank Mission exposed GoG. The Mission members visited villages in the deepest interior, and in the presence of government officials declared that each encroacher and/or landless family was entitled to a minimum of 2 ha of land of their choice. To this GoG responded saying that it did not care for the WB loan and that it was not possible to give land entitlement to the encroachers and their major sons. People were thus stuck in suspense. The Bank continued its tough stance. GoG finally yielded but was not before trying yet another angle.

During the 1987 WB Mission's visit, the government maintained that there was not enough private land that could be offered to the displaced families. However, in the wrap-up meeting, one of the Bank Mission members produced a long list of private owners who had shown willingness to sell land. This list was supplied to the Bank members by ARCH-Vahini. The government had no answer (Patel *et al.*, 1991).

The suspense ended in late December 1987, when GoG issued three GRs dated December 4, 14 and 17. The first granted to each eligible land owning family land of its choice. It also granted the difference between the cost of land purchased and the compensation as *ex-gratia*. This was an improvement over the NWDT provision, in which the difference in cost was to be given as interest-free loan to be repaid in 20 yearly instalments.

The second GR replaced the November 30, 1995 GR on encroachers. It clearly recognised encroachers and their major sons and provided 2 ha of land of choice to each family and some *ex-gratia payment*. The Third GR extended the same provision to landless families. Thus, by December 1987 the Gujarat government had announced what has now come to be known as the most liberal R&R policy package in India and perhaps in Asia. The provisions certainly conformed to the overarching principles of the World Bank's 1980 policy guidelines for projects involving involuntary displacement. By this provision of a minimum of 2 ha of irrigable land of choice, it became feasible to ensure for a tribal family at least the restoration of the standard of living that obtained prior to its involuntary displacement. The Bank has called this process an incremental strategy for policy reforms in R&R in the SSP.

Between the Loan Agreement in 1985 and the final policy declaration in December 1987, the tribal population facing involuntary displacement in the villages of Maharashtra and MP was also organised. A protest movement surfaced there too. Earlier, in March 1984, when ARCH-Vahini had organised a protest rally, tribal families from submergence villages in Maharashtra had joined. However, their total involvement in the protest movement happened only after 1985 (Varghese, 1984 and Parasuraman, 1993). This movement decided to oppose the project itself in August 1988. Therefore, its direct contribution to the policy reforms in Maharashtra and MP was not much. The movement could not succeed even in ensuring provisions similar to those of Gujarat. A comparative profile of the R&R provisions in NWDT Award and their modifications by the three states is given in the Appendix to Chapter 11, this Voulme.

Critique of World Bank's Incremental Strategy

A major and scathing criticism of the World Bank's incremental strategy for policy reforms in R&R of the SSP was advanced by the Independent Review Mission (IRM) in 1992. Their major argument was that the Bank had failed to implement its own policy which it had evolved in 1980 for projects involving involuntary displacement and in 1982 for projects involving involuntary displacement of tribals. Making a serious charge, the Report says, "...absence of human and environmental assessment *ab initio* creates the impression that the demands of engineering carry more weight in the Bank than the needs of people to be affected...." (Morse and Berger, 1992: 355).

Similarly commenting on the incremental strategy the Report says, "the record of resettlement and rehabilitation in India, which has been unsatisfactory in virtually every project with a large resettlement component, should reasonably have prompted the Bank to adopt a less flexible standard for resettlement and rehabilitation of the project affected people. In this context, the Bank's incremental strategy to obtain compliance, made explicit in 1989 greatly undermines prospects for achieving successful resettlement and rehabilitation" (Morse and Berger, 1992: 351).

According to the Bank's Operational Manual of February 1980 for R&R[4], the Bank has to necessarily carry out an appraisal of the R&R component in every project it finances. The basic requirement for this appraisal is the data and resettlement plan which is to be provided by the borrowing country. In the case of SSP, there was neither data on total number of people affected nor a blueprint that explained the resettlement plan. Since the basic data was absent the Bank couldn't have carried out any appraisal and it did not. If it did not carry out an appraisal, it had failed in implementing its own policy. The omission of the process was more serious because the project affected population was tribal.

Yet another charge that IRM made was that the Bank failed to have any consultation with the people who were actually affected. Referring to 1982 Operational Manual on tribal people, which says

4. The policy was first issued as an internal Operational Manual Statement (OMS 2.33) to staff in February 1980.

that the design of an appropriate tribal component depended upon detailed and contemporary knowledge of the peoples to be affected, the IRM held that it was essential for the project appraising authority to discuss it in detail with those to be impacted. The Report further says, "In 1985, the basis for designing, implementing and assessing the overarching principle of at least retaining oustee's standard of living simply was not in place. The numbers of people to be affected were not known; the range of possible impacts had never been detailed, the effects of canal had been ignored the social organisation of the peoples to be displaced was little understood in Gujarat, and not at all elsewhere. There was virtually no sociology with which to formulate a policy that might achieve the Bank's long term objectives. Nor had there been any consultation with those at risk in virtue of which "adequate participation" could be said to be taking place. There were no benchmark data with which to assess success or failure. The resettlement and rehabilitation component of the Sardar Sarovar Projects had never adequately been appraised" (Morse-Berger, 1992: 50).

It may be true that the World Bank did not appraise the R&R component of the SSP when it carried out the techno-economic appraisals of the project before 1983, it appointed Professor Thayer Scudder as a consultant and he visited India in November 1983 accompanied by three Bank staff members (a lawyer, an engineer and an agricultural economist). The Bank has considered this visit as an R&R appraisal visit in the 1985 Staff Appraisal Report (Patel, 1993). Scudder also visited India again in September 1984 when he met the ARCH-Vahini team and discussed the field situation. This was over and above the written material and data supplied to him on the field reality and situation assessment. Prior to meeting Scudder, ARCH-Vahini had written a long letter detailing the field situation and furnishing quantitative data as evidence to the World Bank in August 1983. Since then, all review missions from the Bank made it a point to meet and discuss with ARCH-Vahini and the tribal leaders in the submergence villages of Gujarat, the field reality, know the people's mind and the government's response. The IRM was privy to all this detailed information and hence its contention that the R&R part of the project was never appraised and the affected people were never consulted was not true.

The IRM was also selective in quoting from Scudder's report of November 1983 (Patel 1993: 4). It mentioned only the lack of adequate information and data, inadequate institutional framework, the shortcomings of the NWDT Award and state governments' lack of commitment to honour the Award. It missed out totally on the information and data which Scudder collected, the analysis he did and the recommendations he made. Similarly, his visit during September 1984, when he met with ARCH-Vahini, enabled him to generate a lot of hard data which he used in his recommendations. As stated earlier, he was also a key person in negotiating the R&R component in the Loan Agreement.

In the context of the policy reforms, it would have been more pertinent if the IRM had examined – which it could have very easily – whether Scudder had studied the R&R problems in SSP during his November 1983 visit and whether he studied them within the framework of the Bank's Policy of February 1980. In the absence of any uniform policy across the states, it was obvious that the Bank would start with the NWDT Award. Scudder did appraise and realise that the NWDT Award was inadequate. Based on his discussions with ARCH-Vahini and the tribals, and the data he gathered from ARCH-Vahini and the government, he recommended crucial changes. The state governments were forced to comply. Describing the networking that happened between the grassroots organisations, international NGOs and WB to push and pressurise the state governments into accepting the changes, Anil Patel says, "…in retrospect at least, Thayer Scudder's November Report is a major breakthrough …it forced the state governments to recognise the rights of the encroachers to entitlement … A meaningful and informal partnership amongst Thayer Scudder, Oxfam, Survival International, ARCH-Vahini and some key individuals in the Bank including Tim Lancaster (Executive Director, UK), Michael Cernea, etc. came very soon into existence…" (Patel, 1995: 4).

This network not only ensured that the state governments did not mislead the Bank missions but also became instrumental in formulating a policy that was acceptable to the tribals and went beyond the parameters of the February 1980 Guidelines. According to Patel, "Thayer Scudder, as a key negotiator of the R&R part of the Loan Agreement (LA) of May 1985, built upon the solid foundation of the NWDT Award, a radically new R&R policy. Most importantly, the land entitlement of all the tribals (including the encroachers, who constituted majority), were effectively secured" (Patel, 1995: 4).

It appears that in its enthusiasm to fail WB in its test, IRM overlooked many important achievements of the local-international NGOs network and the Bank. These achievements, which pressurised GoI and the state governments to comply in increments, did indeed promise to fulfil the overarching principle with respect to Standard of Living in WB's February 1980 Guidelines. Further, by inserting a clause on the land of choice and the commitment of GoI to get denuded forest land released if the tribals demanded it, the LA secured a safe and worthy policy for the tribals in the Valley. Thus, the reformed R&R policy package of SSP also conformed to the Bank's 1982 Operational Manual on Tribals[5].

Activists and scholars, both in the country and outside, have been rather impatient in examining the process and sifting evidence. Parasuraman's article (1993) is a case in point. He starts by saying that the period of the early 1980s was full of confusion. Both government and people were confused. I think a better assessment would have helped inform him that in the early 1980s government was out to short-change tribals as described earlier, and people were simply stunned because of the shock and lack of proper and adequate information. There was no question of confusion.

Parasuraman also seems to have misunderstood the sequence of events when he writes, "it was only after the completion of negotiations with the GOI for a loan, that the World Bank realised that R&R aspects had not been appraised at all. Thus, Dr. Scudder, an expert on resettlement issues was deputed by the Bank to analyze the situation ..." (Parasuraman, 1993: 9). The Bank negotiated for the Loan Agreement in late 1984 and it was finally signed in May 1985, whereas Scudder's first visit was in November 1983 and it was followed by a second visit in September 1984. Very often, such misinterpretation of sequences of process and events by both activists and academics has led to faulty assessment and judgment of the ground reality. This in turn led to an unwarranted criticism of the project.

5. In 1982, the World Bank became the first multilateral financial institution to issue a policy on tribal people [Operational Manual Statement (OMS) 2.34]. It was titled Tribal People in Bank-financed Projects.

The IRM had harped only on the non-availability of data without which no plan regarding R&R could be formulated for SSP. I think the IRM overemphasised this aspect to prove that WB failed to do its homework. In the process, it overlooked what was achieved. The state governments which were unwilling to implement even the NWDT Award and MoEF-GoI, which had taken a rigid stand over the issue of the release of denuded forest land for resettlement of tribals, were persuaded into going beyond the NWDT Award. This by itself was no mean achievement. It should be remembered that WB could do it partly because of the tremendous pressure that was built up by the local-international NGO network. Secondly, the critics who thought that the policy in all the three states should have been uniform and the same as Gujarat's 1987 policy, and that the World Bank failed in pushing these reforms through, overlooked the fact that the Bank was dealing with the states of a Federal Union. Both Maharashtra and MP had their own compulsions in not going as far as Gujarat. This aspect has been aptly described by Varghese. The caution exercised by Maharashtra and MP could, therefore, be well understood. And given the rights of states in a federal polity "there are limits beyond which they cannot be pushed. The World Bank tacitly recognizes this. The independent Review did not and thought nothing of derogating from federal principles though both its members, Bradford Morse, an American, and Thomas Berger, a Canadian, come from federal countries. This is not to say that federal principles can ride roughshod over social justice. The NWDT, a Tribunal headed by a Supreme Court Judge, set out certain norms which are being observed" (Varghese, 1994: 138).

And we know that all the three states have gone beyond the NWDT Award. The IRM's charge that the World Bank had not received compliance to its incremental strategy of policy reforms is thus untenable. Unfortunately, the IRM did not stop at that. It recommended the World Bank to 'step back', which caused ultimately more harm than good as we shall see in the discussion on implementation.

Policy Reforms in Maharashtra and Madhya Pradesh

R&R policy reforms in the case of Maharashtra and MP need to be viewed in somewhat different perspective. First of all, both the states had their state level policies for R&R. Secondly, both were yet to harness all their major surface water resources. This is particularly true of MP. Both the governments had argued that the Gujarat could be very liberal in its R&R package because SSP is the last mega-project and hence the government could afford to offer such package. Thirdly, and more importantly, the activist-led people's initiative for improving the R&R package, which began in late 1985, very soon abandoned the cause of reforming R&R policy and implementation and directed all efforts to oppose the SSP itself.

The tribals in both the states had almost same set of problems as their counterparts in Gujarat. The "encroacher farmers" did not have land titles despite their occupation of those lands for long years. The application of the NWDT Award was relevant only for a limited number of families having land titles. Landless families also had no claims of any kind in the Award except getting a house plot at the resettlement site. Thus, the need for reforms and improvements over the NWDT Award was equally acute in Maharashtra and MP.

Between 1985 and 1988 the Maharashtra and MP organisations had similar charters of demands as that of the organisations in Gujarat. In the case of Maharashtra, the tribals also had a clear choice about the resettlement area. They wanted to settle in a degraded forest area that was relatively close to their native lands and, more importantly, had similar setting and environment. The Narmada Dharangrast Samiti (NDS), Dhule, the organisation of tribals in Maharashtra, put up a charter of 38 demands, one of which was the allotment of degraded forest land for resettlement. Even in MP, the tribals were looking for a good alternative agricultural land and had expressed a clear preference for an improved policy of R&R (Patel, 1994 and 1995).

It may be worth recalling at this point that by the time tribals were mobilised and organised in Maharashtra and MP in late 1985, WB was already in the total picture. ARCH-Vahini had already detailed the ground situation to the Bank in August 1983, and international NGO pressure had begun. Scudder had visited twice and had made important provisions in the Loan Agreement that was signed in May 1985. The organisation in Maharashtra in fact raised the issue of the provision made in the Loan Agreement when MoEF-GoI declined to release the forest land demanded by the Maharashtra tribals. Bruce Rich of the Environmental Defence Fund (EDF) of the US termed MoEF's denial as a breach of agreement (Loan Agreement) at a US Congressional hearing on the Narmada Projects sometime in 1988 (ARCH-Vahini, 1991; and Varghese, 1994). Thus, between 1985 and 1988, a case for an urgent need to reform the R&R policy in the SSP was built up at grassroots, local, national and international levels. Apparently, the organisations and the tribals in none of the three states indicated any signs to oppose the project itself.

Interestingly, Parasuraman (1993) has built up a case which reads as though the decision to oppose the dam was the result of the frustrations experienced by the tribals and activities in the valley during 1985-88. He refers to this period as 'Period of Tension' between the people and the government. He has discussed the involvement of activists and their efforts to organise tribals in Maharashtra. These activists, according to him, pressurised the state government to provide and strengthen the basic facilities and social services like health and education. They also mobilised the youth among tribals to raise demands for better R&R. He then says that with activists on their side the tribals submitted specific demands to the government. These demands included right to information on all aspects of dam, implementation of NWDT Awards in all the three states, compensation for encroachers, a comprehensive blueprint of R&R plan, right to exercise the choice to resettle in Gujarat or their own state, whichever they preferred, and extension of R&R package to people affected by colony, canal, sanctuary and compensatory afforestation programme. Nowhere does he mention anything about the intention to oppose the project. However, he then says that between1985-88 the Maharashtra and MP governments did not make any changes in the policy. There were some changes in the policy in Gujarat. Availability of land became uncertain and evidence from resettled villages pointed to increasing misery and prolonged deprivation. Thus people decided to oppose the dam.

Admittedly neither MP nor Maharashtra made any changes in their policies. They were willing to go by the NWDT Award. MP was slow because even the first submergence was five years away in 1985. The Maharashtra government had requested for the release of forest land in Somaval – a village in Taloda taluka in Dhule district – the land which the tribals had been demanding. MoEF did not give permission to release the forest land. It was about this forest land that Bruce Rich of EDF had made representation in the US congressional hearing. But it was wrong of Parasuraman to view the policy changes in Gujarat as trivial. It has been shown how vital and how improved the changes in the R&R policy of Gujarat were. The real reason for trivialising the improved R&R policy in Gujarat could be different. The activists in the Valley and the intellectuals in urban areas perceived the improved policy document of Gujarat as merely a piece of paper (Patel, 1993). They were certain that the government did not intend to, nor was it capable of, implementing the improved policy even in Gujarat. These activists apparently did not consult the tribals about the improved policy of Gujarat and declared that R&R in SSP was impossible.

The issue of non-availability of land was hastily judged in all the three states. In fact, by 1987, some affected tribal families had already identified more than 4000 ha of private lands that were available for sale (ARCH-Vahini, 1988b). The assessment of conditions in resettlement sites was also not taken up with full comprehension of the situation. The evidence about life and living in the resettled villages did not generate confidence because the community was in transition. Secondly, the families lived and tried to manage production at the both, the submergence village and the relocation site. This went on for years. When they finally settled at the relocation sites, their conditions improved. Surprisingly, both studies – one showing increasing misery and the other showing improvement in the same site – were conducted by Tata Institute of Social Sciences (TISS) with which Parasuraman was closely associated. Yet he commented later that life in relocation sites did not inspire any confidence among the affected families.

The real reason for opposition to SSP by Maharashtra and MP organisations lay elsewhere. Two important sets of reasons were advanced. One, it was *a priori* decided that rehabilitation was impossible, and two, that the project was economically not viable and environmentally hazardous. Opposition to SSP became more important for the activists and the intellectuals in the country. The vital breakthrough that was achieved by the activists – the international NGOs and World Bank combine – was relegated to background

(ARCH-Vahini, 1988b & 1991; Patel, 1994 & 1995; and, Varghese, 1993).

Despite the stand taken by the organisations in Maharashtra and MP, the government did introduce reforms in their respective policies, mainly due to WB pressure. As earlier, the final R&R policy packages in the three states differed. The IRM had been critical about non-uniform provisions. However, the subsequent reforms introduced by the Maharashtra and MP governments contain provisions for encroachers and landless families. Thus, the R&R policies in all the three states have to help regain the standard of living in relocation sites. Pushing for reforms by people's organisations and civil society groups could have led to further consolidation of the policies in Maharashtra and MP and the implementation could have been far better. Interestingly, a majority of the affected tribal population was also in favour of having a good land based R&R package instead of opposing the SSP with or without a good R&R policy.

II

The Independent Review Mission submitted its Report to the World Bank in August 1992 and recommended in unequivocal terms that it steps back and conducts a comprehensive review before renewing assistance to the project. This was a watershed in the implementation of the R&R policy in all the three states. Further, the set of issues which were raised before the IRM and after its Report also differ. Hence it may be worth reviewing the implementation of R&R between 1987-93 and 1993 onwards.

ARCH-Vahini in Gujarat expected that the activists in the Valley would take the fight to Maharashtra and MP for reforms in R&R policy on the lines of Gujarat (ARCH-Vahini, 1988a). The activists, however, did a *volte-face* and abandoned the cause altogether. The combined forces of activists and organisations came to be known as the Narmada Bachao Andolan (NBA). It was led by Medha Patkar. NBA opposed the SSP and was critical of the R&R policy and its implementation. This put ARCH-Vahini in a dilemma (ARCH-Vahini, 1988b; Patel, 1994). The tribals in Gujarat villages expected continued support from the activists in getting the reformed policy implemented. They did not trust government officials in this matter. ARCH-Vahini decided to collaborate with the government to implement the reformed policy. Another voluntary organisation, Anand Niketan, led by the late Harivallabh Parikh, who was known for holding Lok Adalats (people's courts), also joined. Most NGOs in Gujarat expressed satisfaction over the new policy and supported it.

Building a Database

We have seen earlier that IRM had raised the issue of lack of data and information while criticising the World Bank for not conducting the R&R appraisal. What it expected of the Bank was that the latter should have prepared an R&R plan on the basis of the detailed data supplied by the project authority of the borrowing country. According to the IRM, this was a principle which the Bank itself had established out of its learning from lending for development projects involving involuntary displacement.

The World Bank could not have extracted the requisite data from the states. Until the reforms in the policy were announced, the governments were thinking of either giving cash compensations to the land owners or allot some government waste land to them on holding basis. This type of conventional resettlement practice would have required data on land holding families. This information was available from the revenue records. The other type of information that the government could have supplied was on houses and house owners since acquisition notices were to be issued. The state governments could not have supplied any other data. It is a different story that the governments did not have an update even on these two aspects. Also, unless the normative criterion of the policy was clearly spelt out, i.e. unless the policy was clear about who among the affected families were entitled for land as compensation, the exact number of families entitled for agricultural lands could not have been known. Thus, the IRM expected data which could not have been generated before the policy decision was made.

At the time of implementation, therefore, the first task to be performed was to identify the persons who had land entitlement. Obviously this could not have been accomplished in one meeting or a sitting. It was realised that the identification of eligible persons was going to be a long drawn out process until persons in all categories such as land owners, co-sharers, encroachers, landless labourers, socially isolated and marooned and the major sons falling in all the above categories were identified. At this point an institutional innovation was introduced. A committee was set up which came to be known as the List Committee. It comprised members from government departments, representatives from two voluntary organisations, ARCH-Vahini and Anand Niketan, and the Sarpanch (head) of the village for which the lists were reviewed. Members of the Legislative Assembly (MLAs) and Taluka and District Panchayat members were also invitees to the List Committee. It met as and when required.

The main function of the Committee was to verify the claims of persons to be identified as a project affected persons (PAPs) and ascertain their category. Unfortunately, the government did not collect any baseline information about the families living in the submergence villages prior to issuing notifications. This made the verification work more difficult. However, the Centre for Social Studies (CSS), Surat, had in 1981, conducted a census of all the 19 villages in Gujarat facing submergence. These household level data served as the benchmark. In addition, the voters' list, birth registrations, school leaving certificates which carry date of birth and ration cards etc. were also checked wherever available to examine the claims. The government of Gujarat's policy had made a provision to give land to major sons (Males above 18 years of age). A need arose for fixing a cut-off date to decide whether a person was major or not, i.e. whether he was 18 or above. This date was set at 1-9-1987.[6] It became necessary to examine more than one evidence to establish the domicile as well as age of the person before confirming his/her eligibility.

Documentary evidence, however, was available for a relatively few persons. ARCH-Vahini between 1980 and 1988 had interacted closely with a large number of villagers in these villages and had first-hand knowledge about the residents. It had also listed all the households by category of land ownership and operation and lineage. It had family trees for almost all the families. This initial listing

6. The government was not willing to set this as the cut-off date. Initially it was proposed to fix the cut-off date as the date of issuing notice to vacate. But the organisations insisted on advancing it to the date of declaring the policy. Government agreed.

exercise also helped a great deal in establishing the identities of persons, and also helped identify those who had land titles or land rights in the submergence villages but had migrated over the years to other villages for socio-economic reasons. This category of people was known as *uparwadias* and they also put up claims. Legally, the land owners had the right to compensation as per the policy. The Committee also deliberated on all such claims and tried to settle them.

The displaced families from Maharashtra villages, who had opted to resettle in Gujarat, also placed their claims before the List Committee in Gujarat. TISS, Mumbai, which was the monitoring agency for Maharashtra, had during its survey prepared family trees for the persons surveyed in the villages of Akkalkuva taluka. These villages would be the first to face submergence in Maharashtra. The TISS listing was also very helpful in determining the domicile and age of the claimants.

It is important to emphasise that without intimate knowledge of the affected population and without a deeper understanding of the tribal culture and practices prevalent in the submergence villages, it would not have been possible to settle the claims in the List Committee. Further, it must also be emphasised that the task could be accomplished to the satisfaction of most of the tribal claimants largely because of the presence of the NGOs and village heads in the Committee as full members. The officials on their own, or even with the formal people's representatives such as Members of Legislative Assembly (MLAs), Panchayat members etc. would have found it extremely difficult, if not impossible, to settle all the claims so satisfactorily.

Before the discussion on this is closed, two aspects merit attention. One, it was difficult for the Committee to settle a number of cases of persons claiming to be major sons. Since the cut-off year was 1987, it became difficult as the years passed to convince the young persons and their wards that in the year 1987, the claimants were minor. Invariably, the young claimants came with their wives, children and infants and questioned how they could have been minors in 1987. For them wives and children were the evidence of their being major, whether now or then. Claims of this type kept coming and the Committee could not satisfy them all. In the villages of Maharashtra many displaced families resisted resettlement over this issue for a long time before they ultimately moved out.

The second aspect relates to a major issue that was raised with respect to the total number of PAPs. Activists raised a major claim on this issue and introduced many more categories to be recognised and identified as PAPs. Based on such claims, a series of varied estimates were generated. For instance, the NBA estimated the number in the following way:

- 150,000 tribals were to be affected in 245 submergence villages of Gujarat, Maharashtra and Madhya Pradesh;
- 170,000 farmers were to be affected due to the canal network;
- Shoolpaneshwar Wild Life Sanctuary was to displace 40,000 tribals;
- Thousands were to be displaced because of Catchment Area Treatment (CAT) and Compensatory Afforestation (CA);
- Thousands of landless labourers would be displaced as a result of transfer of private lands to the tribals.

(ARCH-Vahini, 1993d)

The IRM had used this argument of number of affected persons to point out the World Bank's failure in conducting the R&R appraisal, estimating the total number at 40,000 families or 200,000 to 240,000 people (affected by submergence alone) in 1992 and argued, "needless to say, the financial organizational, logistical problems of resettling the original estimate of 6,603 families (figure given by the NWDT) are substantially increased when the number rises by six times to the current estimate of 40,000 families" (Morse and Berger, 1992: 51).

While this writer is no apologist for the government, it must be stated that the NBA and IRM figures were exaggerated. There is a separate discussion on this issue in a different chapter of this volume. But the salient points can be captured here as well. It is true that because of the dam and the reservoir 245 villages were going to be affected. But this did not imply that all the households and all the agricultural lands in all the villages would come under full submergence. This fact has been neither appreciated fully by the critics of the project nor highlighted fully by the project authorities. A detailed explanation is available in two of the ARCH's publications (1993d and 1994). The points in brief are as follows:

- Estimated figures of displacement in the villages of Gujarat and Maharashtra respectively were 4500 and 3000 families. In MP, the official figure of families affected is 23000. This was accepted by the IRM also. However, out of these, nearly 67 per cent families are not likely to be uprooted from their place of origin either because only their houses were under submergence or they were losing less than 25 per cent of agricultural land or both. In all such cases there was no dislocation. A maximum of 7,600 families had faced displacement. Thus, in all, a little more than 15,000 families or about 75,000 to 80,000 persons were likely to be displaced. There has been revision in numbers after the 1992 estimates. The latest figures are discussed after the points of contention that arose then are covered.
- Displacement in canals was limited to small and marginal land holders on the first few km. length of the main and branch canals. IRM had given an estimate that 1100 farmers would lose their entire holding. If at all there was dislocation, it would be of these 1100 farmers and their families.
- There would not be any displacement in Shoolpaneshwar Wild Life Sanctuary due to Catchment Area Treatment (CAT). In fact, MoEF had, with great reluctance, agreed in principle in August 1994 to GoG's proposal to regularise 60,000 ha of degraded forest land under cultivation and accorded approval to first 18,000 hectares of forest land under cultivation by tribals in the Sanctuary and the CAT land that was under the control of the forest department.
- Contrary to the general belief that the states have been underestimating the number of persons/families displaced, MP has kept on revising the figure upward. And yet charges of underestimation continued up to 1996. Kothari, while reviewing various estimates of the total number of displaced persons, arrived at a figure of 400,000 (Kothari, 1996). Interestingly, he had tried to enumerate amongst the displaced also those dependent on private lands allotted to tribals in the relocation sites. Such displacement is known as secondary displacement. ARCH-Vahini had conducted a survey to find out if such secondary level displacement was taking place in

the relocation sites where private lands were sold. It found hardly any evidence of the kind. It had presented this finding to the IRM, but it chose to remain silent on this issue. It was perhaps a tacit acceptance that secondary displacement was not a problem. The actual figures of the total number of displaced families, tentatively arrived at by the end of 1994, were 4578 and 3103 respectively for Gujarat and Maharashtra. These figures were derived in collaboration with the collaborating NGOs, ARCH-Vahini and Anand Niketan. The formation and the working of the List Committee in Gujarat had thus made a significant contribution in sorting out the issue of numbers and correct identification of the affected families. It was also suggested to both Maharashtra and MP to form such committees and institutionalise the efforts to prepare realistic estimates but neither has not done it. The default on the part of MP on this count has created more problems.

It will now be useful to look at the latest figures that we get from the Sardar Sarovar Punarvasvat Agency, the rehabilitation agency and the NCA. The table below present the figures given by the Agency.

State-wise Distribution of Affected Villages (As of 31.07.2009)

State	*Villages Affected*			*PAFs likely to be resettled*	*PAFs likely to be resettled in Gujarat*	*PAFs already resettled in Gujarat*
	Full	*Partial*	*Total*			
GUJ	3	016	019	04744	04744	04744
MH	0	033	033	04204	00771	00771
MP*	1	191	192	37773	05579	05579
Total	4	240	244	46721	11094	11094

* In addition about 20,000 PAFs have been resettled in MP when the height of dam had reached 121 metres. Those remaining to be resettled number 12 to 15 thousand, all in MP.

PAF – Projected Affected Families

Source: Sardar Sarovar Punarvasvat Agency, Vadodara.

The Narmada Control Authority puts the total figure at a higher level[7] for MP. Figures for Maharashtra and Gujarat are lower. The differences are small but it should be noted that there was no regular verification.

States	*Villages Affected*			*Families Affected*
	Full	*Partial*	*Total*	
MP	1	191	192	39,369
Maharashtra		33	33	4,163
Gujarat	3	16	19	4,737
Total	4	240	244	48,269

http://nca.gov.in/rnr_index.htm access verified again on February 2, 2011.

The issue of numbers is still not settled. It can be assumed that all those PAFs who have to be compensated with land have been resettled. That MP has not accepted any alternative land allotment liability is apparent from a report submitted in July 2006 by an

7. Interestingly, even NCA also does not claim to be always correct. Its website http://nca.gov.in/rnr_elev_table0.htm contains the following disclaimer in *CAPITAL LETTERS*:
DISCLAIMER : DATA AND INFORMATION REGARDING RESETTLEMENT OF PROJECT AFFECTED FAMILIES (PAFs) CONCERNING SARDAR SAROVAR PROJECT (SSP) ARE PROVIDED FOR INFORMATIONAL PURPOSES ONLY, AND ARE NOT INTENDED FOR ANY OTHER PURPOSE. NEITHER GOVERNMENT OF INDIA/STATE GOVERNMENTS NOR ANY OF THE DATA OR CONTENT PROVIDERS SHALL BE LIABLE FOR ANY ERRORS OR DELAYS IN THE CONTENT, OR FOR ANY ACTIONS TAKEN IN RELIANCE THEREON. BY ACCESSING THIS SITE, A USER AGREES NOT TO REDISTRIBUTE THE INFORMATION FOUND THEREIN. GOVERNMENT OF INDIA/STATE GOVERNMENTS SHALL NOT BE LIABLE FOR ANY DAMAGES OR COSTS OF ANY TYPE ARISING OUT OF OR IN ANY WAY CONNECTED WITH USE OF THE INFORMATION OR ANY LOSS TO ANY PERSON CAUSED BY ANY SHORTCOMING, DEFECT OR INACCURACY IN THE INFORMATION AVAILABLE ON WEBSITE.

oversight group.[8] If this assumption is correct then the original estimates of PAFs made by NBA and IRM were exaggerations. IRM was interested in taking the World Bank to task and NBA wanted to prove that rehabilitation was impossible.

The figures given by the Sardar Sarovar Punvarvasvat agency show that by the last quarter of 2009, all the entitled PAFs in the three states had been provided with land. The number is 11,094. The total PAFs range between 48,000 and 54,000. Thus between 38 and 42 thousand families had to be relocated with housing and other facilities. This number is indeed high, but as we have noted Maharashtra and Madhya Pradesh did not develop any institutional arrangement for correct identification and estimation. There could be scope for mistakes and manipulations.

Land Purchase Committee (LPC)

We have seen earlier that a majority of affected tribals wanted to resettle on private agricultural land. The anti-dam movement raised many doubts about the availability of such lands. It was also argued, when the new policy was announced by GoG, that the tribals would have to wander in search of private land at their own expense and risk to health. All this was not to happen. The government again responded to need of the time and formed a Land Purchase Committee headed by a revenue official of the rank of Additional Collector. Its members were officials, selected MLAs, Panchayat presidents of Vadodara and Bharuch districts and those taluka Panchayats from which land was purchased, Sarpanchs from selected submergence villages and representatives of ARCH-Vahini and Anand Niketan. In early 1993 ARCH-Vahini resigned its membership in protest against unacceptable compromises.

It would be worthwhile to describe in some detail the process of land purchase and selection by the tribals. Prior to the announcement of the new R&R policy, some tribals had already explored the availability of private lands. In fact some of them were readying themselves for a fall-back position of accepting compensation and

8. Report of the Sardar Sarovar Relief and Rehabilitation Oversight Group, on 'The Status of Rehabilitation of Project Affected Families in Madhya Pradesh', New Delhi, July 2006 (NCA).

buying private land instead of accepting unacceptable government waste land. This exercise came in handy when it was decided to purchase private land. Also the mechanism of the Land Purchase Committee was already in place. Some tribals also had the knowledge of fertile agricultural land in the plains because they seasonally migrated to work on some of these farms. Thus, the tribals made exploratory visits to the areas, talked to owner farmers to ascertain their minds on selling. These exploratory trips also triggered communication among farmers that the government was willing to buy land. The farmers then made offers to the LPC.

At this stage, it must be understood why the farmers were willing to sell their land which either had irrigation facility or was to get it from the Narmada canals once the project was commissioned. According to the land laws of Gujarat, until 1996, only a farmer could buy agricultural land and that too within an eight km radius of where [s]he lived, so that [s]he could operate that land. These laws had virtually frozen the land market in the state. The economic growth and development in Gujarat suggests that the industry and service sectors have done well and agriculture has been going through ups and downs (Iyengar, 1998). Many farmers did not see good prospects for agriculture. Obviously, some agriculturists wanted to sell land and invest in other more paying enterprises. Secondly, their children had got educated and they were not interested in agriculture. Thirdly, in the rural areas of Vadodara district, the scope for non-farm employment had increased and as a result, agricultural labourers had taken to non-farm labour which was more paying and gave exposure to city/town life. Thus, there was a serious shortage of labour in the fertile plains in central Gujarat. Fourthly, the cotton growing farmers in Vadodara district also suffered a serious setback in the 1980s because of uncontrollable pests and diseases in cotton. Many of them became indebted to credit cooperatives and commercial banks. For these reasons, many farmers wanted to dispose of their land if they could realise proper prices. The SSP project offered such an opportunity.

The LPC members considered the proposals, collected opinion about the quality of land, verified the preference and choice of the displaced groups and then negotiated prices with the farmers who offered land for sale. The LPC also formed sub-committees at times and inspected the land parcels on offered before the LPC negotiated prices. The NGOs helped officials to ascertain the willingness of displaced groups and also to control the upper limit of the land prices offered by persuading displaced groups to look for alternative sites if the farmers offered their land at very high price and did not accept the negotiated prices. Until 1992 end, by which time substantial purchases were made to resettle almost all Gujarat and Maharashtra tribals willing to resettle in Gujarat, the average prices settled for unirrigated and irrigated land were Rs.12,500 and Rs.20,000 per hectare respectively.

The LPC functioned well between 1988 and 1992. Beginning 1993, its discipline got diluted. The norms, established by conventions rather than rules, began to be violated. This happened specifically, when the MP government pressurised the Gujarat government to first buy land, prepare sites and then invite the displaced tribals and other groups to visit them. There was also pressure to buy land faster. On the sellers' part, the farmers also realised that the demand was reasonably high and hence inferior land could also be pushed for sale. Commitment and seriousness of purpose may also have been lost among the lower level officials. All this resulted in some substandard land purchases in the post-1992 period. ARCH-Vahini stopped attending meetings, protested and then resigned. An innovative institutional arrangement was slowly dismantled. LPC continued to exist but lost credibility. This has been one of the critical blows to the healthy process of R&R which had begun in Gujarat. An institutional innovation that had improved cost effectiveness and efficiency gave way to administrative and political pressure and encouraged unfair ways and means.

Process of Land Allotment and Joint Inspection Committee

Government departments arranged vehicles to pick up the tribal family heads from the submergence villages to inspect the agricultural lands in the plains. Special four-wheel-drive vehicles were pressed into service for villages which did not have motorable roads. The tribal family heads were accompanied by the members of the NGO groups. The lands offered for sale were visited and inspected. When the tribal groups selected the land, they submitted an application indicating their choice of a particular site. The government then issued orders. Two-hectare plots were mapped and were demarcated on the site. Once the plotting was done, the groups visited again. During this visit, uncultivable plots were rejected and only those that were accepted were allotted. NGO workers were constantly present in this entire process, which took a long time. Arrangements were made for overnight stay in government guest houses in the project colonies and for food. After the completion of joint visits, the groups were dropped back at their villages. This service was offered not once but as many times as the groups had to visit the possible resettlement sites.

Despite the care and caution exercised by the officials and NGO members, there were cases in which the first allotment was not acceptable to a member of the resettling group. To tackle all such cases a joint inspection committee was formed, comprising officials and representatives of the NGOs. The members made visits, invited experts from the state agriculture department, if necessary, and took decisions to change the plots if they were found unfit for cultivation. Thus, the entire process of land allotment went smoothly. IRM, which was been highly critical of WB, expressed satisfaction about the resettlement in Gujarat while airing serious concern about the rehabilitation, which is a long drawn out process (Morse and Berger, 1992).

Resettlement Experience in Gujarat: Critics' Apprehensions, 1988-1992

A major criticism of R&R in Gujarat has also been that private lands have been allotted and due to this the displaced tribal population has been scattered. Related criticisms include loss of cultural life and natural resources support, allotment of uncultivable, inferior and waterlogged agricultural land etc.

To answer all this criticism, it is probably enough to refer to the IRM report which has categorically stated that the tribals have selected land of their choice and decided to form groups which they wanted to. The social reality, also accepted by Scudder, was that tribal communities in the Narmada Valley were not homogenous or monolithic. The WB Loan Agreement said that the tribal communities might be resettled as a village, hamlet or a group that was formed out of choice. The tribals have indeed exercised their choice freely. It is true that families belonging to one village in the submergence area have settled in more than one village and in that sense they are scattered. However, if one contrasts it with the distances between hamlets of a single village in the hilly areas and the fact that all those distances had to be covered on foot, the present scatter may be considered as compact settlement. The plains also offer a road network and transport including state transport bus services. The women at a number of sites have reported that their frequency of visits to their maternal homes and families has increased (Solanki, 1994).

The dependence of tribals on natural resources including forest has been exaggerated both by the activists and social scientists, particularly in the case of tribals in the Valley. Even the IRM report has cited their expert's report according to which there are hardly any forests. Hence dependence on forest produce, fish etc. was in fact marginal. It is interesting that both the arguments – one about commercialisation of forests leading to massive degradation by vested interests and the other about heavy dependence of tribals on the forest – were simultaneously made. Even the IRM report was not free of this anomaly. Except for fuelwood and some seasonal fruits, flowers and leaves (used to roll bidis – indigenous cigarettes) and occasionally fish, not much was available in these forests to the tribals. A number of them told me that the teak and other house building wood had disappeared so fast during last 30 years that they found it extremely difficult to find wood for house construction.

Regarding the quality of land, I have already mentioned that the LPC as well as the Joint Inspection Committee were in place and working well until 1993. There has been an erosion of quality of work of these institutionalised set-ups which I shall discuss in the next section. Until 1993, cases of allotment of bad quality land were rare and all attempts were made to change those plots. The system was in place. Up to December 1993, a majority of Gujarat and Maharashtra families had been allotted land. The progress after that has been very slow. This time period coincides with the withdrawal of the Project from the World Bank. The review missions stopped coming. There was no pressure anymore. Nobody from outside the country was breathing down their necks. Local resistance could at best be handled and at worst be ignored.

Post-1993 Implementation

The post-1993 period does not speak very well for R&R in the SSP. The allotment of agricultural land and housing plots for most of the families in Gujarat was over. Families from Maharashtra were coming in large numbers and in both Maharashtra and MP the tribals had realised that their leaders were preoccupied with jubilation over the Morse-Berger verdict on the project and that they had to act on their own if they wanted to get resettled well. ARCH-Vahini was receiving feelers from villages of both states. The stage was set for the GoG to seize the initiative and get on with the work. But the administration could not respond as it had done until then, and the post-1993 R&R faced some rough weather.

ARCH-Vahini, which had actively collaborated until then, became critical of the way in which the Gujarat administration and government were handling the resettlement of displaced families, largely from Maharashtra and MP. According to ARCH-Vahini, it appeared that the administration's main interest was in seeing that the affected families relocated and planted themselves in new sites. Beyond that, it probably failed to perceive any responsibility. It seemed to escape them that proper and satisfactory relocation or resettlement was just the first step to proper rehabilitation. This has been one of the major problems with resettlement in the post-1993 period. Between 1993 and 1998 following problems had been highlighted:

1. The drinking water facilities were poor. The major source had been hand pumps. An ARCH-Vahini survey conducted then found 200 non-functional hand pumps. The quality of service rendered by another 150 hand pumps was doubtful. Frequent breakdown of the pumps was the most common complaint.
2. From among those tribal families from Gujarat and Maharashtra allotted agricultural land in the post-1993 period – a process in which ARCH-Vahini field staff collaborated, more than 400 families had been allotted plots which had following problems:
   (i) The plots did not admeasure 2 ha,
   (ii) Major parts of full plots were uncultivable, and

(iii) The plots had not been demarcated.

3. In all, about 500 families from MP had been allotted land plagued by the following problems:
   (i) One site was in a village targeted by dacoits from across the district border.
   (ii) Another site was a village where the cattle herder community was known to harass cultivators rendering cultivation difficult.
   (iii) Agricultural land was rendered uncultivable because of un-removable roots of a grass weed locally known as *Dabh*.
   (iv) About 50 families had been shown good quality land but were subsequently allotted inferior quality of land.
4. The fourth set of problems related to the quality and condition of the physical infrastructure in relocation sites. Approach roads, internal roads and the drainage had not been laid properly. It is said that the quality of workmanship and the material used were sub-standard. In a number of cases the quality of roof tiles supplied was poor. Despite repeated requests, they had not been replaced.
5. Services of regular revenue administration had lagged behind, raising doubts about the intentions of the government regarding the ownership title deeds of the allotted land to the relocated families.

The response of the administration and the government was that they had initiated corrections and the complaints and shortcomings were being attended to. However, this did not convince ARCH-Vahini, and it resigned from the membership of the executive committee of the autonomous rehabilitation agency known as Sardar Sarovar Punarvasvat Agency. In the post-1993 period, the MP government took a stiff stand against agreeing to the increase in the height of the dam. It took the position that the resettlement efforts by GoG had not been satisfactory. Hence, the height of critical section of the dam was not being raised.

The Supreme Court of India directed the SS authorities to make an institutional arrangement for monitoring R&R work in the state. GoG, through a resolution on February 16, 1999, set up a body called the Grievances Redressal Authority (GRA) for Sardar Sarovar PAP. A Grievances Redressal Cell (GRC) was created with the objective of sensitising the SSPA infrastructure and providing prompt redressal to the PAFs. It gave opportunity to PAFs to voice their grievances and highlight the problems they faced. SSPA responded to the problems that cropped up in the process of implementing the R&R programme through different units created to attend to the problems. There is a Land Cell, Engineering Cell, Rehabilitation Cell, Agricultural Cell, Medical Cell and Education Cell. Only a comprehensive independent review would inform us about the current status of R&R in Gujarat. But it can be said that even with problems and grievances its performance is better than that of most projects in the past.

III

Problems and Prospects of R&R in SSP

SSP has been under the world's gaze. We have seen that the R&R programme of the project was initially as bad as it had been in any development project involving involuntary displacement. We also saw that there were many firsts in the project. For the first time, such a massive mobilisation of PAPs took place and for the first time the struggle bore concrete results. For the first time, any state government in the country announced and committed an R&R package which could be said to be fair and just to those who were to get involuntarily displaced. The setting up of a List Committee, the Land Purchase Committee, the GO-NGO collaboration and the process of allotment of alternative agricultural land – everything has happened for the first time. A combined effect of all these first-time initiatives, the resettlement and relocation of the displaced population in the first phase, i.e. 1988-93 was also largely successful. This was also assessed and approved by an International Review Mission. However, problems again plagued SSP's R&R programme until the setting up of the Grievance Redressal Cell.

Two sets of problems have affected the programme. The first concerns the government's understanding of R&R issues. Cernea (1996: 1516) has drawn attention to the processes involved in involuntary resettlement. He says, "What is usually described as 'involuntary resettlement' consists of two distinct, yet closely related social processes: (a) displacement of people and (b) reconstruction of their livelihood; this reconstruction is sometimes called rehabilitation. Each has its own demands, risks, costs logistics and socio-cultural and economic effects." He further says that in theory, the two processes are part of the same continuum; however, the second does not always follow the first. In the case of the SSP, it appears the rehabilitation of the population resettled during 1988-93 was facing rough weather and the resettlement of the post-1993 displaced families was also plagued with problems. The government and the administration was either not aware of or had decided to ignore the seriousness of both the problems. The officials at the lower levels were hardly able to appreciate these fine aspects of involuntary resettlements. Therefore, lacking clear policy pronouncements and procedures for implementation, the implementers were going to create chaos in R&R. Probably, inept handling of serious problems arising in the relocation sites with regard to reconstruction of living and livelihood, and the resettlement efforts in terms of allotment of land to the affected families from MP, resulted in dissatisfaction among the displaced people. This, as a result, prompted the otherwise effective and fruitful GO-NGO collaboration to break.

It appears that the feeling among most persons in the government and the administration was that, with the relocation of families from the submergence villages the rehabilitation was completed. A majority of displaced families in Gujarat got good quality land and a plot to build a house. What more should they expect? The efforts required to set up and roll socio-economic life smoothly at the new sites, the displaced population's integration with the host communities, the requirement and aspirations of the host communities, potential areas of conflict and cooperation between the host and the guest population etc. didn't seem to be worth attending. Many of these aspects did not involve a monetary component, but they involved understanding and commitment on the part of government officials.

The second set of problems relates to the role of the IRM, activists, and NGOs in and outside the country in handling the SSP issue in general and the R&R issue in SSP in particular. The Independent Review Mission came into existence because of the pressure of the national and international NGOs on the World Bank. One can raise a serious question whether it was required. Of course, even if one does not go into rhetoric and say that the sovereignty of the nation was threatened by such international interventions, enough understanding existed within the country to seriously discuss the problems and look for solutions. However, once it became a given, there should not have been any argument against the setting up of the IRM. What is important is that the IRM did not remain objective in "its search for truth" on both the major agenda items: R&R and the environmental impacts of SSP (ARCH-Vahini, 1992a, 1992b, 1992c, 1992d, 1992e, 1993a, 1993b, 1993c). The IRM by its advice to the World Bank to "Step Back" from the project set such a process into motion that the reform processes that were taking place in the project and in the system got derailed. The World Bank, which was the main actor and the keel that could have kept the ship on course, was forced to leave. Both the national and international NGOs celebrated this only to weaken the reform process that had just begun in a developing country.

This exit of the World Bank also weakened and further isolated the already isolated NGOs like ARCH-Vahini which had provided critical support to the project and had collaborated to implement the policy. It also released the pressure on the government. It is probably clear by now, after we have seen how the policy reforms were won, that the government was better persuaded by the Bank than the combined strength of the NGO network. Interestingly, what the NBA and its allies achieved by sending the World Bank packing was a rise in the chances of the failure of R&R in the SSP which they had always predicted. It must be conceded, though, that NBA's dream failed to materialise because other forces such as GRA directed R&R in the right direction.

Had the IRM given critical feedback and recommended continued but critical support to the SSP by WB, the Bank Missions in conjunction with sensitive NGOs such as ARCH-Vahini could have set into motion processes that would have ensured a decent and timely resettlement, in Gujarat as well as in their respective states, of the rest of the families displaced from Maharashtra and MP. It could have also designed a path to take the resettled population to a proper rehabilitation. Even GRA would not have come into existence. Because it did not happen that way both resettlement and rehabilitation were at risk. With the lone worthwhile NGO getting out of the main frame, the prospects of R&R weakened and delays occurred. Even in 2009, incomplete R&R has held up raising the dam height to its design level. The promise of the project remains unfulfilled.

18

And Quiet Flows the Narmada: A Recapitulation of Events Surrounding the Efforts to Harness the Narmada Waters for the Public Weal

K. Srinivas

The Narmada is a long and vivacious river. Her story is ancient lore even as mythology speaks of significant angst on her banks, be it the Aryan conquest of the Dravidian people, the southward travel of Agastya Muni, or the travels of Lord Rama in the Daṇḍakâraṇya. Events on her banks have not only been epochal but also tumultuous. Therefore, it is of no surprise that even efforts to harness her immense potential have engaged in the hermeneutics of intellectual debate, competing claims and ostensibly viable alternatives.

In the fading light of British rule, efforts began to harness the energy of such bounty for the public weal were initiated in the Narmada valley, when less than 4 per cent of the potential of this huge river basin covering nearly 100,000 sq. km was being utilised. While the significant signposts in this process are recounted below, the article focuses more on the challenges to the Sardar Sarovar Dam, and the premises upon which they were raised, fought and finally settled.

Initial Plans

In 1946, the then Governments of Bombay and of Central Provinces and Berar requested the Central Waterways, Irrigation and Navigation Commission (CWINC) to undertake investigations on the Narmada river system from a holistic perspective comprising flood control, irrigation, power and extension of navigation. The study commenced in 1947 and after inspection of all potential sites by engineers and geologists, seven projects were shortlisted and recommended for detailed investigations.

Consequently, in 1948 the Central Ministry of Works, Mines & Power appointed an *Ad-hoc* Committee headed by A.N. Khosla, Chairman, CWINC to study the shortlisted projects and to recommend prioritisation amongst them. This *Ad-hoc* Committee recommended, as an initial step, detailed investigations for the following projects keeping in view the availability of men, materials and resources:

A. Bargi Project,
B. Tawa Projects near Hoshangabad,
C. Punasa (Narmada Sagar) Project, and
D. Broach Project.

Government of India (GoI) sanctioned this proposal in March, 1949.

The Central Water & Power Commission (CW&PC) carried out a study of the hydroelectric potential of the Narmada basin in the year 1955, and the Navagam site was finally decided upon for the construction of the dam, in consultation with the erstwhile Government of Bombay. The CW&PC recommended a two-stage implementation to the then Government of Bombay. In Stage-I, the Full Reservoir Level (FRL) was restricted to 160 ft with provision for wider foundations to enable raising of the dam to 300 ft in Stage-II. A high level canal was envisaged in Stage-II.

The Bombay Government suggested two modifications, (i) that the FRL of the dam be raised from 300 to 320 ft. in Stage-II and (ii) the provision of a power house in the river bed and a power house at the head of the low level canal. This project was then reviewed by a panel of consultants, appointed by the Ministry of Irrigation and Power, who submitted a report in 1960 suggesting that the two stages of the Navagam dam should be combined into one, and the dam be constructed to its final FRL 320 ft. The consultants also stated for the first time that there was scope for extending irrigation from the high level canal towards the Rann of Kachchh.

With the formation of the State of Gujarat on 1st May, 1960, the Narmada Project stood transferred to that State. Accordingly, the Government of Gujarat (GoG) gave an administrative approval to Stage-I of the Narmada Project in February 1961. The Project was inaugurated by the late Pandit Jawaharlal Nehru on 5th April, 1961.

GoG undertook surveys for the high level canal in 1961. The submergence area survey of the reservoir enabled assessment of the storage capacity of the Navagam reservoir if its height should be raised beyond FRL 320 ft. The studies indicated that a reservoir with FRL 460+ ft. would enable realisation of optimum benefits from the river by utilising the untapped flow below Punasa dam and would make it possible to extend irrigation to a further area of over 2 million acres. Accordingly, explorations were done for locating a more suitable site in the narrower gorge portion. Finally, in November 1963, site No. 3 was found to be most suitable on the basis of the recommendations of the Geological Survey of India, foundation investigations and availability of construction materials at the proposed site.

In November 1963, the Union Minister of Irrigation and Power held a meeting with the Chief Ministers of Gujarat and Madhya Pradesh (MP) at Bhopal. The discussions resulted in what is known as the Bhopal Agreement. Its salient features were:

(a) That the Navagam Dam should be built to FRL 425 by the GoG and its entire benefits were to be enjoyed by the State of Gujarat.

(b) Punasa Dam (MP) should be built to FRL 850. The costs and benefits of Punasa Power Project shall be shared in the ratio 1:2 between the Governments of Gujarat and MP. Out of the power available to MP half of the quantum was to be given to the State of Maharashtra for a period of 25 years for which the State of Maharashtra was to provide a loan to the extent of one-third the cost of Punasa Dam. The loan to be given by the State of Maharashtra was to be returned within a period of 25 years.

(c) Bargi Project was to be implemented by the State of MP. The dam was to be built to FRL 1365 in Stage-I and FRL 1390 in stage-II. GoG were to give a total loan assistance of Rs. 100 million for the same.

In pursuance of the Bhopal Agreement, the GoG prepared a brief project report envisaging the Navagam Dam FRL 425 ft. and submitted it to the CW&PC on 14th February, 1964. Madhya Pradesh, however, did not ratify the Bhopal Agreement, and a stalemate resulted. To cut the Gordian knot, a High Level Committee headed by A.N. Khosla, the then Governor of Orissa, was constituted on 5th September, 1964 by the Government of India. The Terms of Reference of this Committee were as under:

(i) Drawing up of a Master Plan for the optimum and integrated development of the Narmada water resources.
(ii) The phasing of its implementation for maximum development of the resources and other benefits.
(iii) The examination, in particular of Navagam and alternative projects, if any, and determining the optimum reservoir level or levels.
(iv) Making recommendations of any other ancillary matters.

Within a year, in September 1965, the Khosla Committee submitted a unanimous report to GoI, and recommended a Master Plan for Narmada Water Development. In Chapter XI of the report, the Committee outlined its approach to the Narmada Development plan. An extract from this Chapter is reproduced below:

> "11.1 In the meeting from 14th to 18th December, 1964 at which the State representatives were also present, the Committee laid down the following basic guidelines in drawing up the Master Plan for the optimum and integrated development of the Narmada water resources:
>
> 1. National interest should have overriding priority. The Plan should, therefore, provide for maximum benefits in respect of irrigation, power generation, flood control, navigation etc. irrespective of State boundaries;
> 2. Rights and interests of State concerned should be fully safeguarded subject to (1) above;
> 3. Requirements of irrigation should have priority over those of power; subject to the provision that suitable apportionment of water between irrigation and power

may have to be considered, should it be found that with full development of irrigation, power production is unduly affected;

4. Irrigation should be extended to the maximum area within physical limits of command, irrespective of State boundaries, subject to availability of water; and in particular, to the arid areas along the International border with Pakistan both in Gujarat and Rajasthan to encourage sturdy peasants to settle in these border areas (later events have confirmed the imperative need for this); and
5. All available water should be utilized to the maximum extent possible for irrigation and power generation and, when no irrigation is possible, for power generation. The quantity going waste to the sea without doing irrigation or generating power should be kept to the unavoidable minimum."

The Master Plan recommended envisaged 12 major projects to be taken up in Madhya Pradesh and one at Navagam in Gujarat. In respect of the latter the Committee recommended as follows:

1. The terminal dam should be located at Navagam.
2. The optimum FRL of the Navagam dam worked out to RL 500 ft.
3. The FSL (Full Supply Level) of the Navagam canal at off-take to be RL 300 ft.
4. The installed capacity at the river bed power station and canal power station should be 1000 mw and 240 mw respectively, with one standby unit in each power station (thus the total installed capacity at Navagam would be 1400 mw).

The benefits of the Navagam Dam as assessed by the Khosla Committee were as follows:

"(i) Irrigation of 1.58 lakh hectares (39.4 lakh acres) in Gujarat and 0.4 lakh hectares (1.00 lakh acres) in Rajasthan. In addition, the Narmada waters when fed into the existing Mahi

Canal system would release Mahi water to be diverted on higher contours enabling additional irrigation of 1.6 to 2.0 lakh hectares (4 to 5 lakh acres) approximately in Gujarat and 3.04 lakh hectares (7.5 lakh acres) in Rajasthan.

(ii) Hydro-power generation of 951 MW at 60 per cent LF in the mean year of development and 511 MW on ultimate development of irrigation in Gujarat, Madhya Pradesh, Maharashtra and Rajasthan."

The Khosla Committee stressed an important point in favour of the high Navagam Dam, namely, additional storage. They emphasised that this additional storage would permit greater carryover capacity, increased power production and assured optimum irrigation and flood control and would minimise the wastage of water to the sea. The Committee also observed that instead of a higher Navagam Dam as proposed, if Harinphal or Jalsindhi dams were raised to the same FRL as at Navagam, the submergence would continue to remain about the same because the cultivated and inhabited areas lie mostly above Harinphal while in the intervening 113 km gorge between Harinphal and Navagam, there was very little habitation or cultivated areas.

Inter-State Water Dispute and Constitution of Tribunal

The Khosla Committee report could not be implemented on account of disagreement among the States. On 6th July, 1968 the State of Gujarat represented to GoI under Section 3 of the Inter-State Water Disputes Act, 1956 that a dispute had arisen between the State of Gujarat and the States of Madhya Pradesh and Maharashtra over the use, distribution and control of the waters of the Inter-State River Narmada.

The substance of the allegation was that executive action had been taken by Maharashtra and MP which had prejudicially affected the State of Gujarat and its inhabitants. The State of Gujarat objected to the proposal of the State of MP to construct Maheshwar and Harinphal Dams over the river Narmada in its lower reach and also to the agreement reached between the States of MP and Maharashtra to jointly construct the Jalsindhi Dam across the Narmada on its course between the two States. The main reason for the objection was that if these projects were implemented, they would prejudicially affect the rights and interests of Gujarat by compelling it to restrict the height of the dam at Navagam to FRL 210 ft. or less. Reducing the height of the dam would be to the permanent detriment of irrigation and power benefits that would accrue to the inhabitants of Gujarat and this would also make it impossible for Gujarat to reclaim the desert area in the Ranns of Kachchh.

According to the State of Gujarat, the principal matters in dispute were as under:

(i) The right of the state to control and use the waters of the Narmada river on well-accepted principles applicable to the use of waters of inter-state rivers;
(ii) The right of the State of Gujarat to object to the arrangement between the State of MP and the State of Maharashtra for the development of Jalsindhi Dam;
(iii) The right of the State of Gujarat to raise the Navagam dam to an optimum height commensurate with the efficient use of Narmada waters including its control for providing requisite cushion for flood control; and
(iv) The consequential right of submergence of area in the States of Madhya Pradesh and Maharashtra and areas in the Gujarat State.

Acting under Section 4 of the Inter-State Water Disputes Act, 1956, GoI constituted a Tribunal headed by Justice V. Ramaswamy, a retired Judge of the Supreme Court. GoI made a reference of the water dispute in the following terms:

> "In exercise of the powers conferred by sub-section (1) of Section 5 of the Inter-State Water Disputes Act, 1956 (33 of 1956), the Central Government hereby refers to the Narmada Water Disputes Tribunal for adjudication of the water dispute regarding the Inter-State river, Narmada, and the river-valley thereof, emerging from letter No. MIP-5565/C-10527-K dated 6th July, 1968, from the Gujarat."

On 16th October, 1969, GoI made another reference of certain further issues raised by the State of Rajasthan. The State of Madhya Pradesh filed a Demurrer before the Tribunal stating that the constitution of the Tribunal and reference to it were *ultra vires* of the Act. The Tribunal framed 24 issues which included issues relating to Gujarat's right to construct a high dam with FRL 530 feet and a canal with FSL 300 feet or thereabouts. Issues 1(a), 1(b), 1(A), 2, 3, and 19 were tried as preliminary issues of law and by its decision dated 23rd February, 1972, the said issues were decided against the respondents.

The Tribunal held that the Notification of the Central Government dated 16th October, 1969 referring the matters raised by the State of Rajasthan by its complaint was *ultra vires* of the Act, but the constitution of the Tribunal and referring to it the water dispute regarding the river Narmada was not *ultra vires* of the Act, and the Tribunal had jurisdiction to decide the dispute referred to it at the instance of State of Gujarat.

It further held that the proposed construction of the Navagam project involving consequent submergence of portions of territories of Maharashtra and MP could form the subject matter of a "water dispute" within the meaning of Section 2(c) of the 1956 Act. It also held that it had the jurisdiction to give appropriate direction to MP and Maharashtra to take steps by way of acquisition or otherwise for making submerged land available to Gujarat in order to enable it to execute the Navagam project; and that the Tribunal had the jurisdiction to give consequent directions to Gujarat and other party states regarding payment of compensation to Maharashtra and MP, for giving them a share in the beneficial use of Navagam dam, and for rehabilitation of displaced persons.

First Petition before the Honourable Supreme Court

Against the aforesaid judgment of the Tribunal on the preliminary issues, the States of MP and Rajasthan filed Special Leave petitions to the Hon. Supreme Court and obtained a stay of the proceedings before the Tribunal to a limited extent. The Court directed that the proceedings before the Tribunal should be stayed but discovery, inspection and other miscellaneous proceedings before the Tribunal may proceed. The State of Rajasthan was directed to participate in these interlocutory proceedings.

On 30th June, 1974, the Chief Ministers of MP, Maharashtra, Gujarat and Rajasthan had entered into an agreement to settle the matters in dispute under the guidance of the Prime Minister of India. A formal agreement dated 12th July, 1974 was arrived at between the Chief Ministers of MP, Maharashtra, Rajasthan and the Adviser to the Governor of Gujarat. It was agreed that,

- On the assumption of 75 per cent dependability (i.e. availability for 75 per cent of the years), water available per year was to be assessed at 28 million acre feet.
- The net available quantity of water for use in MP and Gujarat, for the purpose of allocation amongst States, was determined as 27.25 million acre feet.
- The height of the Navagam Dam was to be fixed by the Tribunal after taking into consideration various contentions and submissions of the parties.
- Madhya Pradesh and Rajasthan would withdraw appeals filed in the Supreme Court.
- After the withdrawal of the appeals by the States of Madhya Pradesh and Rajasthan, the Tribunal proceeded to decide the remaining issues between the parties.

Approach for Adjudicating Narmada Water Dispute

- National interest should have over-riding priority;
- Rights and interests of state concerned should be fully safeguarded subject to (i) above;
- Requirements of irrigation should have priority over those of power;
- Irrigation should be extended to the maximum area irrespective of State boundaries;
- All available water should be utilised to the maximum extent possible for irrigation and power generation;
- The quantity going waste to the sea – to be kept to the unavoidable minimum.

Declaration and Notification of the NWDT Award

On 16th August, 1978, the Tribunal declared its Award under Section 5(2) read with Section 5(4) of the Inter-State Water Disputes Act,

1956. Thereafter, the Union of India and the States of Gujarat, MP, Maharashtra and Rajasthan filed references under Section 5(3) of the Inter-State Water Disputes Act, 1956, and they were numbered 1, 2, 3, 4 & 5 of 1978. The Tribunal heard these references, and final orders were passed on 7th December, 1979, and published in a Gazette Extraordinary by the Government of India on 12th December, 1979.

Salient Features of the Award

A. Height of the Sardar Sarovar Dam

The height of the Sardar Sarovar Dam was determined at FRL 455 ft (138.68 m) and Maximum Water level at 460 ft (140.21 m). It was determined that FRL 436 ft.+ was required for irrigation use alone. For power generation throughout the year, it would be necessary to provide all the live storage above Minimum Draw Down level (MDDL) for which an FRL of 453 ft.+ with MDDL 362 ft.+ would obtain gross capacity of 7.44 MAF. Therefore, the Tribunal held that FRL of the Sardar Sarovar Dam should be 455 ft.+ providing gross storage of 7.70 MAF. It permitted the State of Gujarat to undertake and complete the construction of the dam.

B. Geological and Seismological Aspects of Dam Site

The Tribunal accepted the recommendation of CWPC that there should be a seismic coefficient of 0.10*g* for the dam.

C. Relief and Rehabilitation

The final Award contained directions regarding submergence, land acquisition and rehabilitation of displaced persons. The award defined the meaning of the land, oustee and family. The Gujarat Government was to bear all costs of rehabilitation incurred by the Governments of MP and Maharashtra including compensation, charges and expenses incurred in respect of Land Acquisition. A scheme for rehabilitation of the 'oustees', including aspects such as civic amenities to be provided was also laid out. The Award also provided that if the State of Gujarat was unable to resettle the oustees or the oustees were unwilling to migrate to other states, then the oustees were to be re-settled in the home State, and all expenses incurred thereto were to be borne by Gujarat. A mandatory provision for rehabilitation in Clause XI sub-clause IV(6)(ii) stated that *no submergence of any area would take place unless the oustees were rehabilitated.*

D. Allocation of the Narmada Waters

The Tribunal determined the utilisable quantum of water of the Narmada at Sardar Sarovar Dam site on the basis of 75 per cent dependability at 28 MAF. On that basis it made the following allocation amongst the states from the utilisable quantum of Narmada water:

Madhya Pradesh	:	18.25 MAF
Gujarat	:	9.00 MAF
Rajasthan	:	0.50 MAF
Maharashtra	:	0.25 MAF

E. Period of Non Reviewability of Certain Award Terms

The Award provided for review of the specific clauses pertaining to Apportionment of the Utilisable Quantum, Excess waters and sharing of distress after a period of 45 years from the date of the publication. However, the determination of '75% dependable flow as equivalent to 28 MAF' is non-reviewable along with other governing levels for the dam and off-taking canal.

Thus was evolved a water utilisation plan that envisaged the construction of 30 major dams, 135 medium dam projects and more than 3000 minor dams. The major terminal dam, Sardar Sarovar being in Gujarat, the remaining 29 were in MP. Of them three are major dams – Narmada Sagar (rechristened as Indira Sagar), Omkareshwar and Maheshwar. Rajasthan was to construct a canal in its territory to utilise its share of 0.5 MAF.

Institutional Mechanism – The Narmada Control Authority

The Tribunal in its Award directed that an inter-state administrative authority – Narmada Control Authority (NCA), be constituted for the purpose of securing compliance with and implementation of the decision and directions of the Tribunal.

The Sardar Sarovar Construction Advisory Committee (SSCAC) was also to be constituted for ensuring efficient, economical and early execution of the project, to be chaired by the Secretary, Ministry of Water Resources (MoWR).

The Tribunal also directed constitution of a Review Committee with the Union Minister for Water Resources as its Chairperson and the Chief Ministers of MP, Maharashtra, Gujarat and Rajasthan as members. The Review Committee could review the decisions of NCA and the SSCAC.

The NCA was to be a high powered committee with Secretary, MoWR, GoI as its Chairperson, and Secretaries in the Ministries of Power, Environment & Forests, Welfare, and Chief Secretaries of the concerned four States as members. In addition, there were a number of technical persons like Chief Engineers as members.

NCA was empowered to constitute one or more sub-committees and assign to them such of the functions and delegate such of its powers as it thought fit. Accordingly, NCA constituted the following discipline based sub-groups:

(i) Resettlement and Rehabilitation sub-group under the Chairmanship of Secretary, Ministry of Welfare;
(ii) Rehabilitation Committee under Secretary, Ministry of Welfare, to supervise the rehabilitation process by undertaking visits to R&R sites and submergence villages.
(iii) Environment Sub-group under the Chairmanship of Secretary, MoEF;
(iv) Hydromet Sub-group under the Chairmanship of Member (Civil), NCA;
(v) Power Sub-group under the Chairmanship of Member (Power) NCA;
(vi) Narmada main Canal Sub-committee under the chairmanship of Executive Member, NCA.

Post-Award Clearances

After considerable consultations between the Ministry of Water Resources and the Ministry of Environment, on 24th June, 1987 the

Ministry of Environment and Forests, Government of India accorded environmental clearance subject to the following principal conditions:

(i) The NCA would ensure that environmental safeguard measures were planned and implemented *pari-passu* with progress of work on project.

(ii) The Catchment Area Treatment programme and the Rehabilitation plans to be so drawn as to be completed ahead of reservoir filling.

Environment Sub-Group: In November 1987 for monitoring and implementation of various environmental activities effectively, NCA created an independent Environment Sub-Group. This Sub-Group is chaired by the Secretary, MoEF, GoI, and has various other independent experts as members.

Thereafter, the Planning Commission, accorded investment approval for an estimated cost of Rs. 64.06 billion on 5th October 1988.

Financial Closure of World Bank Funding and its Aftermath

Consultations with the World Bank started in 1978 for a loan to meet the financial obligations of implementing the project. WB sent a Reconnaissance Mission to visit the project site and carried out the necessary inspection. In May 1985, the Narmada Dam and Power Project and Narmada Water Delivery and Drainage Project were sanctioned by WB under International Development Agency Credit No. 1552. An agreement in this respect was signed with the Bank on 10th May, 1985 and credit was to be made available from 6th January, 1986.

According to WB, central to the Bank's credit and loan agreements was the objective requiring that all oustees, including those described as landless, be enabled as a result of resettlement and rehabilitation measures taken on their behalf, to "improve or at least regain the standard of living they were enjoying prior to their displacement". Following its comprehensive resettlement policy on oustees generally (1990) and specific resettlement policy on tribals (1991), the Bank established for the first time an Independent Review (IR), a panel to evaluate and examine the SSP.

The ToRs by which the IR should evaluate the project were set forth in March 1991, and Bradford Morse, former head of UNDP, was appointed Chairman of the IR in June 1991. An international aid organisation had never before established an investigatory body with a mandate as sweeping as that of the IR, no matter how controversial. Under the agreements, the Bank had acknowledged only those persons whose villages would be affected by submergence as "project-affected" individuals entitled to resettlement and rehabilitation. The ToRs of the IR recognised such persons "displaced/affected by the reservoir and infrastructure." However, the then President of the World Bank, Barber Conable, asked the IR to consider, under the ToRs the status of resettlement and compensation for "canal-affected persons", also. It is interesting to note here that it was almost in the same days that the Supreme Court of India had refused to stay SSP (August 9, 1991) when it was not convinced by a plea for a stay on the further execution of the SSP "until such time as the basic issues raised in the petition are settled and the concurrence of the tribal people is obtained to the acquisition of land and to the plan for their alternative resettlement if any based on full facts, clear understanding and free will of the concerned people". On the environmental side, the ToRs required consideration of the measures being taken to ameliorate the impact of "all aspects of the Projects".

With characteristic openness, the Review Mission was welcomed in the state. The political leadership, civil service, NGOs and the press took keen interest in the Review. There were considerable expectations that the SSP would be subjected to a constructive review. In particular there was widespread anticipation that the rehabilitation and environmental studies would benefit from constructive proposals for improvement in light of the new developments that had taken place in the interim. The Mission acknowledged the cooperation of the Governments of India, Gujarat, Maharashtra and MP, of NGOs and of people affected by SSP.

In June 1992, this Mission came out with its Report calling upon WB to "step back" from SSP on the grounds that rehabilitation was "impossible", human rights, especially of the tribal population were being – and must necessarily be – infringed, and that on account of lack of water in the Narmada the project would in any event "not perform as planned". This was contradictory to the Bank's own assessment that 'the Sardar Sarovar Plan represents a break with past approaches to the planning, design and construction and operation of irrigation projects in India' and again, 'within a period of three years an impressive array of high quality studies and designs have resulted, including a comprehensive framework plan by the Narmada Planning Group and production by the Irrigation Department of high quality designs, specifications and procurement documents'.

Calling the IRM Report "disappointing", GoG, in August 1992, presented "a point by point rebuttal of the intellectual bankruptcy of the so called Independent Review". Government of Gujarat also desired that the partnership in progress (with WB) must continue. Stating that "There is no alternative to the SSP, the design and scope of which has been determined after careful optimisation studies" and that "Its promise of a better tomorrow is real and realisable", the Government called for "No Cancelling Tomorrow" (Appendix-I). The IRM Report was elaborately critiqued by many and was not even accepted by WB, whether in regard to R&R or ecological outcomes. In fact these were specifically rejected in a Bank press release dated 22 June, 1992.

Rejecting the IRM Report outright, GoI decided to unilaterally withdraw from the funding agreement with the World Bank and to go ahead with the project by raising funds from its own resources and from the public. Public participation was ensured by devising investment instruments like Fixed Deposit Schemes, Deep Discount Bonds etc. These financial instruments were unique and have been much studied and written about (including in other parts of this publication) and are not being repeated here.

B.D. Sharma's Intervention

In November 1990 the former Commissioner of Scheduled Castes and Scheduled Tribes, B.D. Sharma wrote a letter to the Supreme Court seeking effective steps for proper rehabilitation of the oustees of Sardar Sarovar Dam. This letter was entertained and treated, under Article 32 of the Constitution, as Writ Petition No. 1201 of 1990. On 9th August, 1991, the Supreme Court directed that an NCA Committee headed by Secretary (Welfare), to monitor the rehabilitation aspects of SSP be activated.

The Narmada Bachao Andolan

Differing perceptions are inherent to a healthy democracy and even icons like Bhakra Dam were no exceptions. Despite the overwhelming consensus that prevailed in the Nehruvian era, Pt.Nehru himself came across as being ambivalent on the issue of 'Large *Vs* Small' – on November 17, 1958, while addressing the 29th annual meeting of the Central Board of Irrigation and Power: "*For some time past, however, I have been beginning to think that we are suffering from what we may call, disease of gigantism. We want to show that we can build big dams and do big things. This is a dangerous outlook developing in India...The idea of having big undertakings and doing big tasks for the sake of showing that we can do big things is not a good outlook at all*" (Thakkar, 2005); while on the other hand, on October 22, 1963, during the dedication of 'Bhakra' he described it as a 'New temple of resurgent India and symbol of India's progress'. The Bhakra Dam has been widely credited with being responsible for the India's green revolution, and yet even this does not prevent the opposite perception to coexist; according to which Bhakra Dam is "a most ordinary project, an ordinary dam much like any other large dam — with all its flaws and blemishes" (Manthan Adhyayan Kendra, 2005).

This divergence of opinion in the case of the Sardar Sarovar project was not, however, dispassionately academic. It was ideological, acrimonious, and driven by violent emotions on all sides. The principal 'dissenter' was the Narmada Bachao Andolan, and its methods captured much media attention. The movement led to much realignment of policies, and improvement in the delivery methodologies and strategies; and the resultant post- R&R situation received widespread public approval for the entire SSP. The constructive intervention of the Supreme Court and the supervision of the state rehabilitation agencies helped improve the lot of the affected people vastly. However, the 'Spirit of Accommodation' did not moderate democratic values of freedom to differ and agitate, nor were they tempered by intellectual humility. Ideological determinism seemed over-arching and intemperate opposition by activists continued with grandstanding tactics.

In ancient Rome, triumphant Generals in ceremonial processions honouring them, were reminded 'Memento Mori, Memento Mori – remember you too are mortal' by a slave riding behind them. Instrumentalities of State and Civil Society too must be constantly reminded of this truism.

The Five Member Expert Group

The NBA agitation against construction of the Sardar Sarovar Dam led the Government of India, on 3rd August, 1993 to constitute a Five Member Group headed by Jayant Patil, Member, Planning Commission and Vasant Gowariker, Ramaswmy R. Iyer, L.C. Jain and V.C. Kulandaiswamy as its members, to engage in dialogue with NBA on issues relating to SSP. The group had three months to submit its report.

During this time, construction of the dam continued and on 22nd February, 1994 MoWR allowed closure of the construction sluices to enable raising the dam. On 23rd February, 1994 the ten construction sluices were plugged permanently.

The Second Petition before the Supreme Court of India (Writ Petition 319)

In April, 1994 the NBA filed a writ petition before the Supreme Court of India, *inter alia* praying that the Union of India should be restrained from proceeding with the construction of the dam and they should be ordered to open the aforesaid sluices. The Court, finding the plea impractical did not allow it. However, on 15th November, 1994, it called for the report of the Five Member Group and the Government of India was also directed to give its response to the said report.

On 13th December, 1994, the Supreme Court directed that the Report of the Five Member Group be made public, that responses of the states be filed and that it be considered by NCA. The NCA discussed the report on 2nd January 1995. MP differed on issues of height and hydrology. Thereafter GoI and the Governments of Gujarat and MP filed separate responses on the Report before the Supreme Court.

On 24th January, 1995, the Supreme Court directed the Five Member Group to submit a detailed further report on the issues of:

(a) Height
(b) Hydrology
(c) Resettlement and Rehabilitation and environmental matters

Dr. Patil, who had headed the Five Member Group, expressed his unwillingness to continue on the ground of ill health. Hence, the remaining four members continued their work. On 17th April, 1985 they submitted a report that was not unanimous. Instead the group were equally divided. While Kulandaiswamy and Gowariker supported 75 per cent dependable flow of 27 MAF for design purpose, Iyer and Jain favoured 23 MAF. On the height issue Gowariker felt that it was impractical to revisit the tribunal decision to keep the FRL at 455 ft. The other three members, considering the magnitude of the R&R issues, opined that the idea of phased construction offered a practical solution. It did not prevent the FRL from being raised to 455 ft. in due course if the necessary conditions were satisfied; and it enabled GoMP to take stock of the position at 436 ft. and call a halt if necessary. They recommended phased construction, where at each phase advance completion of R&R should be fulfilled before proceeding to the next phase (i.e. the installation of the next tier of gates).

The members left open the possibility of further construction when the FRL 436 ft. was reached or halting at that stage. With regard to the environment it observed that this subject had been by and large covered in the first FMG report.

Supreme Court Judgment of October 2000

Lifting the stay on further raising the dam, the apex court gave a historic judgment in October 2000 and put in place a regulatory mechanism that left no ground to apprehend that the human and environmental dimensions of the project could or would be ignored or jeopardised in any manner. During these proceedings the Court, however, noted that NBA had been opposing the dam since 1986. It could not, therefore, in 1994 challenge the conditional environmental clearance given in 1987 after considerable public investment in the project. It barred the NBA's plea to reopen questions such as the dam height, hydrology, environmental studies and seismicity and held its objections to be belated and procedurally bad.

It was hoped that with this landmark judgment the Apex Court had cleared the last roadblock in the implementation of SSP. But it was not so. On November 16, 2000, the World Commission on Dams (WCD) released its final report in London (at a WB-sponsored workshop held at Gland in April 1997 a decision was taken to create such a Commission which would be independent). The WCD Report, amongst many other anti-dam findings, concluded that "large dams have failed to produce as much electricity, provide as much water, or control as much flood damage as their sponsors claim". This was no surprise, because seven out of twelve commissioners of WCD were known for their anti-large-dams stand, and it also included one woman commissioner whose plea against the SSP was disposed of by the Supreme Court. The Report incidentally (or obviously!) referred to SSP in a veiled manner.

During 2000-06, stage-wise clearances were obtained from the NCA to raise the dam height up to 121.92 m. Even during this period, several attempts were made to suspend the construction work of the dam at different levels, which led to one such decision at the level of MoWR, GoI. Tremendous public support to SSP, however, succeeded in overcoming all these obstacles, which reflected the true value of this project.

Last, but Not the Least

Borne out of the deliberations of a constitutional body, following the principles of 'Equality of Right' and 'Equitable Utilisation' of the whole course of an inter-state River, SSP has reached its present state overcoming various hurdles – legal, rehabilitation-related, environmental, financial or technological. In the history of human endeavour, this project undoubtedly enjoys a special place, not only because of its awesome dimensions and benefits but also because of the kind of support it received from the beneficiaries and the public at large. An analysis of its evolution opens up a new window to think about the developmental issues in a democratic society. Removing all doubts, it reconfirms and strengthens our belief that the Indian legal framework adequately provides safeguards for the interests of the people.

As we go into print, the 458 km long lined irrigation canal –

Narmada Main Canal has been completed along with 28 branch canals and a total distribution network of more than 19,000 km in Gujarat. The Main Canal is further extended by 74 km in Rajasthan and Narmada water has been flowing in the dry lands of Barmer and Jalore districts since March 2008 to irrigate about 60,000 hectares. In MP too, the Man (15,000 ha) and Jobat (9,848 ha) projects have been completed and other major projects like Omkareshwar (146,800 ha) and Indira Sagar (123,000 ha) are progressing rapidly. As regards Hydropower projects in MP, Indira Sagar (1000 MW), Bargi (River Bed and Canal Head), Tawa and Morand have been commissioned and work on canal head power houses of Indira Sagar and Omakareshwar are in progress.

The Sardar Sarovar Dam awaits the final environmental clearance for the installation of the gates, which would take its height to the FRL 138.68 m (455 ft), and this is on the anvil.

So, now, quiet flows the Narmada, disbursing its bounty.

Appendix I

No Cancelling Tomorrow: A Critique of the report of the Independent Review Mission[1]

The independent Review Mission's (IRM) report calls upon the World Bank to "step back" from the Sardar Sarovar Projects on the grounds that rehabilitation is "impossible", human rights, especially of the tribal population are being and must necessarily be infringed, and that on account of lack of water in the Narmada the project will any event "not perform as planned". This is not mere pessimism but outright rejection based on misperception, unsubstantial conjecture and fallacies partly related to issues far beyond its terms of reference as set out by the World Bank.

IRM Condemns Hopes of Millions

The IRM's findings, would condemn millions of potential beneficiaries, including the so-called project affected persons to continued deprivation and immiserisation just when opportunity beckons after long decades of controversy, adjudication, investigation and mobilisation to effect a positive transformation in their lives.

IRM Disregards Human Right to Development

These long-suffering tribal and other marginalized communities mires in poverty are in effect being cruelly told to cancel tomorrow. This is unacceptable and would more truly than the IRM seek to argue in reverse, be a gross denial of human rights embodied in the right to development and life to dignity. Worse, and ironically enough in terms of the origin of the IRM, the World Bank—the International Bank for, mark it, reconstruction and Development—is being told to abandon its charter. The bank has rightly rejected this gratuitous advice and pledged continued support to the Sardar Sarovar project and Narmada Valley Development generally even while striving to mitigate human, environment and other costs. This coincides exactly with the

1. Government of Gujarat (1992): Sardar Sarovar Project, No Cancelling Tomorrow: A Critique of the Report of the Independent Review Mission, Unpublished, Gandhinagar, August.

commitment of the three beneficiary states Gujarat, Maharashtra and Madhya Pradesh, and the Government of India. There remains a strong basis for continuing partnership between the World Bank and India in the great enterprise of Narmada Valley development.

Aswan Dam: UNEP Observations

Two decades and more ago, the Aswan High Dam on the Nile in Egypt was the environmentalist's bête noire as the Narmada Valley projects are today. All manner of dreadful consequences were forecast and it was said that nothing could go right with it. Twenty years after commissioning the project has been evaluated. In a foreword to "Environmentally Sound Water Management" edited by N.C. Thanh and Asif K. Biswas (1990), this is what the executive Director of the U.N. Environment Programme, Mostafa Kamal Tolba has to say: "There is no argument: the dam's contribution to substantially increasing food production and hydropower generation have made an enormous difference to the quality of life of the Egyptian people....The adverse impacts were all anticipated and plans set for addressing them, but international political situations delayed the implementation. So the overall costs of the negative impacts and their redressing skyrocketed like any other cost in the world today. What is thus necessary is a balanced environment development philosophy within which environmentally-sound water development could occur." And again: "Water development Projects...have both positive and negative impacts. What needs to be better understood are the positive impacts of sustainable water management."

UNCED Observation: Disasters from Water Shortage

At the end of the U.N. International Drinking Water supply and Sanitation Decade (1981-90), the Red Cross organization of the world have banded together to prevent looming "disasters" from water shortages in many parts of the world, South Asia, Africa, west Asia, China and parts of North America among them. In a joint preface to "water: The international Crisis" by Robin Clarke, publishes by Earthsacan (1991) Maurice Strong, Secretary-General of the Just concluded UNCED meeting in Rio, the Swedish Minister for Environment and Chairman of the Swedish Minister for Environment and the Chairman of the Swedish Red Cross speack of the danger of "water blindness" in discussing the future of growing numbers of vulnerable people. And then: We should ask how much water there is and how we can best benefit from it. Much of the world is in for a bad time. But in a period of so many opportunities it is in for a bad time. But in a period of so many opportunities it is not worthy of the international community or any group or individual to succumb or surrender to the "Malthusian thesis" with growing population struggling to survive when water could make such a difference to food supplies and better health.

Life Giving SSP cannot wait

It is in this international context that the IRM has called on the World Bank to step back and India to desist from going ahead with a life-giving project on wholly untenable grounds. The idea that the tribal conditions or the environment in the Narmada Valley will be improved, or even only salvaged, by putting Sardar Sarovar on hold pending further studies or "consultation"—with whom? about what? Other than that which is already being done-is fallacious. Like the greenhouse effect, current subsistence lifestyles to eke out mere survival is rapidly degrading the environment and eroding the tribals' natural resources base compelling more and more of them to undertake distress migration, first seasonally and then permanently, to urban work-sites, there to live preciously in noisome shanty towns. The evidences are everywhere to see. The Narmada tribals whom the IRM is so solicitous to protect through further studies and improvident delay, are fleeing the Valley—not because of the dam or submergence (which is largely in the future) but because of the scourge of hunger, diseases and lack of opportunity, with ever-growing numbers pressing on the land and vanishing forest. These helpless actors in a Malthusian drama are moving from rural to urban habitations and from hunter-gathering/agricultural to non-farm avocations.

It is against this background that the planned relocation of project-affected persons in the Narmada Valley should be viewed. To be displaced from ancestral homes, kinship grounds, familiar environments and social networks is indeed traumatic and so resettlement and rehabilitation must be infused with compassion and caring. There is no difference of option about this or about the objective of ensuring that the PAPs are better off than previously or at least as well of at the end of the day. The Narmada Valley R&R package, and Gujarat terms in particular, are very liberal and will indubitably lift all those who accept it above the poverty line. To equate transition problems of resettlement with a permanent incapacity to ensure proper rehabilitation in due course is unfair and unfounded. Notwithstanding any amount of study and consultation there are liable to be teething troubles in any large and complex human undertaking. These can be and are being overcome and it is the IRM's own findings that the R&R package and its interpretation have steadily improved. Logic would dictate reinforcing that trend rather than crying halt and throwing everything out of gear.

No Violation of Human Rights—Wrong Assumption of IRM

The IRM appears to have adopted the vocabulary of the Narmada activists opposed to the dam is alleging human rights violations and forcible eviction. It has in fact embroidered these themes in speaking of the tribal's as "Indigenous people" a common international fallacy these days—the IRM is perhaps inadvertently conjuring up vision of American Indians or Australian aborigines being overwhelmed by outsiders. Nothing can be further than the truth. The tribal Indian is no more or less an Indian than any other Indian. The use of the word "Indigenous" in the Indian context is wrong and misleading.

Tribal Culture not Properly Appreciated by IRM

The IRM has sought to disparage the resettlement of Sardar Sarovar tribals from the submergence villages in scattered clusters as inimical to their ethos and culture. It appears to find something almost sinister in some of the terminology used, such as forest-dwellers in a given context, as though violating their pristine status as isolated communities. It would also appear in some places to deny the processes of assimilation and acculturation long in progress and better understood as Sanskritisation, through it admits this elsewhere. Altogether the IRM gets locked into a powerful muddle by inventing a new history, sociology and cultural anthropology for India and from false premises, it leaps to erroneous conclusions.

ILO Convention not flouted

Considerable store is set on ILO Convention 107, adopted in 1957 and ratified by India concerning the protection and integration of indigenous and other tribal and semi-tribal populations in independent countries. This stipulates (Article 12) that "the population concerned shall not be removed without their free consent from their habitual territories except in accordance with national laws and regulations for reasons relating to national security, or in the interest of national economic development and no infringement of the ILO Convention can be made out as lands, alternative employment or cash compensation are being paid in all cases with primacy given to land. Long before the ILO Convention, India had provided constitutional and legal guarantees against tribal land alienation and institutionalized other measures to protect tribal life and culture while yet encouraging these communities to develop and modernize at their own pace.

Supreme Court of India refuses to stay SSP

That these are not empty promises is evident from the fact that a mere letter from Mr. B.D. Sharma, then commissioner for Scheduled Castes and Tribes and now a leading activist against the Dam, to the Supreme Court of India was converted into a plaint under the rubric of public interest litigation. The Court was not convinced by Mr. Sharma's plea for a stay on the further execution of the SSP project "until such time as the basic issues raised in the petition are settled and the concurrence of the tribal people is obtained to the acquisition of land and to the plan for their alternative resettlement if any based on full facts, clear understanding and free will of the concerned people". The stay was refused (August 9, 1991). The "basic issue" Mr. Sharma raised related to "the tribal people's right to life with dignity, their access to and command over natural resources in their habit and the serious dissonance between the law and the social customs of the people and also their economic system."

R&R Sub-Groups

The Supreme Court did however noted the commitments for R&R

made by the concerned authorities and the schedule of submergence notified by the project officials and required that a high level committee report to it once every month on developments and progress in this regard. That the highest court of the land should so promptly and readily take such a direct interest in this matter is sufficient guarantee that any real infringement of human rights will surely not go unnoticed. The action of the Supreme Court is unprecedented and it is receiving monthly reports as required.

Manibeli experience – Dhule District Court Refuses Stay Order

Forcible eviction has been alleged by the IRM. The reference is to Manibel, the first village to come under submergence and which happens to fall in Maharashtra. The activists have made Manibeli a test case and have resolves to drown themselves in the rising waters of the Narmada (Jal Samadhi) rather than allow relocation in Gujarat and Maharashtra where alternative land and house sites have been provided. The activists went to Dhule district court to seek an injunction against enforcement of the statutory eviction notice at least six months before likely submergence. The Court found that of the 267 PAPs represented by the plaintiffs and on whom eviction notices had been served, 153 had already accepted lands in resettlement sites in Gujarat and Maharashtra and their properties in Manibeli had accordingly been long since acquired. However under the Narmada Tribunal Award PAPs are permitted to continue occupation of their ancestral homes and lands, even while enjoying ownership of the new sites given in compensation, until six months before the likely date of submergence when they are to be served the prescribed eviction order. It is at this stage that the plaintiff petitions the court for a stay which was properly refused. The court found that these PAPs had declined to exercise their option for allocation of new lands of their choice although the project authorities were ready and able to provide irrigable lands, house sites and the scheduled civic amenities. As for the remaining 114 PAPs, the court found the plaintiff incompetent to represent them. It is the basis of such subterfuge and defiance that the stuff of alleged “forcible” eviction is made.

Smooth Transition

Efforts have been made to integrate the new settlers into the host community and smooth the transition for both. If tribal PAPs have in some cases settled in small and scattered clusters this has been of their own choosing after a close examination of alternatives. Nor have they been planted in totally alien cultural surroundings as all the settlement zones up to date lie within or at worst, at the fringe of the existing or extended socio-economic universe of the tribes concerned.

Common Tribal Belt

Gujarat is committed to resettling in its territory of every family desirous of moving to this State from submergence areas in Maharashtra and Madhya Pradesh. Some have already done so and more are likely to follow. They too will not be cultural strangers in Gujarat as the Bhil/tribal ethnographic belt extends along and around the Satpuda range and Narmada Valley and spreads over Madhya Pradesh, Maharashtra and, Gujarat and Rajasthan. And in each case there has been a history or age-old cultural inter-penetration with local non-tribal communities.

Consultation of PAPs

The IRM echoes the activists' allegation that the tribals "have not been consulted". This is untrue as there has been continuous interaction between political leaders, officials and other on behalf of the 'state' and the local communities, including the tribals, both directly as well as through the mediation of NGOs some of whom have been working in the area even before the commencement of the project. That apart the affected communities are not without local, state and national representation through elected legislators. Gujarat has 26 tribal representatives in a legislative assembly of 182 Members. Some of these are from the submergence and command area. Indeed, during the early 1980s, Mr. Amar Singh Choudhary, a tribal himself (though not from the Narmada region), was Chief Minister of Gujarat and there have always been tribal Ministers in the State Cabinet who could be relied upon zealously go protect real tribal interests. Nothing is this suggests a willful, gross, continuing disregard of human rights. The charge is grotesque.

Activists talking in the Name of Tribals

On the other hand, it is instructive to look at what is said about the activists claiming to speak for the tribals and "the people" generally by Ms. Amita Baviskar, a doctoral scholar who is quoted by the IRM in another context. The Lokayan Bulletin issue of May-August 1991 carries her articles. "The Researcher as Pilgrim" and introduces her as "a supporter of the Narmada Bachao Andolan," easily the best-known of the many activist formations.

Activists Manipulating Tribal Symbols

Amita Baviskar asks "what role does the Andolan accord to the adivasi in its theoretical and practical scheme?" Her findings: I maintain that adivasis (tribals) as well as non-adivasis from the Valley—do not initiate any programmes of Andolan action. Strategic and tactical decision is taken by act....so the structure of the Andolan is Pyramidal..." Again, "....the Andolan takes resources to the manipulation, (and we use the word advisedly) of adivasi symbols. The *shapath* – pledge not to move from the Valley – has been repeated several times in Andolan's public meetings – in the name of Narmada and *Juvar* Mata. Activists argue that these *shapaths* are indigenous and that we are making creative use of adivasi culture. But authentic *shapaths* or *mantras* occur in a rituals context and are taken by believers in the religion. In andolan programmes, on the other hand, we have instant shapaths recited over microphones shorn of their religious meaning: Such decontextualization only serves to make the programme seem gimmicky to adivasis who are asked to pledge their resolve over and over again. However, adivasis go along with such strategies so long as they believe it serves their interest in the cause of the Andolan." In the Dhule case mentioned earlier, activists went to the district country on behalf of PAPs, 114 of whom they had no authority to represent as the Court observed.

Activists Opposing "Development"

But return to Amita Bhaviskar on another issue that places the opposition to the Narmada Valley dams in perspective. "The Andolan resists the dam not only because it benefits (the) urban middle class and capitalists (both industrialists and farmers) interests but because it embodies "development", i.e. a process based on values antithetical to adivasis culture. We (have) based our belief on the Andolan's press for Jan Vikas (People's Development) by Jan Andolan (Peoples struggle). The notion of vikas—egalitarian, need-based, sustainable development—is posited as a counter to Vinashkari "development" destruction brought about by modernization and industrialization..... The dam is a target of protest because it is both a symbol and a part of "development".

Romantic Image of Tribals

Return to Amita Baviskar just once more. This time on the image evoked by the word "tribal". In many minds, the picture painted is essentially a romantic one of forest and people living in harmony with nature, practicing some sort of primitive communism. As will be seen, this image as problematic as the earlier, notorious notion of the "adivasi" as "noble savage." Further: "The romantic image of the "Tribal" does not match the reality of adivasi existence as variously experienced in the Narmada Valley....while the Andolan in Dhar and Khargone (district in Madhya Pradesh) is dominated by *Patidar* land owners, it is also supported by adivasi land-owners. Adivasi labour, like their harijan counterparts, are sympathetic to the Andolan but participate only marginally in its programmes...Agriculture in these two districts is capital-intensive, employing the full complement of HYVs, fertilizers, pesticides and irrigation—a far cry from harmony with nature."

Depleted Resource Base for Tribals

"In Jabua and Dhulia (districts), the so called adivasi areas, 'the Andolan comprises only adivasis.... And although they (the adivasis) depend on the forest and river to make ends meet, their resource use cannot be termed sustainable". Finally, "somehow the reality of (the adivasi's) depleted resource base remains unperceived. Adivasis seem to believe that the forest is inexhaustible and this faith in the regenerative powers of nature paradoxically facilitates its destruction. This perception has changed only recently...."

Ground Realities not Appreciated by IRM

Amita Baviskar is quoted at length only because she is a friend of the Andolan and scholar cited by the IRM. But she is by no mean the sole source for these perceptions. Myth making is an ancient hobby. Ground realities are different. The activists, using a romanticized image of the tribal, are not just fighting "development". These are the real terms of debate and they are free to argue their case. But this is the hidden line the IRM has bought and propagated. Is this the purpose for which the IRM was appointed—to oppose modernization, fertilizers, irrigation, industrialization, development? Is the world's Bank being called upon to "step back" while its charter is rewritten by neo-Luddites?

Gujarat's Liberal Package of R&R

The IRM has referred to the disparities between Gujarat's most advanced R&R package and the somewhat more modest terms offered by Maharashtra and Madhya Pradesh in terms of major sons, encroachers and landless in the Sardar Sarovar submergence zone. It must however be appreciated that all three submergence states have variously improved upon what was at time a path-breaking minimum package prescribed by the Narmada Tribunal in 1979. Gujarat has gone farthest in extending the two hectares irrigable land allocation to all landholders and at later stages to major sons, landless agriculturists and encroachers. Each PAP family also gets a homestead plot of 5400 square feet apart from transitional allowances, cash assistance and access to a number of civic amenities such as electricity, tubewells and standpipes, schools, dispensaries, communication facilities, etc., few, if any, of which are available in the submergence area. Fuel wood lots, fodder banks and ration shops are being organized in steadily developing programmes aided by social workers and NGOs and monitored by the project authorities. Yes, there have been some delays and mistakes, some hard cases and the usual complement of transient problems all of which have been or are being corrected. Land sites have been exchanged in certain cases. None of this suggests human rights violation.

Gujarat Setting New Trends

If PAPs from Maharashtra and Madhya Pradesh are opting for resettlement in Gujarat, it is in some cases on account of proximity and in other because of the more attractive R&R package for major sons, landless and encroachers. Maharashtra and Madhya Pradesh have not gone as far as Gujarat as they are concerned with setting a precedent for the many other large and medium water resource projects as well as other development programmes in progress or yet to be taken up entailing substantial land market required for resettlement with landowners willing to sell their lands and invest the proceeds in the secondary and tertiary sectors. Maharashtra and Madhya Pradesh are not so fortunate placed yet.

IRM's Wrong Appreciation of R&R Policies of States

Nevertheless, both States which have legislation governing rehabilitation (a state subject under the federal provisions of the constitution), have announced improvements from time to time a process that has not necessarily come to an end. At the time the Narmada Tribunal gave its Award in 1979, there was an assurance that degraded forest land would be available for R&R. But with the passage of the National Forest Conservation Act in 1980 this avenue was barred. Many difficulties have essentially stemmed from this change. However, Gujarat could find out a novel solution to purchase agriculture land from the private owners through the District Land Purchase Committee. Maharashtra, which had made some prior commitments has after considerable discussions been permitted to open part of the forest land in Dhule district for resettlement. It has also announced that encroachers regularized up to 1978 will be entitled to two hectares on resettlement, including major sons, while all other encroachers, will be given an acre (0.4 ha) each. Madhya Pradesh has also moved cautiously in the direction of liberalization. The landless too are being offered preference in employment and income generating schemes. There too earlier encroachers have been regularized. In the district of Khandwa and Khargone in the Narmada Valley alone 5000 square kms of forest land have been encroached upon (M.N. Buch, Lokayan, May-August, 1991).

Encroachment is taking place by tribals and non-tribals for cultivating and grazing with the inexorable pressure of population. Alternative avenues of employment must be found through increased labour intensity of agriculture with irrigation and more agro-processing and industrialization, which require power. The Narmada valley project are designed to provide that irrigation and energy. The IRM's theory of pristine tribal rights to all forests, thereby conjuring away altogether the notion of encroachment cannot be accepted Maharashtra and Madhya Pradesh legitimately fear that if all encroachment is regularized, this will offer carte blanche to anybody, everywhere to invade the forest, the forest conservation act notwithstanding.

However, the bottom line is that any encroacher or landless PAP and his major sons can opt for R&R in Gujarat. The IRM argues that this means shifting of M.P. and Maharashtra PAP to Gujarat. It does. But listen to the IRM on Gujarat's "scattered" R&R. Here the worry is that the tribal PAPs are free to make a choice between "one parcel of land and another. The parcel of land offered in Gujarat are proven cultivable lands and not Government waste lands. This focuses their social and cultural preferences.... Are they being forced, in effect, to sacrifice social and cultural needs as a result of an overriding economic apprehension, or indeed, through a wish to maximise economic opportunity?... The policy itself, with a central emphasis on two hectares of land, the conception of development that prevails in the institutions that design and implement policies, and the social scientific difficulties that arise when it comes to measuring non-economic variables – all these factors converge to limit the entire issue to narrow economic concerns."

It's a no-win situation with reference to R&R in Maharashtra and Madhya Pradesh the "central emphasis" to the extent of alleging cultural rape and human rights violation, is on "two hectares of land". In Gujarat, the actual allotment of two hectares of land for R&R is seen as abit and a "narrowing of the issue"! No comment is necessary.

Assessment of "Canal Affected Persons"

When it comes to the so-called canal affected persons in Gujarat, who are described as CAPs and fully deserving of PAP entitlements, "tow hectares" is again emphasized. The fact is that the median holding in the Sardar Sarovar command is 0.4 ha, though it is larger in the dry arid zones through which the canal will pass in the second phase, beyond the Mahi river, Those receiving canal irrigation are by and large beneficiaries though some many lose modest parcels of land.

There will be hard cases too and the project authorities and the Gujarat Government have promised to look into these cases and consider assistance to CAPs to enable them to purchase some land or assistance for income – generation schemes and give preferential employment.

The Sardar Sarovar Nigam has studied the canal land requirements in some detail. From the main canal to the field outlets just under 74,000 ha. Will be required (in a command area of 1.8 m ha). Just under 170,000 land holders will be affected. Of these almost 146,000 will lose less than 25 per cent of their holding and another 15,766 between 25 and 50 per cent. Only 758 will lose the entire holding and 1992 over 75 per cent of their land. Another 4997 will lose between 50 and 75 per cent of their holding. It is only the last two categories that will need something over and above cash compensation under the land acquisition act. This is being looked into. Some canal affected persons also hold lands elsewhere in the command which would also mitigate their loss.

The IRM observes "Ironically, a parcel of irrigable land would be more appropriate compensation for the canal-affected people than for the tribal villagers for the submergence areas canal communities are already accustomed to farming in the Gujarat plains, are not dependent on river and forest resources, and can be expected to adjust well and quickly to relocation to the vicinity of their existing villages." The IRM's bias in favour of tribal isolation is evident.

Conclusions on R&R

The IRM has presumptuously concluded that the kind of R&R stipulated under World Bank covenenants and international human rights norms are virtually unattainable. Even if all the Maharashtra and Madhya Pradesh PAPs were to be resettled in Gujarat, true rehabilitation would be impossible. The fact is that all bar 700 Gujarat PAPs have been resettled and are being rehabilitated and the remaining numbers should be relocated within the next 12 months.

As currently estimated of the 2464 Maharashtra PAP families 625 have so far expressed a desire to relocate in Gujarat and two third of them have already shifted to Gujarat. The other PAPs wish to remain in Maharashtra. Some 2700 ha of Taloda forest have already been released and another 1000 ha may be dereserved. This was the precise area that the Andolan had earlier agitated to secure for R&R on behalf of the tribals. The World Bank also pressed for de-reservation here. Now that this land has been allotted, it is curious that objections are being raised by the Andolan and the tribals to the very site they sought.

The IRM has raised issues about Taloda forest as an appropriate relocation site and about possible secondary displacement. It has done so on inadequate premises. If two problems are raised about every solution then one is inevitably trapped in a vicious circle. A problem-solving approach is required and any problems that may arise in Taloda are capable of reasonable solution.

The number of PAPs in Madhya Pradesh is much larger, some 23180, families in all. Of these 11,250 have indicated that they would like to be relocated in Gujarat and 818 families have already relocated in the state. The Gujarat R&R land bank now has 1600 ha reserved for further Madhya Pradesh PAPs with a further 3000 ha on offer. Together, these will take care of 2300 families or all those willing to resettle in Gujarat until the submergence year 1996-97, with something to spare.

The real problem in Madhya Pradesh will arise in 1997-98 when as many as 21,057 families will require relocate. Of these, 9137 families are willing to come to Gujarat which should have no difficulty in finding land for them by that time. The remaining 11.920 families will have to be resettled in Madhya Pradesh itself. The bulk of these PAPs will be from 193 partially submerged villages out of a total of 245 affected villages. R&R for this number pose fewer problems as only the lower portions of the villages will be submerged and that too mostly for the period july to September/October. Homestead plots can be found at higher locations in the same villages in a large number of cases. The land losses can be made good to a considerable extent by granting the original landholders rights of drawdown cultivation in the nutrient-rich lands that surface as the annual floods recede. Some additional compensation to improve their remaining lands or homes or to invest otherwise may persuade them to remain in their original surroundings.

Madhya Pradesh started later than the other in building up an R&R organisation because it is affected by submergence much later. It has sufficient time now to do so and has already begun to locate land for resettlement of PAPs within the state.

The IRM's conclusion that R&R is going to be impossible is not founded on fact but on conjecture and a jaundiced conviction that there can be no real rehabilitation in the circumstances.

The notion that there should have been a total master plan for R&R at the state exhibits impractical idealism in the case of a very large project, spread over a huge area stretching across three states. People will not be hustled into making instant decisions and any elaborate plans made so far in advance could only be notional and likely to go awry. The task is being done pari passu with the submergence schedule but within a known framework. Certainly, the task of R&R is challenging and will not be accomplished without meticulous organisation, training of personnel, surveys, timely consultation, and preparation and monitoring. The magnitude of the undertaking is not being minimised. But it can and will be accomplished with all round good will and cooperation to everybody's final satisfaction. The Kevadia and so-called dyke village problems are relatively minor (numerically) and manageable.

So much for R&R, let us turn to the environment impact assessment.

Environmental Impacts

The IRM is equally damning of the Sardar Sarovar Project on environmental grounds. But its arguments are perhaps weaker than in regard to R&R.

Hydrological Premises

For a start it questions the hydrological premises of the project and makes the totally unsubstantiated assertion that 28 million acre feet of water at 75 per cent dependability, which is the basis of the Narmada Tribunal's Award is most uncertain and that on this count alone the project design and related environmental assessments must fail. The basis of this extraordinary statement in the face of expert opinion to the contrary is not revealed.

Premises on Back Water Effects

The IRM's dissertation on another technical matter, namely backwater effects, is again without foundation. It's assertion of severe backwater effects requiring larger than planned R&R efforts is needlessly alarmist and can only aggravate misconceptions about the projects which have attracted the "hostility" it laments. No backwater effect of any significance is likely to manifest itself over the next several decades. With catchment area treatment extending into the upper Narmada Valley and the construction of the Maheshwar, Omkareshar, and Narmada Sagar dams on the main stream of the river, and other storages on tributaries above the SSP, sedimentation must in fact decline.

Assessment of Downstream Effects

The IRM is once more tilting with windmills in its despairing assessment of downstream effects on fisheries, salinity, and availability of industrial water and so on. The fact is that river flows will not be fully diverted into the main canal until 30 years after the system comes into operation. Till then the river bed power house will discharge its tail-race waters back into the stream. Moreover, the lower Narmada below Sardar Sarovar is flanked by the Mahi, Karjan and Orson irrigation Commands. About 30 percent of the water released into this area is expected to be regenerated, some part of it replenishing the river as return flows. In addition to this, the dam will continue to spill excess flows during the floods between July and September.

The least that can be said is that the kind of worst case scenario of downstream effects that the IRM implicitly portrays as imminent is in fact quite some time away. The impacts are likely to be far more modest. There is more than sufficient time for making more realistic downstream assessments in the light of actual observations and for instituting ameliorative measures as required.

Planned Fisheries Development Programme

The hilsa's migratory run could probably be adversely affected. But, as experience at Farakka (on the Bhagirathi-Hooghly arm of the Ganga) shows, the hilsa can adjust to shorter runs in course of time. Coastal fisheries adjacent to the mouth of the Narmada could similarly suffer some deleterious effects. But, as at Ukai (on the Tapti River) reservoir fishing can be very productive. The Sardar Sarovar canal will also feed innumerable tanks and fish ponds in the command which can be developed into highly profitable shrimp fisheries. The net gain to fisheries can therefore be enormous both in terms of protein and employment. What it needs is a sound fisheries development and training programme. This is planned.

World-class Technological Sophistication of SSP

Poor water management, over irrigation and faulty crop planning can lead to water logging and salinity in the absence of proper drainage. The anxieties of the IRM on this score however appear to be exaggerated as much careful planning and technological sophistication have gone into the design of this part of the project. As usual, the IRM claims to have made its own studies of the irritability characteristics of various soil types found in the command and come to categorical conclusions. It cites the poor water management record of various earlier irrigation projects in India. It finds it difficult to concede that anybody can learn from experience and is judgmental in pronouncing that none of the technological innovations of the SSP will work. So the project is damned though, sets new standards and equally damned on other counts for allegedly failing to set or conform to higher standards.

Drinking Water Supply

The drinking water programme for Saurashtra and Kuchchh is derided as unplanned and in any case more or less unimplementable. An allocation of 0.86 m ha has been made for domestic water supply and off take points are being provided. The Gujarat Water Supply and Sewerage Board is preparing a separate scheme and project report. The quantum of water allocated is estimated to take care of the needs of a growing population up to 2025. Should and this for any reason appear likely to fall short of requirements, adjustments in sectoral allocations can be made or savings affected through switching to drip irrigation for horticultural and other crops in some of the more arid zones of North Gujarat, Saurashtra or Kuchchh apart from other modes of augmentation that may be available later.

Health Impacts

The health impacts, and especially of malaria, have evoked strong comment from the IRM. There was some rise in the incidence of malarial in 1987-89. This could have been brought about by the influx of construction labour at the dam-site. There has otherwise too been a reported increase of malarial in other parts of Gujarat during this same period which witnessed relatively good rains following three preceding years of drought. The IRM has quoted the findings of Dr. N.I. Kalra which the Gujarat government disputes partly on the ground that large conclusions have been drawn from narrowly-focused data. Prevention is certainly better than cure and prophylactic measures are being intensified.

Aforestation and Catchment Area Treatment: Double Standards

The forestation and catchment area treatment programmes are proceeding apace. The IRM however is critical of the fact that Gujarat has undertaken compensatory a forestation in Kuchchh, which it describes as "marginal forest land" Instead of acquiring land in the command for a forestation as much as for R&R. This is a very curious argument. Why or how is Kuchchh marginal, and to whom? The ecological value of a forest is not necessarily diminished by its location. In Madhya Pradesh and Maharashtra on the other hand, the IRM twists the states for taking "the easy way out" by resorting to densification degraded forest lands within the Narmada Valley. Once again the IRM would like to have it both ways. The further point about the need for community participation in a forestation and catchment area treatment is well taken. Efforts are being made in this direction.

Social Forestry in Gujarat

The acknowledged success of Gujarat's aforestation programmes elsewhere in the state testifies to its ability to harness popular support in these endeavours. There is no reason to expect that the Sardar Sarovar catchment area and compensatory a forestation programmes will be any different. Usufruct rights in the catchment area are already going to local tribal communities in the form of grass. The project will green a large part of Gujarat.

Studies *pari-passu* with Construction Justified

It is noteworthy that the IRM has come to very firm, magisterial conclusions on a number of complex environmental issues in a remarkably short span of time only part of which was spent in the field. Yet is strongly denies the possibility of Indian scholars and scientists being able to come up with sufficient broad and relevant data to provide a framework for designing the Sardar Sarovar Project. It keeps harping on data inadequacies which beyond a point, if collected in minute detail and far ahead of operational use could end up as a costly exercise in futility. It has come down heavily on the Indian authorities and the World Bank for completing these studies *pari-passu* with construction. On the contrary, it is perfectly compatible with sound planning to sequence operational (not design) studies to harmonise with implementation schedules. The Bank has not erred in going along with such an arrangement on practical considerations as further studies are being completed in time to mesh with operational schedules.

Detailed basin wide studies of the Narmada Valley as a whole may not be available, but framework studies do exist and the various projects in the Valley are unfolding in accordance with an overall vision and design. As far as the Sardar Sarovar Project is concerned, Gujarat's Narmada Study Group has done a lot of good inter-disciplinary techno-economic and other studies which continue to be developed and refined. The notion that the project has got lost in a blind alley is mistaken.

Unbalanced Views Regarding Cost: Benefit

The IRM has not really stopped to examine the manifold positive benefits, direct and indirect, expected from the Sardar Sarovar project except almost dismissively, in passing. And there too, it has discounted the possibility of the project performing as intended based on its own rather negative calculus. It is convinced that Sardar Sarovar must fail whether or not the Narmada Sagar dam is constructed. These prognostications are not based on fact so much as on fancy.

Yet, the gains are going to be enormous, and costs have little meaning unless matched with benefits, it is surprising that the IRM took such and unbalanced view. It is not only in material but in human, social and ecological terms that the Sardar Sarovar Project promises a highly positive outcome. The irrigated agriculture it provides will greatly supplement the nation's larder and help combat hunger, malnutrition and poverty. The provision of drinking water will quench the thirst of 25 million people by 2025. Thousands of village tanks that run dry in the summer will be replenished with Narmada waters through the massive canal network. Together, all these will end uncertainty, distress migration and the servitude of millions of women and girl-children to a lifetime of walking to fetch water, fuel and fodder. The health impact should be dramatic. Assured water availability will mitigate gastric disorders, skin diseases and other water-related ailments which constitute the largest cause of morbidity in India.

The R&R package will lift all PAPs above the poverty line and give their children and altogether new start in life. The Sardar Sarovar Dam will prevent flooding in Bharuch town and 210 downstream villages. Catchment area treatment and afforestation will vastly improve the environment. The project is planting 40 trees for every-one lost in submergence and more if project plantations outside the catchment and command areas are taken into account, such as coastal mangroves. All this vegetation will generate oxygen and provide a considerable carbon sink – the SSP's contribution towards fighting global warming.

There is a considerable potential for fisheries development. Five wild life and bird sanctuaries are planned and the animals and birds will thrive with assured water supplies. Power generation and the provision of industrial water will further industrialisation, agro-processing and utilisation of groundwater which will be recharged over wide areas. All this will add to employment and save huge amounts on drought and food relief and on ferrying water over long distances every summer. In the three years drought of 1985-87, Gujarat spent Rs. 1500 crores in famine relief alone. The production loss to the community was equally onerous.

The potential for navigation along the Narmada is another future benefit that awaits development up to Hoshangabad in Madhya Pradesh, if not further upstream. The commissioning of the Narmadasagar complex of dams will create an extensive inland waterway. It has been well said that the Narmada Valley dams offer the country, the people of the Valley and, not least, the PAPs a great development opportunity. Not to see this and merely look at the costs is to lose the vision of the future. Carping criticism will achieve nothing.

The World Bank has so far given sanctioned to $ 450 million towards construction of the Sardar Sarovar Dam and the first phase of the canal and drainage programme. Some $ 250m has already been drawn. The Japanese government offered 28 million Yen for river-bed power house which it has since withdrawn after orders were placed on equipment suppliers on environmental grounds. Japanese missions have since visited the project and held intensive discussions. Hopefully this assistance will be soon restored. There have been exploratory talks with the World Bank for two more loans: one for the second phase of the canal and drainage works of the order of $ 450 m and the second for $ 90 m for Narmada Basin Environment Development Plan.

Invasive Usurpation of Jurisdiction by IRM Unwarranted

There are many insights and suggestions in the IRM Report that are useful and will be considered with care. Unfortunately, there are other areas where the IRM goes far beyond assessing the implementation of the Sardar Sarovar project with regard to R&R and environmental impacts. It has transgressed into areas of national sovereignty in the matter of cultural norms, federalism, national integration, the judicial process and development patterns and priorities. This invasive usurpation of jurisdiction though undoubtedly well-meant, is unwarranted and unacceptable.

Delay means Denial of Benefits to Millions

Withholding further assistance in the light of the IRM Report could delay the project and thereby give rise to some of the very negative forecasts of the critics in the manner of a self-fulfilling prophecy.

Mid-course corrections and improvements are routine. But "stepping back" would mean unrequited delay which can be deadly form of denial.

No Cancelling Tomorrow

The Narmada Valley projects should not be seen in isolation but as part of a larger project India which the World Bank and other donor nations have consistently supported with sympathy and understanding. This partnership in progress must continue. It is not without reason that the Sardar Sarovar project, and integral component of a larger Narmada Valley development programme, has been described as the lifeline of Gujarat. There is no alternative to the SSP, the design and scope of which has been determined after careful optimisation studies. It does not exclude smaller water harvesting schemes which would only complement it. Its promise of a better tomorrow is real and realisable. Let none seek to cancel that tomorrow.

Rehabilitation in Gujarat—At a Glance (Upto July-1992)

Sl. No.	*Item*	*Unit*	*Achievement*			
			Guj.	*Mah.*	*M.P.*	*Total*
1.	Total No. of PAPs		4500	2462	23180	30142
2.	Allotment of Land	(A) No. of Site	175	16	19	210
		(B) No. of PAPs	3813	424	821	5058
		(C) Area	7600	865	1642	1017
3.	Allotment of Resi. Plot	(A) No. of Site	81	4	12	97
		(B) No. of PAPs	3217	288	618	4123
4.	Subsistence Allowance	(A) No. of PAPs	3769	264	486	4519
		(B) Rs. in lakhs	139.98	9.96	12.43	161.37
5.	Rehabilitation Grant	(A) No. of PAPs	1940			1940
		(B) Rs. in lakhs	25.71			25.71
6.	Payment of Ex-Gratia amount	(A) No. of PAPs	3038			3038
		(B) Rs. in lakhs	1717.23			1717.23
7.	Productive Assets					
	(a) Relocation Sites	No.	116	4	11	131
	(b) PAP families	No.	2745	191	379	3306
	(c) Amt. of subsidy	Rs. in lakhs	1.08	0.03		1.11
8.	Insurance claimed	(A) No. of Persons	22	1		23
		(B) Rs. in lakhs	1.08	0.03		1.11
9.	Employment Provided to PAPs	No.	360		13	373

(Contd.)

Rehabilitation in Gujarat—At a Glance (Upto July-1992) (*Contd.*)

Sl. No.	*Item*	*Unit*	*Achievement*			
			Guj.	*Mah.*	*M.P.*	*Total*
10.	Civic Amenities Provide	No.				
	(a) Primary School	No.	13	1	1	15
	(b) Tin Shed	No.	3205	223	855	4283
	(c) Well	No.	24	3	6	33
	(d) Hand pump	No.	108	14	14	136
	(e) Approach Road	No.	19.4	2	1.15	22.55
	(f) Internal Road	No.	52.21	4.33	12.3	68.84
	(g) Tree Platforms	No.	61	7	12	80
	(h) Electrification	No.	37	2	4	43
	(i) Plinth	No.	100	44	40	184
	(j) Temple	No.		1		1

Note: Guj.—Gujarat, Mah.—Maharashtra, M.P.—Madhya Pradesh.

Resettlement and Rehabilitation (R&R) Activities of Sardar Sarovar Project

Position as on 31st May, 1992- Submergence Schedule (as approved by SSCAC as per Construction Schedule of Dec.89)—(due to pondage, i.e. storing of water behind Dam and backwater effect

Sl. No.	State	Unit	Submergence Year (July-June)							
			Upto 91-92	During 92-93	During 93-94	During 94-95	During 95-96	During 96-97	During 97-98	Total
(1)	Gujarat	No. of Submergence villages	5	1	4	9	0	0	0	19
		No. of Families	929	543	1405	1623	0	0	0	4500
(2)	Maharashtra	No. of Submergence villages	1	4	10	11	0	0	7	33
		No. of Families affected in these villages	0	238	411	900	763	0	152	2464@
		No. willing to resettle in Gujarat	0	238	350	37	0	0	0	625@
		No. willing to resettle in Maharashtra	0	0	61	863	763	0	152	1839@
(3)	Madhya Pradesh	No. of Submergence villages	0	0	0	1	16	13	163	193
		No. of Families affected in these villages	0	0	0	45	1432	646	21057	23180**
		No. willing to resettle in Gujarat	0	0	0	45	1428	640	9137	11250@
		No. willing to resettle in M.P.	0	0	0	0	4	6	11920	11930@
	Total	No. of Submergence villages	5	2	8	20	27	13	170	245
		No. of Families affected in these villages	929	781	1816	2568	2195	646	21209	30144

$ - Dyke village: @ - Tentative: ** For 18 villages approximate figures are taken. (Circulated in the meeting of Sub-Group on resettlement and rehabilitation held at New Delhi on 9.7.92)

Sardar Sarovar Project

The Sardar Sarovar project is an inter-state multipurpose joint venture of four states—Gujarat, Madhya Pradesh, Maharashtra and Rajasthan—with a terminal major dam on river Narmada in Gujarat. It is the largest water resources development project in India and possibly in the world. The main features of the project are:

(I) Decision of Tribunal	
(a) Allocation of water (at 75% dependable yield)	MAF
Madhya Pradesh	18.25
Gujarat	9.00
Maharashtra	0.25
Rajasthan	0.50
Total	28.00
(b) Height of Dam	FRL 455'
(c) FSL to Canal	RL 300'
(d) Power Allocation	
M.P.	57%
Maharashtra	27%
Gujarat	16%
(II) DAM	
Length	1210 m.
Max. Height above deepest foundation Level	163 m.
High Flood Level	140.21 m.
Full Reservoir Level	136.65 m.
Live storage Capacity	0.58 m. Ha.m.
Spillway Gates	7 Nos. 60' × 60' 23 Nos. 60' × 55'
(III) Main Canal and Distribution System	
Length of Main Canal up to Gujarat-Rajasthan Border	460 km.
Design discharge Capacity	
(a) In head reach	1133 cumecs 40,000 cusecs
(b) At Gujarat- Rajasthan Border	1133 cumecs 2,500 cusecs
Bed width of Main Canal in head reach	73.1 m.
Nos. of Branches	42
Length of Distribution System	75,000 km.
(IV) Estimated Cost 87/88 price level	Rs. 6406 crores
Total Expenditure incurred up to June, 1992	Rs.2000 crores

(*Contd.*)

(*Contd.*)

(V) *Submergence*

	Villages Affected			*Families Affected*	*Families to be rehabilitated including major sons*	*Total land under Submergence* (in ha.)
	Full	*Partial*	*Total*			
M.P.	0	193*	193*	7500	23180	20828
Mah.	0	33	33	1350	2464	6570
Guj.	3	16	16	1900	4500	7469
Total	3	242	245	10750	30144	34867

* In 79 villages more than 10% Agriculture land will set submerged. In 89 villages less than 10% Agriculture land or only houses will be submerged under FRL due to back water of 1 in 100 years flood. In 25 villages only Government waste will be submerged.

(VI) *Benefits*

1. Annual Irrigation

Gujarat	18 lac.ha
Rajasthan	0.75 lac.ha
Maharashtra	0.375 lac.ha
Total	19.125 lac.ha

2. Power Generation

River bed power house	1200 MW
Canal Head Power House	250 MW
Total	1450 MW

3. Drinking water supply

Gujarat	Urban Centers	Villages	
Saurashtra	90	4877	(All urban
Kachchh	10	948	Centres and villages)
North Gujarat			
Mehsana	32	1977	(All no source villages)
Banaskantha Ahmedabad	3	413	
Panchmahals District	135	8215	

Rajasthan: Large number of village in Barmer and Jallor districts will get drinking water supply benefits.

4. Flood Control: Riverine reaches measuring 30,000 ha. 210 villages including Bharuch City and 4 lakh. Population
5. Fisheries Development, Recreational Facilities, Development of Wildlife Sanctuaries, water supply to industries.
6. Secondary Benefits: Employment Generation, Agro Industries, Gains due to Compensatory, forest and tree plantation, prevention of loss of cattle and large scale migrations of communities along with cattle during drought years. Increase in groundwater potential etc.

19

The Story of the Sardar Sarovar Project: A Development Model with a Difference

Pravin Sheth

The River Narmada rises as a small stream at Amarkantak in the Maikala hills in the north-eastern corner of the state of Madhya Pradesh (MP), and flows as a long winding course before merging into the Arabian Sea near Bharuch in Gujarat. From its very source, it takes twists and turns, cascading in rapids through rocky terrain, becoming a roaring waterfall at Dhuandhar just before it flows through the narrow gorge formed by the Marble Rocks at Dhuandhar. From here on it flows through thick forests and through fertile land, occasionally squeezing through gorges or rushing down rapids. Past important places of pilgrimage such as Onkareshwar, Maheshwar and Siddheshwar in its middle reaches, to Hanfeshwar and Shoolpaneshwar, and finally becomes a serene mass of flowing water near Garudeshwar. This meandering river has become the promise of the Sardar Sarovar Project (SSP) itself. SSP has thrown up contentious issues of resettlement and rehabilitation of displaced families, environmental degradation, human rights, escalating costs, and its overall feasibility, before it could materialise as a huge dam, a sizeable reservoir and a network of canals that feed thousands of villages and towns and fields in a vast command area.

Through all the tribulations faced in its making, SSP scaled some significant highs. It is the largest manmade lake in the world. It is shaping as the largest canal network in the world and will be India's biggest water management project. The dam will also be ranked as the third largest in the world with a height of 138 metres; it will be

India's third highest concrete dam. It will be the world's second largest gravity dam, next to Grand Coulee in the USA, in terms of the mass of concrete used.

Delays in Implementation: The Political Drama

SSP became the playground of inter-state politics during the first two decades of its conception. On 1st January, 1974, Prime Minister Indira Gandhi invited Chimanbhai Patel, Chief Minister of Gujarat and P.C. Sethi, Chief Minister of MP to New Delhi where, in their presence and with their concurrence, she was to declare an award regarding the height of the dam. Chimanbhai received a last-minute call from her asking him not to come to New Delhi. The anti-Chimanbhai faction in the Gujarat Congress had persuaded her that if the height was fixed Chimanbhai would go ahead with the construction of the Dam and would become stronger – a thing not expected to be palatable to Mrs. Gandhi!

Following this retraction of invitation by Mrs. Gandhi the CM, expressing his concern over the project, told me of a possibility that the height of the dam as per the award could be less than that demanded by the Government of Gujarat, and hence Mrs. Gandhi's retraction of the invitation to Delhi. He said that he had decided to accept her decision so that the project would not be further delayed. He said that he wanted his farmers to receive water for irrigation for their fields, enabling them to reap the benefits of the project at the earliest.

He paused, and added that he would request the Prime Minister to lay its foundation stone and that he would do it himself if she was not able to preside over the event. He wished to name it as Sardar Sarovar Dam. But the torrent of actions of the Navnirman Movement in the winter of 1974 proved stronger than the Narmada waters. They were to drag away the leader with his dream. Politics was to repeatedly strangle SSP's progress.

Episodes of Domestic Resistance

The Narmada valley project was destined to have a long and chequered history of more than five decades. Its critics and naysayers raised shrill questions ranging from the issue of the futility of building big dams *per se* to those of proper resettlement and rehabilitation of the thousands of persons who would be displaced, availability of land, to ecological degradation, extinction of 'rare' species of plants and wild life, limited life of the dam due to siltation of the lake, possibility of earthquakes, the bugbear of malaria afflicting people in the command area and so on. A novel range of domestic actors entered the fray, altering the trajectory of the Narmada Projects in profound ways.

During 1980-87, Anil Patel, leader of pro-tribal NGO ARCH-Vahini, successfully fought against the Gujarat government in the High Court and the Supreme Court for the rights of the tribals and the displaced Project Affected Persons (PAPs). Responding to severe indictment of the SC and its directive to Gujarat (applicable to other state governments as well), Chief Minister Amarsinh Chaudhary came out with a liberal policy of Resettlement and Rehabilitation(R&R) in December 1987. This led to a progressive change in the profile of SSP. Sensing the Gujarat government's commitment to the modified policy, ARCH-Vahini members became increasingly active in working with villagers and the Government of Gujarat to implement the policy, while Medha Patkar and her NGO Narmada Bachao Andolan (NBA) adopted its negative agitation of 'No dam at any cost.' It raised logical as well as highly skewed arguments against SSP on R&R, environmental protection and other issues enumerated above.

In the mid-1980s, NBA and its international NGO allies for human rights raised their voices to preserve the cultural identity of tribals in the submerged villages. While meeting with the first chairman of SSNNL, Sanat Mehta, known for his sensitivity to the needs of the disadvantaged, I noticed on his table some ethnographic maps of the resettled villages. The maps showed that the displaced tribal communities had settled on plots of their preference to form culturally homogeneous neighbourhoods – Rathwas with Rathwas; Tadvis with Tadvis; Bhils with Bhils and Bhiladas with Bhiladas. In a way, the new villages/sites replicated the layouts of their now submerged settlements. Such virtual transplantation of resettlement patterns, sensitively pursued by Mehta in consultation with pro-tribal leaders like Anil Patel and Harivallabh Parikh, flew in the face of the accusations of 'ethnocide' and 'genocide' made by the human rights groups. In the early 1990s, a young lecturer of Gujarati origin from

Orlando, FL who had come to Gujarat Vidyapith with his group of American students under an exchange programme peremptorily inquired about ethnocide in the Narmada. In response to this, he was told about the R&R maps on Sanat Mehta's table. Endorsement of this came from the tribals themselves whom the group visited at the R&R sites.

The Centre was not very helpful in facilitating Gujarat's case except during the prime ministerial tenures of Rajiv Gandhi, Chandra Shekhar, Deve Gowda and A.B. Vajpayee. Much time was lost over the issue of the height of the dam. Persuaded by his Parliamentary Secretary Ahmed Patel, Rajiv Gandhi gave the much-delayed environmental clearance to SSP. Chandra Shekhar silently supported Gujarat during the Ferkuwa face-off despite pressure from some heavyweight celebrities to accede to the demands of Medha Patkar and Baba Amte.[1]

On the height issue, Gujarat had to grapple with the uncompromising and obstructive attitudes of the state governments of MP and Maharashtra and their procrastination in paying PAP resettlement costs for those who chose sites like Gutal in Gujarat.

In the late 1990s, Babubhai J. Patel, Minister for Narmada Development in the Chimanbhai Patel cabinet, appealed to Baba Amte to review his stand against the SSP. Chunibhai Vaidya, Sooryakant Parikh, and his secretary and PA, respectively, Hasmukh Patel and Prabhakar Khamar and this author went to meet Baba at Kasravad, on the Narmada near Badwani in MP. Babubhai, tried to assure Baba that rehabilitation would be done in a humane and just manner.

Both men warmly shook hands. Then Baba pointed towards the river and said, 'If the water from the dam reaches here, I will embrace death by drowning in it.' Babubhai's mission had failed. Thereafter the height of the dam was raised to 80 m, and more a little later. Baba eventually returned to his *karmabhoomi* Anandwan in Maharashtra.

Medha Patkar/NBA succeeded in creating a powerful anti-project front and managed to get support from national and international celebrities including Nobel Prize winners and resourceful international

1. Information based on personal communication with Rajiv Gandhi's Parliamentary Secretary Ahmed Patel and Deepak Babaria, then Secretary of GPCC in-charge of Environmental Cell in 1987, Ahmedabad, Gujarat.

NGOs such as EDF (USA), Friends of the Earth, and Inter-River Network, etc. Initially, with Baba Amte, she could project the image of an agitation on Gandhian lines, with her threat of *jalsamarpan* when the village Manibeli was to be submerged with the increase in dam height. But both at Ferkuwa earlier and at Manibeli, when she claimed that she would embrace death if the authorities did not meet her demands, she retracted from her extreme stance.

Again, her claim to build a mass struggle on Gandhian lines was proved hollow when she resorted to persuading her supporters in MP to threaten and prevent surveyors from entering their villages to record land ownership data. She even collaborated with anti-SSP political leaders such as V.C. Shukla, the Union Minister for Water Resources, and MP Chief Minister Digvijay Singh, both known for their apathy to Gujarat.

NBA Resistance Proves Ineffective

In December 1990, NBA launched its carefully planned a 'Sangharsh Yatra', and by inviting foreign environmentalists, aimed at making it 'perhaps the most explosive event in the history of the movement.' NBA claimed that SSP was a Government project and that the Patel government would resort to force to thwart their onward march from Badwani in MP to Gujarat. But the pro-dam NGOs of Gujarat and Chimanbhai Patel government countered it with a 'Shanti Yatra'. Farmers, tribals and a section of informed citizens peacefully countered the NBA at Ferkuwa on the Madhya Pradesh-Gujarat border. The month-long face-off was marked by massive rallies on the Gujarat side. Even before starting the 'Sangharsh Yatra,' the anti-dam marchers had to withdraw from Ferkuwa without gaining anything substantial.

Although Baba Amte and Medha Patkar had claimed that the police and the government of Gujarat would "resort to force, repression and violence," during the the entire month-long historic confrontation nothing of the sort happened. Instead the world saw the people of Gujarat silently demonstrating their commitment to safeguard the lifeline of Gujarat. Meanwhile the English-language press generally gave excessive prominence to the anti-dam views and programmes with misleading information.

Ironically, NBA twice wrote the epitaph of SSP. The first was when the World Bank withdrew funding for Gujarat's project of $170 million on 26 March, 1993. Elated by this, Patkar and friends celebrated their victory for 'drying up the SSP'. They scripted the second epitaph when the Project got smothered at the Supreme Court during its long-winding procedure between 1995 and 2000.

This perceived victory and epitaph were erased by the grit and spirit of a determined and persevering Gujarat. At the end, Gujarat could demonstrate its firm resolution to meet the powerful combination of anti-SSP forces at national and international level and the massive media propaganda against SSP. It became the symbol of Gujarat's spirit and grit to take on such forces.

Box 1: Not a Government Project

Chunibhai Vaidya and Krishnaprasad Patel, both of them aged activists, the CM's wife and NGO leader. Urmilaben Patel, Majoor Mahajan leader Hiralal Bhagwati and Suren Choksi with their fragile health, and others sat and slept on the Ferkuwa bridge for four weeks during the cold month of December. This drove home to foreign as well as Indian media that the dam was not the concern of the state government alone, as Medha Patkar and Baba Amte would have them believe, but the cherished goal of the people of Gujarat.

After the setback at Ferkuwa, in 1991 NBA tried influencing villagers of Manibeli in Maharashtra to resist from shifting. Their slogan was 'We will drown but will not move out.' However, the pro-Dam NGOs led by *khadi*-clad activists, Krishnaprasad Patel and Chunibhai Vaidya succeeded in persuading most of the villagers to vacate Manibeli peacefully with their belongings. During these momentous agitations, Patkar concentrated her energies and intellectual resources to down the Project. At Ferkuwa, she undertook a 'fast unto death' until Gujarat conceded her demands; she then converted it into an 'indefinite fast'. Still they had to retreat. At Manibeli she also issued an ultimatum that if the project was not reviewed NBA activists would resort to *jalsamarpan* on 6th August, 1993. But on that date Patkar folded up her programme, marking the denouement of a high-profile drama on a development theme. She was questioned about her claim to being a

Gandhian. Not a genuine Gandhian *satyagrahi* who is ready to die for a cause, she rather proved herself as a conventional strategist, not an activist of international stature so far as Gujarat's Narmada Project was concerned. Ironically, her moves, tactical and strategic, resulted in forcing even mutually antagonistic forces of Gujarat to join hands. This produced four unanimous resolutions in the legislative assembly supporting the cause of development – an unprecedented phenomenon in the legislative history of any Indian state.

Thus it was proved that in a country like India, when the objectives of developmental projects are genuinely in the public interest and for inclusive growth, agitations are controlled. A liberal democracy must be guarded by vigilant citizens who work for a just society against the two extremes of authoritarianism on the one hand and anarchy on the other. Gujarat has remained sensitive to the citizens' right to dissent, protest and to human rights in general. On the other hand, groups like NBA should eschew their air of self-righteousness and their attempt to force decisions on the other side of the dispute. Patkar could have avoided wearing the moral trappings of self-sacrifice like *jalsamarpan* when Gujarat had a proven commitment to implement the R&R policy and to redeem the need of millions of its water-starved people. Gujarat has, time and again, shown a strong, reasoned and popularly backed position on SSP.

Box 2: Arundhati Roy and Narmada Sangharsh Samiti

In the late 1990s, Arundhati Roy rekindled the flagging interest of critics of SSP with her 'Rally for Valley'. The tortuous delay that had paralysed the project for more than five years (May 1995-October 2000), had also led to a decline in public interest in Gujarat. It was during this that the Narmada Sangharsh Samiti led by Chunibhai Vaidya organised a padayatra from Sardar Patel's house in Karamsad to Gandhi Ashram in Ahmedabad, in order to keep the SSP going. Krishnaprasad Patel, Ramesh Shah, and I accompanied him. On the way we addressed public meetings at Vallabh Vidyanagar, Anand and Vasna. At Vasna we were joined by Siddharth C. Patel and Haren Pandya, eminent MLAs of the Congress and BJP respectively. The media did not give it as much coverage as was given to Arundhati.

Box 3: A Case of *Suppressio Veri*

At the instigation of Medha Patkar, some PAPs from the villages to be submerged filed complaints before the Grievances Reprisal Committee (GRC) constituted under the Supreme Court. It was to submit its report on whether the rehabilitation was being carried out according to the orders of the Court. Later, under the persuasion of Chunibhai Vaidya, the PAPs admitted that they had filed false complaints. Vaidya collected relevant affidavits from the PAPs and the complaints were withdrawn. During the hearing of this case, the advocate for NBA read out the GRC report, but he tried to skip the portion that stated the fact of misguiding the PAPs who had later changed their affidavits in favour of SSP. Thus, he tried to conceal these facts from the Supreme Court. Chief Justice Anand, who was also following the text, noticed the omission. He insisted that the NBA advocate read that portion before proceeding further – a rebuke of the advocate in the Court itself when proceedings were on – a case of *suppressio veri*.

The media, as usual, did not adequately report this matter of public interest. A few weeks later, the hearing was again fixed when Chunibhai Vaidya was also present. He made some quick calculations and exclaimed, right there in court, "Does it mean that Gujarat will have to bear, unnecessarily, the burden of Rs 6 crore per day?" Maganbhai Barot, Vaidya's advocate (who worked *pro bono* for the Sangharsh Samiti) was afraid that such sudden outbursts could be construed as 'improper interference' in the court's proceedings, whereby Vaidya might invite contempt of court, which it fortunately did not. However, this incident was to create a good effect on the wise and learned judges of this highest court of the land.

Both Patkar and Roy used unacceptable words against the verdict of the Supreme Court in October 2000, even when it was NBA which had moved the court against SSP in 1995, causing intolerable delay in further implementation of SSP and raising the height of the Dam, thereby leading to an enormous burden of escalated costs on the progress of the Project. The court rebuked both of them for their statements against its decision, though the media that had otherwise projected them in favourable light termed it as undemocratic.

Patkar and Roy indirectly, and ostensibly in the people's interest, challenged the status of the Supreme Court, and its judgment as being against the people, calling it anti-democratic, and further suggested starting the agitation anew.

Box 4: Contempt of Court Judgment

"The respondent (Roy) has not shown any repentance or remorse....However, showing the magnanimity of law by keeping in mind that the respondent is a woman, and hoping that better sense and wisdom shall dawn upon the respondent in the future to serve the cause of art and literature by her creative skill and imagination, we feel that the ends of justice would be met if she is sentenced to symbolic imprisonment besides paying a fine of Rs 2,000."

(Supreme Court Judgment of 6 March, 2002)

The entire chain of events was a demonstration of the opinionated temperament of the critics of SSP who claimed to conduct their movement in a Gandhian spirit. Did they want to insist on their rules of the game in a democratic system or only manipulate the constitutional institutions in their ideological mould? The years 2000-02 marked the decline in their influence and credibility and the deflation of their image.

Supporters Push the Project Ahead

When the Project was paralysed in the legal arena during 1995-2000, Chunibhai Vaidya and his Narmada Sangharsh Samiti, with legal assistance by Maganbhai Barot, tried to push the procedure in the Supreme Court. NGOs staged a *dharna* to persuade the court to expedite the case. Chief Minister Shankersinh Vaghela prevailed upon the NGOs and business groups to plead the case of SSP before Prime Minister Deve Gowda and President S.D. Sharma. As a member of the delegation, this author argued that in view of the infiltration from Pakistan across the Kachchh-Sindh border, if the project was stretched to irrigate areas in that area, together with a wall with built-in bunkers for the Border Security Force, it would help safeguarding the border.

Also, the availability of canal water would help our retired army personnel to settle in the greened, irrigated border region of Kachchh which could work as a sort of human wall to guard against the threat from Pakistan. This was an added reason for using the Dam for national security. Y.K. Alagh, Minister of Science and Technology in the Deve Gowda Government and a member of our delegation actively supported by the Chief Minister, also welcomed this point of view. While President Sharma was unhelpful and appeared rather cool to the delegation that visited him to plead for SSP, Deve Gowda gave unhesitating support to accelerate the project.

After the Supreme Court judgment in 2000, the focused efforts of Chief Minister Narendra Modi, and a proper representation before Prime Minister A.B. Vajpayee, who played a mediator's role under the judgment, succeeded in eliciting the cooperation of MP with Gujarat on the issue of environment. Meanwhile Patkar, by starting a fast in February 2006, made the issue of rehabilitation of 177 displaced persons more acute, so that the Government of Maharashtra would not give consent to Gujarat. As a result, unanimity between the states as required by the judgment could not be achieved. Narmada Control Authority (NCA) met several times to work out a unanimous solution, but the matter remained unresolved. In the mean time Narendra Modi, without getting frustrated, persevered and succeeded. He secured a historic unanimity among four states on 16 March, 2006.

When Patkar was exerting pressure by going on fast, Union Minister for Water Resources Saifuddin Soz adopted a ploy to further delay in the implementation of SSP. Chief Minister Modi went on hunger strike to protest against the decision of Soz to suspend increasing the height of the dam from 110 to 121.92 m. This led to public opinion rally around Modi's demand and to build pressure on the Centre to relent.

Actually, SSP has benefited because of agile and forward-pushing political leaders who could overcome the powerful combined forces that created mammoth roadblocks and the obstructive tactics of highly focused anti-Dam activists like Medha Patkar.

The Narmada Project was initiated in 1979 during the regime of Babubhai J. Patel and its initial infrastructure was put in place by his Irrigation Minister Keshubhai Patel. When Gujarat lost the case against the pro-tribal NGO ARCH-Vahini led by Anil Patel, in the highest court on the issue of fair rehabilitation of the PAPs, Chief Minister Amarsinh Chaudhary offered them a progressive package of giving 5 acres of land with assured water supply. Thus the government won the cooperation of Anil Patel and other NGOs to make it an ever improving development project. It was revived by Chief Minister Chimanbhai Patel (1983-1993) when NBA made fierce attempts to block it. A forceful front of NGOs and public spirited intellectuals countered NBA with logical counterarguments; confronted NBA at Ferkuwa and Manibeli; joined delegations to meet Prime Ministers and the media in New Delhi and Mumbai; and representing the case of SSP before the World Bank Team and other groups who visited Gandhinagar to look into the controversy and the arguments from both sides of the divide. World Bank observed in its Project Completion Report:

> On the implementation side, Gujarat developed a unique mechanism for acquiring replacement for agricultural land, at market price through a Land Purchase Committee (with one representative of the potential PAPs). Well developed R&R units with central monitoring cells were established. Specific lessons on means and procedures required for satisfactory R&R implementation were drawn.
>
> (World Bank Project Completion Report (Sardar Sarovar), 1995).

Babubhai J. Patel, as Minister for Narmada Development, paid frequent visits to Kevadia Colony to expedite the building of the Dam to 80 ft in accordance with the permission of the Supreme Court. In spite of severe backache, he would endure the six-hour journey from Gandhinagar to Kevadia Colony. As in the case of Amarsinh Chaudhari and SSNL chairman Sanat Mehta (1987-1989), there was remarkable rapport between Modi and Babubhai Patel. Similar was the coordination between them and C.C. Patel, SSNL chairman. CM Keshubhai Patel and SSNL's agile chairman Jaynarayan Vyas managed to decide on and construct a bypass tunnel that helped Gujarat circumvent the obstructive policy of MP. This helped Gujarat in utilising Narmada waters even when the dam had not reached the desired height.

As the first chairman of SSNL, Sanat Mehta, having experienced different hues of public life, understood the importance of NGOs and succeeded in harnessing the cooperation of a few pro-tribal, pro-justice NGOs like ARCH-Vahini, which was working for health and in the interest of tribal communities, and Harivallabh Parikh whose Anand Niketan, located in the interior of the tribal area, was working for the uplift of Adivasis. Sanat Mehta set up his own NGO, Shramik Vikas Sansthan and collaborated with others like, Sampoorna Punarvasavat Samiti, Narmada Abhiyan and Gujarat Lok Samiti, etc. in the early 1990s. Chimanbhai Patel and Jaynarayan Vyas were perceptive enough to appreciate the moral and political backing that such NGOs could offer and which the concerned governments lacked when it came to issues of the environment and the interests of the deprived. Sanat Mehta and the NGOs stood strong against NBA and its allies and negated Patkar's propaganda to label SSP as the project of the Gujarat government. C.C. Patel encouraged preparation of professional reports at a time when NBA and its foreign allies had flooded the battleground with massive anti-Narmada literature bearing names of celebrities and environmental groups. SSP served as a vibrant symbol of the cooperation between the Government and people – an unusual phenomenon in India's political development. Urmilaben Patel, Vidyut Joshi, Sudarshan Iyengar, Ramesh Shah, and other activists representing different NGOs, met with or protested before the highest authorities at the Centre and the states against policymaking delays. Leaders of the business community like Gujarat Chamber of Commerce and Industry, Khedut Samaj and the Swaminarayan sect also played a vital role.

So also, a delegation of independent intellectuals and NGOs comprising Y.K. Alagh, C.C. Patel, Urmilaben Patel, Sharad Sabnis, Vidyut Joshi and I attended the Global Forum, UNCED at Rio de Janeiro in 1992 among others to articulate the case of SSP with a focus on its good performance on issues of environment and human rights. Chunibhai Vaidya and Narmada Sangharsh Samiti, with limited resources, managed to prepare a documented case and approached the Supreme Court and exposed the pleadings of NBA, almost single-handedly in late 1990s. Along with them Sanat Mehta, Anil Patel and Jaynarayan Vyas played notable roles in clearing many misconceptions skilfully spread by NBA and its allies with effective links and reach to the media.

Jaynarayan Vyas worked at the international level to erase misconception about SSP by presenting the plight of rural women from drought-prone areas like the North Gujarat areas followed by interface between the experts in International Water Resources Conferences at Hague, Amsterdam and Tokyo. P.K. Laheri, chairman of SSNNL under Modi, meticulously handled the matter of speedily raising the dam height to 110 m. Reaching this height helped SSNL to start generating hydropower for the first time that soothed the attitude of MP.

Soon thereafter, the clearance of the Supreme Court raised the height to 121.92 m with active support of CM Modi, and the construction of the Dam up to 121.92 metres was completed by December 2006 in record time. They also tried to facilitate the MP government in the latter's huge task of R&R in that state. MP, however, is found wanting in pushing R&R as scheduled and to the satisfaction of the Supreme Court.

Multiple Roles of Narendra Modi

The Chief Minister (CM) Narendra Modi has shown grit and sagacity in dealing with the one-sided attitude of the Centre in the case of the Narmada Project. He obtained the permission of the NCA for raising the height of the Dam from 95 metres to 100 metres and then further to 110 meters. Ready to provide technical backing to this decision, Laheri, with a dynamic team of engineers from the Nigam, made the under-bed power turbine operational, making energy available to a sceptical MP. Modi's success in obtaining NCA permission, conforming to the norms set by the Supreme Court and taking the height to 110 metres bespeaks an extraordinary feat of leadership. He also arranged funds for hesitant neighbour states to help them speed up R&R in their area. All along, he vigilantly focused on calibrating such an arrangement with perseverance. He used his skill in executing this with the support of Prime Minister A.B. Vajpayee.

Modi could effectively delink an unwritten alliance between MP and Patkar, a task which, during the decade of 1990s, had seemed impossible to achieve. The year 2006 saw him playing a spirited role in persuading the Centre to resume its move on the project. He galvanised the concerned machinery monitored by P.K. Laheri to counter the obstructionist anti-SSP tactics of the Patkar-led combine supported by a section of the media and some ministers at the Centre in 2006. He put Saifuddin Soz on the defensive by exposing his role in the case and observed a fast in Ahmedabad that eclipsed the fast undertaken by Patkar and others in New Delhi, thus proving that it was not the monopoly of Patkar-Baba-NBA to play a moralistic role! With this he managed to force the Centre to concede his demands.

Soon after the Dam reached the height that allowed SSNL to let Narmada waters flow to Kachchh, Modi asked Laheri to arrange to convey the waters to Bhuj by a pipeline network which was already in place (18 May, 2003). Modi himself was present in Kachchh, the farthest point the Narmada water had reached, where the local women awaited its arrival. A huge crowd danced and celebrated this occasion. Medha Patkar's oft-repeated caution that the people of Kachchh would never receive Narmada waters was given a convincing rebuff. Modi's audio-visual demonstrated the lies of NBA and rekindled the hope of people in Saurashtra that the Project would one day prove to be a boon for them as well.

Meanwhile, the Gujarat Government completed the conditions for further raising the dam height to 121.92 m in 2009, and is waiting for the green signal to move on. It was only during Modi's monitoring of SSP that the dam was raised by a dramatic 26 m.

Empowerment and R&R Reform

Vinod Babbar, IAS, Managing Director of the R&R Agency, proved a competent administrator sensitive to the demands of the displaced tribals in the crucial transitional phase of SSP during 1989-1996. The late Shobha Asthana, IAS, who succeeded him at R&R Agency, also worked in the same spirit to attend to the problems of PAPs and succeeded in neutralising the studied efforts of anti-SSP lobby to stir discontent among the PAPs. Bhagyesh Jha, who was working during this crucial period under Sanat Mehta and Chimanbhai Patel, helped the NGOs by providing useful information regarding plots for rehabilitation and pending problems. It was an unusual constellation of some CMs, SSNNL chairmen, NGOs and bureaucrats, supported by the people of Gujarat, which enabled Gujarat to present a common front that countered the powerful combine of anti-SSP forces, considered almost invincible on issues of development. C.C. Patel, a retired technocrat and former Chairman, SSNL, was sensitive to complaints of the tribal PAPs regarding their rehabilitation problems. He even visited in the monsoon the houses built for them on new sites. He observed that the black soil there became sticky in the rains, obstructing entry to the houses built for the Adivasis. The disturbed chairman, on his return to Gandhinagar, related this experience with a touch of sadness and vowed to ensure that such plots were not allotted to the PAPs. SSNL always tried to follow up on complaints regarding mistakes in allotment of land. Following C.C. Patel's instructions, the Nigam would take quick steps to improve the use of such plots with tractors or by offering alternative plots to the complainants.

"Meanwhile, the Gujarat government had fully implemented further phase of R&R programme by rehabilitating all its project-affected families (PAFs) to claim for raising the Dam height to 138.68 metres. But due to slow progress of this programme in the two neighbouring states of MP and Maharashtra, Gujarat has not been able to raise the height further for nearly two years" (*Indian Express*, 2009).

Widespread doubts about proper use of Rs 200 crore given by GoG to MP for purchasing land to resettle agriculturist PAFs in that state affecting some 2,000 PAFs have resulted in the MP High Court-appointed committee to inquire into the alleged scandal. It is another matter, that the delay in further raising the dam height causes both the states to lose electricity. "While MP loses Rs. 800 crore annually due to inadequate power being generated by river bed and canal head units, Maharashtra stands to lose Rs. 550 crore" (*Indian Express*, 2009).

All along, MP has proved either an adversary or a very weak partner in fighting the forces working against the Narmada projects – SSP and those being planned and implemented in MP by its governments. The Chief Minister of MP, P.C. Sethi raised objections against Gujarat's plan to put the Dam at higher level and it was supported by Maharashtra. This led to mediation by Prime Minister Indira Gandhi who avoided granting the award and that caused a decade of delay before the matter was entrusted to the Khosla Commission. The Narmada Development Tribunal Award finally came in 1979. After concurrence of all the respective states, Chief Minister

Babubhai Patel and his Irrigation Minister Keshubhai Patel worked hard to make up for the lost years. But the project again was entangled in the controversy around environmental viability.

Thereafter Medha Patkar and NBA entered, raising questions about availability of land for PAPs in Gujarat. When the land became available, the question about its quality was raised. But good land, too, began to be available. Some near Dabhoi was so fertile and close to the reservoir that PAPs scrambled to get it during 1990-91. After initial hesitation and doubts created by NBA, the villagers of Vadgam, Gadher, Malu, Parveta, etc. began to agree to leave their old villages or habitat and resettle in the new villages with better infrastructure. NBA worked up discontent among them regarding the initial problems that such rehabilitation generally entails in its early phase. SSNNL, R&R Agency and the NGO Sankat Nivaran Samiti tried to sort out such genuine complaints with guidance from ARCH-Vahini and Anand Niketan. Even PAPs from MP began to come to Gujarat and satisfactorily settled in the new village Gutal.

In the 1990s MP Chief Minister Digvijay Singh and NBA were seen to be in a working arrangement to halt the progress of SSP on one excuse or another. Singh even joined her, through a separate but virtually supporting application to the Supreme Court in 1995 and succeeded to checkmate the advancement of SSP. Meanwhile the Union minister of Water Resources V.C. Shukla played his delaying tactics by appointing a Five Member Group to look into Patkar's complaints, and "continue the review discussions in June 1993 on all issues related to Sardar Sarovar Project,"[2] and similar ploys.

When Medha Patkar started her fast in Mumbai, V.C. Shukla, then Union Minister for Water Resources in Narasimha Rao's Government at the Centre, assuaged her and constituted a review committee of experts to look into the issue of the height of the dam and displaced persons. The Government of Gujarat was not consulted in the matter. Thereafter, when a meeting was held in presence of Prime Minister Narasimha Rao, V.C. Shukla gave a set of papers regarding a proposed understanding with Patkar to Gujarat Chief Minister Chimanbhai Patel and told him, "Go through it."

2. Ministry of Water Resources, Office Memorandum, No. 6/4/93-pp. 5 August, 1993.

Chimanbhai promptly pushed it back to Shukla in the PM's presence, and shot back, "It is your problem. Gujarat is not interested in such a deal!"

MP was found to be weak in implementation of the R&R programme even in the Independent Review by the Morse Committee. The review report stated, "While the development of resettlement policy in Gujarat was indeed an achievement, R&R policy in the other two states – Maharashtra and Madhya Pradesh – needed urgent improvement, and implementation needed to be further strengthened in all the three states." It also advised to make "efforts to replicate Gujarat's R&R policy and standards in the other two states" (Morse & Berger, 1992). The Morse Committee was critical of the progress of SSP and working of the states, and advised WB to 'step back'— from Gujarat's ambitious Project.

Another issue that attracted significant attention is the quality of the R&R package offered to oustees with regard to resettlement sites in MP and Maharashtra. For instance, "Dharampuri had been shown to the GoM (Group of Ministers) as a success story by the Madhya Pradesh Government and it turned out to be the worst example of doing anything by way of settlement when there was apparently no difficulty in respect of resources. The people there showed the GoM two dry water pumps and a heap of stones to indicate a day before its arrival that roads would be built soon" (*The Hindu*, 2006). Earlier, the Daud Committee that surveyed rehabilitation sites in Maharashtra found that, "excepting the settlement village of Simamli in Gujarat, which offers a little satisfaction, rest of the resettlement villages from Maharashtra in particular, visited by us, lack almost all the basic facilities required for habitation, especially quality and availability of suitable agricultural land" (Daud Committee, 2001). Similarly, the Shunglu Committee appointed by Prime Minister Manmohan Singh observed, "The quality of land available in the Land Bank was by and large average; it was not irrigable and cultivable.... "The offer of government land (in MP) in place of land acquired did not meet the requirements of PAFs and was the most significant reason for their opting for SRP" (Shunglu Committee, 2006).

After field research, Neeta Deshpande comments, "In the Adivasi villages on the banks of the reservoir, they live dangerous lives surviving extreme, unthinkable odds. In the most fertile plains of

Nimad in Madhya Pradesh, they fear the certainty of an impoverished future. And for those who are actually resettled, the outcome is mired with insurmountable difficulties: uncultivable land, tin shacks to live in, flooding in the monsoons, brackish water, poorer diets, fragmented communities, and broken promises" (Deshpande, 2009).

The performance of Gujarat needs to be evaluated in this over all perspective also. It does not imply, however, that Gujarat should not continue to expand the scope of improvements on this front. By 2009-end, SSP has evolved, in an escalating process of reforming and integrating some valid points of criticism given by NBA and its allies. It has also adopted some useful recommendations of WB and such agencies. A few findings of the recent Evaluation and Monitoring Report 21 of the Centre for Social Studies, Surat are relevant here:

"In so far as access to basic amenities and services is concerned, it can be stated that the PAFs are by and large better off compared to the situation of their submerging villages." (Sah, 1994). The same E&M report states, "The services provided by the state in the fields of education and health care are highly appreciated by the PAFs. Crop diversification and market links are developing and yields have substantially improved compared to submerging villages. The proportion of population engaged as casual or agricultural labour has reduced" (Sah,1994c). Similarly, its Report 24 finds that "at current prices the total annual expenditure of the households has increased about two and half times when compared to expenditure level of submerging villages. The improved purchasing power of household after relocation is undisputed even at constant prices" (Sah, 1996d).

The Task Ahead

The project is still evolving. In the process it needs to integrate some of the critical inputs in its ongoing reform framework. For example, the plight of those who are displaced from Gora, the original village where Nehru first laid the foundation stone of Narmada Dam still nurse some complaints that need looking into. So also, those who have lost their fields because of the canals, have neither land to irrigate nor a decent job to earn their livelihood. Even if such CAPs (canal-affected-people) may not have legal rights for compensation, their situation needs to be taken care of. "In the shadow of the Narmada Dam, those displaced by the canal are demanding justice" (Deshpande, 2009).

Many areas located far from the Main Canal, where villagers have been promised Narmada water, are still waiting to quench their thirst and irrigate their fields. The canals for this are not yet constructed and there are no links to the points where they have been mapped and promised. In the absence of infrastructure to take Narmada waters to such needy villages, Narmada water is diverted to the River Sabarmati and it helps embellish the Sabarmati Riverfront Project. Such a scenario provides a chance to highlight the contrast of feeding or beautifying an urban riverfront project while the people in distant rural areas still keenly wait for irrigation and drinking water. Pradeep Mallik who has studied the rehabilitation process of SSP, says, "Nearly 50 years later and Rs. 29,000 crore down the dam on the Narmada and the canal across the state, not even one-third of the network has been completed. According to a quick calculation, work worth nearly Rs. 22,000 crore remains to be completed at the current price level. If the state government spends Rs. 2,000 crore on the project every year, it will take more than 12 years to wrap it up" (Mallik, 2010). The SSNNL authorities were quick to respond to Mallik's comments. A senior officer assured him that the points he had raised would be addressed. Such is the agile people-oriented response process under its new chairman, D. Rajgopalan.

- The anomalies are now being removed by planning an extensive network of pipelines to carry Narmada waters to such villages. Required funding for the completion of the canal network is also being planned with the support of the Centre. Firm steps are being taken to acquire land for extending/constructing new canals. Legal hurdles are being removed for acquiring such land. Sufficient funding is also assured for compensation to farmers for such acquisition.
- The practice of drawing water illegally from small canals and feeding lines using improvised devices are claimed to have recently ceased. It is alleged that sugar-cane factories sprang up in the area around the Main Canal up to the River Mahi. Such sugar-cane growers should be made to pay extra

tax for using the canal water which is meant for providing drinking water.

- Farmers in the command area should be educated for prudent use of the canal water as it reaches them, by developing model farming, green patches, free water lifting, workshops for adopting drip irrigation etc. Difficulties faced by small farmers in availing of water due to water tariff or taxation should be resolved well in time. Facilities for easy loans/ subsidies need be arranged for the farmers who cannot afford to buy material for drip irrigation.
- Projects to develop drip irrigation and model greenery parks in one tribal area, one arid area and one each in North Gujarat and Saurashtra, will serve as demonstration centres for those in the command area who are hesitant to adopt irrigation-centred farming or plantations. Agro-clinics should be started to train them about water use in different areas from wetlands to waste lands.
- A normative policy to prioritise the use of water for distant fields or dry zones waiting for canal water should be implemented.
- At present, there are no water users' associations or farmers' cooperatives in the command area. There is a need for NGOs, related agency/board and government to collectively work out a framework to offer an innovative model of democratic water management.

Social and Economic Development Benefits

The project has started augmenting the revenue of the state. With increased irrigation provided by the SSP, a significant increase in agricultural production has been recorded, such as rice, cotton and sugar-cane. Multiple cropping instead of only one crop during the monsoon and diversified crops have helped the farmers increase their agricultural income. A survey done by Nanak Bhatt has shown that farmers from submerged villages who used to earn Rs. 8,000-10,000 are now earning Rs. 76,000-86,000 and more after resettlement in new villages. A plot of 5 acres of good land at a stretch, assured water supply, training by NGOs with help from the government's agricultural extension service agency in cultivating diverse crops like cotton, wheat etc. instead of only rough grains like maize and *jowar* as in the hilly terrain of the old villages have transformed the farmers' lives.

The anti-project forces like NBA have aggravated the problems of SSNNL and GOG by continually hindering the completion of SSP with their successful delaying tactics. This has escalated the cost of construction, land availability and rehabilitation of increasing numbers of PAPs over the original estimate, as more members became adults, claiming additional chunk of 5 acre plot and other amenities. Liberalising the definition of who is an oustee and the type of R&R package has resulted in added financial and other burdens on the riparian states. From 6,147 families (NWDT, 1979), the number of the displaced increased to 13,335 families (DoEF, mid-1980s), to 40,245 families (FMG, mid-1990s) to 40,827 (SC, 2000). This number, by 2003, was a whopping 100,000 families (Scudder), and was increasing. Only Gujarat – its people, NGOS, public-spirited intellectual-activists and SSNNL and some dynamic CMs with resourceful leadership and political will have been able to face such insurmountable and ever new hurdles have made SSP a reality in its present form.

Overall per capita income level is going up. Pre-project expenditure on electricity to be used for irrigation was estimated to go down by 33 per cent with SSP. However, with water level in wells and tube-wells going lower and lower – sometimes 213 to 305 m deep, the cost using power to draw water from tubewells had become prohibitive. This is one of the reasons why farmers resorted to power theft and the government suffered from power loss and reduction in revenue from tariff. It also caused an environment of political conflict like farmers' agitations – particularly, since the 1980s. With cheaper and controlled power supply in the wake of SSP, such conditions will ease and a healthy rural scene is expected to emerge. Moreover, the annual public expenditure of about Rs. 2 billion to supply water by tankers will be saved. Panchayats will also receive power for drinking water without paying for it.

Many additional sources of income are estimated as multiplier effects. Schools, health centres, infrastructure like *pucca* roads transportation services to modern living will expose the once-hill/forest dwellers to the modern world. All this will generate more income-earning activities like jobs in transport services, auto-repairs, construction works related to roads and houses etc. Migration to urban areas due to paucity of rural livelihoods will decrease. This will help reduce the burden on urban infrastructure. Rural bazaars, service sector and supplementary income opportunities will also be encouraged. All this will help the hitherto forlorn or neglected communities to come closer to the mainstream without losing their identity or ethnic culture.

The pressures of daily chores like fetching water from a distance or for collecting firewood by going to a distant *vagada* or forest area will ease and they will be spared the tedious daily routine that saps their energy and wastes their time, especially women. All such trends, together and cumulatively, will lead to a better tomorrow.

SSP actually contributes to the nation's economic development. Widely misunderstood and not adequately supported by the Centre and national media, the project would contribute to food security and energy self-sufficiency of the country.

In the late 1980s, international environmental activists had almost succeeded in their propaganda that Sardar Sarovar Dam was the 'world's most ecologically disastrous dam.' Rather, whatever adverse ecological effects were envisaged, they were well taken care of by SSNNL, NCA and the relevant environmental agencies. Another widely prevailing misconception was the fear spread about "ethnocide/ethno-genocide" or the loss of the "pristine tribal culture and their identity." This myth is also exploded by pro-Dam NGOs, SSNNL and WB. A widespread myth is that the tribal people to be resettled from the submergence area are living in pristine forests in a traditional manner as hunters in harmony with environment. This is not correct. As satellite imagery shows and field observation confirms, the tribals live in mostly barren, stony, steep and increasingly degraded hillsides with a small and decreasing part of their income derived from the forest land. Many of them welcome the opportunity to improve their lot in the more fertile and more sustainable command area and while any move has considerable social costs for a family, many have quite widespread tribal connections in resettlement areas in Gujarat which is expected to help their adjustment...with the least opposition to the Dam.

In fact the progress scenario needs to be visualised in a larger perspective. Known for their pragmatism, farmers have started using the canal waters for better farming with improved agricultural techniques. Farmers in Chansma and Kalala have started using drip irrigation in 100 hectares which maximises the benefits of using the canal waters that have reached their fields. Sixty farmers in Nayka have also shifted to drip irrigation and cultivate superior variety of potatoes. Farmers near Sanand are waiting to do the same for growing tomatoes. Farmers near Kadi are growing thin skinned lemons that yield more income. Krushipandit Jivraj Farm (agro-forest) at Surendrabaug ('the first Eucalyptus District of the world') near Bhavnagar has already attracted national attention for successful implementation of drip irrigation and rich, productive intensive planting of trees like Subabul and Bekum Neems. During the celebration of Umiya Mata Mahotsava (a festival in honour of the revered Goddess of Patidar clans, which, in 2009, was held at November-end)), at Unjha, 100,000 Patel farmers took an oath that they would shift to drip irrigation. "This is welcome news when you consider that only four districts in the state have received normal rainfall this year," says Anand Agricultural University vice chancellor MC Varshneya. "Apart from saving water, drip irrigation improves yield and lowers labour cost."[3]

"More than two-third of the state's cultivable land is semi-arid or arid. This year (2009) we have limited water to cover the 18 lakh hectares under the Sardar Sarovar Project Command Area. The answer to the problem is drip irrigation. "I hope more farmers are inspired by Unjha," says R.B. Malaviya, Executive Director of Agriculture, SSNNL. Such success stories abound in Gujarat and will be replicated elsewhere in the Narmada command. The same week, Narendra Modi, at a ceremony for releasing Narmada canal waters at Dholi Dhaja Dam to quench the thirst of people of the arid area of Surendranagar, Vadhvan, Joravarnagar, said, "*Pani* is more precious than *Parasmani* (Philosopher's stone)." The CM exhorted them to adopt drip irrigation that will save precious water, and to use water with prudence.

Gujarat has brought into practice the enlightened views emphasised in international forums or by thinkers. World Commission on Water, 2000 has proposed for a 'universal acknowledgement that

3. From a note furnished by functionaries of Sardar Sarovar Narmada Nigam Limited.

basic supply of water to allow a healthy lifestyle is a fundamental human right.' Thereafter, the Hague Declaration (March 2000) also recognised that access to safe and sufficient water and sanitation are basic human needs and are essential to health and well-being, and to empower people, especially women, through a participatory process of water management.

Viewed as a whole, SSP may turn out to be the best of its kind in India. The democratic process in India has enabled the grassroots activists from both sides of the great divide to provide the maximum corrective inputs in the ever 'revised' project plan. Critics of the project, through commendable homework and national and international networking of environmental NGOs have pressurised the Government of Gujarat and SSNNL to respond and correct the flaws or lacunae in the implementation of R&R. Though resorting to hyped arguments and rhetoric, Medha Patkar, respected for her commitment to tribals' rights and well-being, has sustained her movement by her perseverance. But the pro-project NGOs and intellectuals have matched her arguments on strength of data and overall performance of SSP and successfully checkmated her and NBA's moves to stop the construction of the dam. Gujarat Government and SSNL, persuaded by pro-project NGOs, have tried to respond to some well-founded critical suggestions and promoted the process of renewal of the Project where necessary or desirable in the interest of the oustees and environment.

Millions of villagers displaced during the last six decades in India have had to migrate to cities and in the process, they stay either in the wilderness in temporary hutments or in ghettos in the cities. The rights of indigenous people also include their right to develop and join the mainstream. A Study by Vyas (2001) also shows that the tribal population in fact welcomes an opportunity to join the mainstream of development.

In a way, when environmentalists and other anti-Dam forces seem to be losing their ground they should take satisfaction in having gained. They have generated public consciousness and official recognition that will prevent the proverbial 'going back on their promises' kind of behaviour in future development/infrastructure-related projects anywhere in the country. A benign rehabilitation policy was nationally debated and is being put in place for future projects.

The *Jan Sunavai* – public hearing and feedback before a new project is planned – has been now accepted by governments at all levels.

Involving potential PAPs through their representatives in identifying their new land (out of three alternative sites) was a progressive step of SSNNL, if one recalls how such a consultative process was missed in case of previous dam projects such as Bhakra, Hirakud, Ukai and Nagarjunasagar. PAPs representing or articulating their demands for just resettlement for more than two decades in Gujarat, Maharashtra and MP have served to correct these past lapses, and to instil confidence in the aggrieved poor to effectively use democratic processes.

A novel approach of the authorities in Gujarat to encourage formations of the local population – water users' cooperatives or farmers' associations to look after proper distribution of canal waters – is also a case of inducting components of people's participation in the project operation in its end-process. If the idea of villagers' cooperatives to manage their water-use and crop-planning in the command area takes concrete shape, even in the long run, it will be Gujarat's novel contribution in cultivating a model of decentralized democratic development.

The distribution of good cultivable land, with assured irrigation and a package of inputs and infrastructure of social development are making the living patterns of PAPs in the new villages a well planned operation. Thinking at the level of the Nigam's planners and executives (C.C. Patel, M.B. Joshi) to devise mechanisms for equitable distribution of water and its fair management by the users' groups or NGOs is another component of distributive development – though still in the process. So also, increased income – multi-fold in many cases – and rising of standard of living of the PAFs after generations are examples of distributive development.

"SSP has the potential to be the lifeline of the naturally under-privileged people and provide them with their right to development. While the issue of R&R is an important consideration, equally important is the acute deprivation of a far larger number of some 20 million people in command area who need earliest regional transfer of Narmada waters" (Gupta, 2001: 125-147).

One subtle aspect of this development dynamics, as noticed by perceptive activists, is the growth of awareness among tribals – so far mute and helpless – about their rights, increase in their capacity to struggle against a powerful State and develop confidence to confront the problems which the PAPS-turned-settlers have faced, or have to face on certain points of in course of becoming mainstream citizens. Witness, for instance, the PAPs of Vadgam, Gadher or Manibeli of late 1984-85, in Ferukadi, Golagamdi, Ambawadi or Simalia in the late 1980s who have 'politically' transformed into vocal and persistent settlers during 1990-95. It is not a behavioural metamorphosis but slow learning curves of democracy that pro- and anti-project groups have helped them to negotiate and to transform themselves in the process. With the support of pro- and anti-project NGOs, Supreme Court, Grievances Redressal Committees and responsive government/Nigam, the resettled PAPs/PAFs have demonstrated confidence in resisting what they felt were unjust elements in the resettlement process. This is an immeasurable aspect of participative development.

The most critical question is: Which alternative is preferable? Planned displacement of a few thousand people which is carried out in a democratic way, with social security, sustainable living with standard income-generating package and infrastructural facilities, or the forced migration of millions of people and cattle under the pressures of water-scarce conditions – turning into a phenomenon of eco-refugees? The question comes down to: What do tribals in particular actually prefer? Equitable distribution of water by regional transfer with local participation, or the dreaded scenario of water riots that even the World Bank has projected for not a very far-off future? Gujarat has shown how macro-politics can be effectively countered by microeconomics – with the determination of people/ NGOs, government, Nigam and leadership at chief minister level.

Abhayanvayi to *Ubhayanvayi*

Almost every aspect of this unique project has been fully debated with feedbacks and response at policy and operational levels, checked and rechecked. At every level, it has been politicised – State, Centre, NGOs, courts, Investigation/monitoring committees, NCA and the like. SSP seeks to dam the India's fifth largest river – and, the only unpolluted major river – called *abhayanvayi* (one that liberates from fear); and truly so, for the once-mute tribals living along its long, winding riparian hills and forests have now become fearless. *Abhayanvayi* has made the poor fearless.

It should now become '*Ubhayanvayi*' – one that bridges the chasms between and bonds the different – even polar-sides of political and ideological spectrum, the pro- and anti-project forces. The politics of contention and conflict should be replaced by politics of dialogue and convergence. Confrontation should give place to consensus and focus on developing a new 'centre' in place of misplaced polarities. At another level, Narmada Project has served as '*ubhayanvayi*' to join Gujarat with its Diaspora. SSP, even in its still evolving form and progressive contents, has carved a unique place in the heart of Gujarat, and in the history of India's development. *Swarnim* Gujarat has all the reasons to celebrate this model of eco-development – a development with a difference.

References

Acharya, S., S.P. Punalekar, N.K. Gourah and N.V.V. Char, (1992): *Report of Fact-finding Team Appointed by Ministry of Water Resources on Manibeli,* Government of India. New Delhi.

Ahluwalia, M.S. (2000): "Economic Performance of States in Post-Reforms Period," *Economic and Political Weekly*, May 6-12.

Ahmedabad Municipal Corporation (AMC) (2010): *Information on Reforms Achieved for Consideration under The Awards for Excellence In Municipal Reforms and Service Delivery Improvements*, Report Submitted to Ministry of Urban Development (MoUD), Government of India.

Alagh, Y.K. (2010): "A Sardar Sarovar riddle", *The Financial Express*, Ahmedabad Edition, April 19.

Alagh, Y.K., Mahesh Pathak, and D.T. Buch (1995): *Narmada and Environment—An Assessment.* Har Nand Publications. Delhi.

Alagh, Y.K., R.D. Desai, G.S. Guha, and S.P. Kashyap (1995): *Economic Dimensions of the Sardar Sarovar Project.* Har Anand Publications. SPIESR. Ahmedabad.

Alexander, K.C., R.R. Prasad and M.P. Jahagirdar (1991): *Tribals, Rehabilitation and Development,* Rawat Publications. Jaipur.

Allen, R.G., L.S. Willardson and H. Frederiksen (1998): "Water Use Definitions and Their Use for Assessing the Impacts of Water Conservation" in J.M. de Jager; L.P.Vermes and R. Rageb (eds.), Proceedings ICID Workshop on *Sustainable Irrigation in Areas of Water Scarcity and Drought*, Oxford, England. September, 11-12.

Alvares, C. and R. Billorey (1988): *Damming the Narmada*. Natraj Publishers. Dehra Dun.

Alvares, Claude and Ramesh Billorey (1987): "Damming the Narmada: The Politics Behind the Destruction," *The Ecologist*, Vol. 17, No. 2.

— (1988): *Damming the Narmada: India's Greatest Planned Environmental Disaster,* Natraj Publishers. Dehradun.

— (1988): *Damming the Narmada,* Third World Network. Penang, Malaysia.

Amarasinghe, Upali *et al*. (2005): *Spatial Variation in Water Supply and Demand across River Basins of India,* Research Report; 83, International Water Management Institute, Colombo.

ARCH-Vahini (1988a): *Displacement in Sardar Sarovar (Narmada) Project: A Gujarat Experience*, Mangrol, Rajapipla. April.

— (1988b): *Sardar Sarovar Oustees – Which Way to Go: An Activist's Dilemma*, Mangrol, Rajpipla. September.

— (1991): "Resettlement and Rehabilitation in SSP: Are the Critics Right?" (*Mimeo*).

— (1992a): *Is R&R Impossible in SSP: The Role of NBA (India), EDF (USA) and MOEF (India): A Submission to Morse Mission*. Mangrol, Rajpipla. February.

— (1992b): *A Critique of Memorandum Submitted to the World Bank by the EDF and Other NGOs* on *March 20, 1991*, Mangrol, Rajpipla. September.

— (1992c): *Letter to Mr. Lewis Preston, President, World Bank on Morse Report,* Mangrol, Rajapipla.

— (1992d): *Letter to David Jones, Associate Director, Oxfam U.K. on Morse Report*, Mangrol, Rajpipla. September.

— (1992e): *The Independent Review: A Search for Truth?* Mangrol, Rajpipla. November.

— (1993a): *Morse-Berger Review of Sardar Sarovar Project: A Counterview*, Mangrol, Rajpipla. January.

— (1993b): *The SSP Withdrawn from the World Bank: ARCH-Vahini's Views*, Mangrol, Rajpipla. April.

— (1993c): *The Sardar Sarovar Project: An Intellectual Fashion*, Mangrol, Rajpipla. July.

— (1993d): *Human Rights in the Sardar Sarovar Project: A Submission to Fact Finding Committee of PUCL*, Mangrol, Rajpipla. October.

— (1994): *A Note of PAPs in M.P. who are Entitled for Agricultural Land for Research,* Mangrol, Rajpipla. March.

— (1995): *R&R in SSP: Are the Critics Right?* in Pathak, Mahesh (ed.) *Development Imperatives and Environmental Concerns*. Agro Economic Research Centre. Vallabh Vidyanagar.

ASCE (1987): *Planning, Operation, Rehabilitation and Automation of Irrigation Water Delivery System,* Proceeding of ASCE Symposium, Portland, Oregon.

— (1993): "Special Issue of Canal System Modelling", *Journal of Irrigation and Drainage,* No. 119 (4).

Barot, J.M. (1996): *Occurrence of Salinity in Gujurat in Reaching the Unreached: Challenges for the 21st Century*, 22nd WEDC Conference, New Delhi, India.

Baviskar, Amita (1991): "The Researcher as a Pilgrim," *Lokayan Bulletin*, Volume 9 (3-4). March.

Berg, Elliot and Mary M. Shirley (1987): *Divestiture in Developing Countries,* World Bank Discussion Paper No. 11, Washington, DC.

Bhaskaracharya Institute for Space Applications and Geo-informatics (2009): *Geographical Information System for Sardar Sarovar Project (SSP) Command Area Phase I: Project Proposal*, Gandhinagar.

Bhatia, Bela (1992): "Lush Fields and Parched Throats – Political Economy of Groundwater in Gujarat", *Economic and Political Weekly*, December 19-26.

— (1997): "Forced Eviction in Narmada Valley" in Jean Dreze; M. Samson and S. Singh (eds.) *The Dam and the Nation,* Oxford University Press. Delhi.

Border Environment Cooperation Commission (2003): "Hidalgo County Irrigation District

Brewer, J., S. Kolavalli; A. H. Kalro; G. Naik; S. Ramnarayan; K.V. Raju; and R. Sakthivadivel (1999): *Irrigation Management Transfer in India: Policies, Processes, and Performance,* A collaborative study by the Indian Institute of Management, Ahmedabad, India and the International Irrigation Management Institute, Colombo, Sri Lanka. Oxford & IBH Publishing Co. Pvt. Ltd, New Delhi.

Bribiesca, Sanchez (1967): *9th International Commission on Large Dams.* Turkey.

Bruns, Bryan Randolph and Ruth Meinzen-Dick (2000): *Negotiating Water Rights,* Sage, New Delhi.

Bureau of Indian Standards (1986): *Criteria for Earthquake-Resistant Design of Structures* in *Indian Standard IS 1893: 1984.* New Delhi.

CAG Report (2000 and 2001): Comptroller and Auditor General Report 2002, Government of India.

Census of India (2001): *Housing and Amenities Table, Gujarat.*

Central Board of Irrigation and Power (1993): *Canal Automation*, Publication No. 238, New Delhi.

Central inland Capture Fisheries Research Institute (1993): *Studies on Fish Conservation in Narmada Sagar, Sardar Sarovar and it's Downstream*, Report submitted to Narmada Control Authority (NCA), Indore, Government of India.

Central Soil and Water Conservation Research and Training Institute (n.d.): Interim Report on *Rainfall Runoff and Sediment Yield Measurement in Catchment Area of Sardar Sarovar Project lying within Gujarat.* Vasad.

CEPT - CRISIL (2006): *Public Private Partnership: A Study of Potential Benefits in the Sardar Sarovar Narmada Project* (Unpublished), Submitted to Sardar Sarovar Narmada Nigam Limited, Gandhinagar.

Cernea, Michael (1995): *Keynote Address: Understanding and Preventing Impoverishment from Displacement* in *International Conference on Development Induced Impoverishment.* University of Oxford. Oxford.

— (1991): "Involving Settlement: Social Research, Policy and Planning" in Michael M. Cernea (ed.) *Putting People First: Sociological Variables in Rural Development*, Oxford University Press, Second Edition. New York.

— (1996): "Public Policy Responses to Development-induced Population Displacements," *Economic and Political Weekly*, Vol. 31 (24).

Chakravorty, Ujjayant and Chieko Umetsu (2003): "Basinwide Water Management: A Spatial Model", *Journal of Environmental Economics and Management*, 45 (2003): 1-23.

Chambers, Robert (1988): *Managing Canal Irrigation: Practical Analysis from South Asia,* Oxford and IBH, New Delhi, India.

Chandra, A. and C. Sudhir (2010): *A Study of Kolhapur Lift Irrigation Co-operatives*, MTS Report, Institute of Rural Management, Anand.

Choudhury, N. and V. Kher (2006): "Public-Private Partnership in Surface Water Irrigation: A Case of Kolhapur", IWMI Tata Water Policy Program, Anand (unpublished research).

CICFRI (n.d.): *Studies on fish Conservation in Narmada Sagar, Sardar Sarovar and its Downstream.* Barackpore.

CMIE (2010): *India Harvest Database*, Centre for Monitoring Indian Economy.

Commissioner of Fisheries, Govt. of Maharashtra's letter No. Fish/T/011403/ 2168 to Secretary, Narmada Ghati Development Department, Government of MP.

Commissioner of Health, Medical Services and Education (1993): *Environmental Impact Assessment Studies on water Related Diseases in SSP command Area Including the Area Downstream of the Narmada Dam.* Government of Gujarat. Gandhinagar.

Cook, Paul and Colin Kirkpatrick (eds.) (1988): *Privatization in Less Developed Countries,* St.Martin's Press, New York.

Cooter, Robert D. (1997): "The Rule of State Law and the Rule-of-law State: The Economic Analysis of the Legal Foundations of Development" in Micheal Bruno and Boris Pleskovic (eds.) *Annual World Bank Conference on Development Economics 1996*, World Bank, Washington, D.C.

Cowan, L. Gray (1990): *Privatization in the Developing World,* Greenwood Press, New York.

Cullet, Philippe (2007): *The Sardar Sarovar Dam Project: Selected Documents*, Ashgate Publishing Company. Hampshire. U.K.

Dalua, A.K. (1993): *Environmental Impact of Large Reservoir Projects on Human Settlement*, Ashish Publishing House. New Delhi.

Dam Safety Panel (1997): *Report of the Thirty Fourth Meeting of Dam Safety Panel, Sardar Sarovar (Narmada) Project,.* Sardar Sarovar Narmada Nigam Limited. *Gujarat.*

DANIDA (1996): *Community Participation in Rural Drinking Water and Sanitation Projects: A Comparative Study of Five Projects in India*, RWS & Sanitation Group, SA, UNDP-WB & IIMA.

Das, Biswaroop (1983): "Studies on Rehabilitation of Submerging Villages – Vadgam" *(Mimeo)*. Centre for Social Studies. Surat.

— (1986): "Submerging Tribal Belt in Gujarat: Spatial Variation and Stratification," *State and Society*. 6(1).

— and A.S. Charan (1983): "Cost of Submergence," *(Mimeo)*. Centre for Social Studies. Surat.

Datta, Samar K. (ed) (2003): *Cooperatives in Agriculture – A Millennium Study* (Volume 24), Academic Foundation, Delhi.

— *et al* (forthcoming): *Developing India's Strategies Response to the Global Debate over Fisheries Subsidies*, Chapter 4, Allied Publishers, Delhi.

Daud Committe (2001): *Report of the Committee to Assist the Resettlement and Rehabilitation of the Sardar Sarovar Project-Affected Persons*, Government of Maharashtra.

Dave Kapil (2010): "Half of Gujarat will live in cities". *Daily News and Analysis*, June, 2.

Davies, Susanna and Naomi Hossain (1997): "*Livelihood Adaptation, Public Action and Civil Society,*" *Working Paper 57*, IDS, Brighton.

De Fraiture, Charlotte and Chris Perry (2002): "Why is irrigation water demand inelastic at low price ranges?" Paper presented at the conference on *Irrigation Water Policies: Micro and Macro Considerations*, Agadir, Morocco, June, 15-17.

Department of Botany (1983): *The Sardar Sarovar Narmada Project Studies on Ecology and Environment.* M.S. University. Baroda.

— (1992): *Eco Environmental and Wildlife Management Studies on SSP Submergence Area in Gujarat.* M.S. University. Baroda.

Department of Earthquake Engineering (1982): *Design Earthquake Parameters for Navagam Dam Site in Gujarat.* University of Roorkee.

Department of Environment and Forests (1986): *Environmental Aspects of Narmada Sagar and Sardar Sarovar Multi-purpose Project*, URL: http://www.ielrc.org/content/c8601.pdf. December.

Desai Vaidya, Prerana (2010): "Impact Of Sardar Sarovar Project: A Field Evaluation." *Seminar on Evolution and Growth of Sardar Sarovar, the Lifeline of Gujarat,* CEPT University, Ahmedabad.

Desai, G.M., Gurdev Singh and D.C. Sah (1979): "Impact of Scarcity on Farm Economy and Significance of Relief Operations," *(Mimeo)*. Indian Institute of Management. Ahmedabad.

Desai, N.B. (1990): "*Influence of Sedimentation on Service Life of Sardar Sarovar,*" *Indian Journal and Power and River Valley Development, Special Number on Sardar Sarovar Narmada Project,* September (pp. 249-253)

Desai, S.J. and M.B. Joshi (2008): *Narmada Water Plays Its Role – Capturing Initial Trends from Gujarat,* Presentation, Sardar Sarovar Narmada Nigam Ltd., Gandhinagar.

Deshpande, N. (2009): *Displaced by the Dam: Living on the Edge,* URL (http://www.indiatogether.org/2009/oct/hrt-sardar.htm), October.

Dharmadhikary, S. (2005): *Unravelling Bhakhra: Assessing the Temple of Resurgent India,* Manthan Adhyayan Kendra, Bhopal.

Dhawan, B.D. (ed.) (1990): *Big Dams: Claims, Counter Claims,* Commonwealth Publishers, New Delhi.

Dholakia, Ravindra H. (1985): *Regional Disparity in Economic Growth in India.* Himalaya Publishing House. Bombay.

—(1994): *Spatial Dimensions of Acceleration in Economic Growth in India* in Economic and Political Weekly. Volume 29, Issue 35 (pp. 2303-09)

—(2010): *Has Agriculture in Gujarat Shifted to High Growth Path?* in Dholakia, Ravindra H. and Samar K. Datta (eds.): *High Growth Trajectory and Structural Changes in Gujarat Agriculture,* Macmillan Publishers.

—and Samar K. Datta (eds.) (2010): *High Growth Trajectory and Structural Changes in Gujarat Agriculture,* Macmillan Publishers.

—and Ajay N. Oza (1999): *Microeconomics for Management Students*, 2nd Edition. Oxford University Press. Delhi.

Dixon J.A., L.M. Talbot and G.J.M. Le Moigne (1989): *Dams and the Environment: Considerations in World Bank Projects*, World Bank Technical Paper Number 110, Washington D.C.

Domanski, Tomasz (1992): "Development of Small Private Companies and Their Marketing Activities," in *Journal of Business Research,* Vol. 24, pp. 57-65.

Falkenmark, Malin and Carl Widstrand (1992): "Population and Water Resources: A Delicate Balance." *Population Bulletin,* Population Reference Bureau: Washington, DC.

FAO (2008): *Report of the Lake Victoria Fisheries Organization and FAO - National Stakeholders' Workshops on Fishing Effort and Capacity on Lake Victoria (2006),* FAO Fisheries and Aquaculture Report No. 817, Food and Agriculture Organization of the United Nations, Rome.

Fernandez, A.P. (1990): "No Dam is an Island" in A P Fernandez (ed.) *Workshop on Rehabilitation of Persons Displaced by Development Projects: A Report,* Institute for Social and Economic Change and MYRADA. Bangalore.

Final Order and Decision of the Narmada Water Dispute Tribunal (NWDT) Gazetted on December 12, 1979, Source: nca.gov.in/forms_pdf/nwdt_finalorder.pdf

Fisher, William F. (2001): "Diverting the Water: Revisiting the Sardar Sarovar Project", *International Journal of Water Resources Development*, 17 (3): 303-314.

Fitch, Lyle C. (1988): "The Rocky Road to Privatization," in *American Journal of Economics and Sociology*, Vol.47, No. 1, pp. 1-14.

Frederick, K D (1992): *Balancing Water Demand with Supplies: The Role of Management in a World of Increasing Scarcity*, Technical Paper 189, Washington DC: World Bank.

GoI (1987): *National Water Policy: 1987,* Government of India, Ministry of Water Resources. New Delhi.

GoI (2002): *National Water Policy: 2002*, Government of India, Ministry of Water Resources, New Delhi.

Goldsmith, Edward, and Nicholas Hildyard (1984): *The Social and Environmental Effects of Large Dams,* Sierra Club Books. San Francisco.

GOPA Consultants (1991): *Narmada River Basin Development Project* – Report submitted to SSSNL, Government of Gujarat. http://www.nvda.nic.in/env.htm

Government of Alberta (2003): "Alberta's irrigation infrastructure," http://www1.agric.gov.ab.ca/$department/deptdocs.nsf/all/irr7197/$FILE/irrigationinalta-part2.pdf (accessed on 11th November, 2009).

Government of Gujarat (1987): *Compilation of Government Resolutions for Rehabilitation and Resettlement Policy for Oustees of Sardar Sarovar Project*, Narmada Development Department. Gandhinagar.

— (1992): *No Cancelling Tomorrow: A Critique of the Report of the Independence Mission of Gujarat*, Gandhinagar.

— (1992): *Comment on the Report of the Independent Review Mission on SSP.* Gandhinagar.

— (1997): *Water Supply & Sanitation, 9th Five-Year Plan (1997-2002)*, Narmada Water Resource & Water Supply Department, Government of Gujarat, Gandhinagar.

— (1998): *Infrastructure 2010: Master Plan on Drinking Water Supply*, Narmada Water Resources and Water Supply Department, Gandhinagar.

— (2000): *Drinking Water Scarcity 2000-2001*, Narmada Water Resources & Water Supply Department, Gandhinagar.

— (2002): *10th Five-Year Plan (2002-2007): Water Supply & Sanitation,* Narmada Water Resource & Water Supply Department, Government of Gujarat, Gandhinagar.

— (2007): *Gujarat Water Users' Participatory Irrigation Management Act*, Government of Gujarat, Gandhinagar.

— (2009): *First Approximation report to increase groundwater resources through artificial recharge in Gujarat State*, Narmada Water Resources, Water Supply and Kalpsar Department, Gandhinagar.

— (2010): *Report of the Expert Group for Strategy for the Accelerated Development of Sardar Sarovar Project Command Area,* Sardar Sarovar Narmada Nigam Ltd., Gandhinagar (unpublished report).

Government of India (1978): *The Report of the Narmada Water Dispute Tribunal with its Decision*, Vols. I and II. New Delhi.

— (1990): *Approach to Eighth Five Year Plan*, Planning Commission. New Delhi.

— (1992a): *Eighth Five Year Plan (1992-97), Vol. II: Sectoral Programmes of Development*, Planning Commission. New Delhi.

— (1992b): *Report of the Five Member Group Set up by the Ministry of Water Resources to Discuss Various Issues Relating to the Sardar Sarovar Project*, Vols. I and II. New Delhi.

— (1987): *National Water Policy, 1987*, Government of India, Ministry of Water Resources.

— (2002): *National Water Policy, 2002, Government of India, Ministry of Water Resources.*

— (1995): *Report of the Committee on Private Sector Participation in Major and Medium Irrigation Projects*, Ministry of Water Resources, New Delhi, India.

— (2002): *National Water Policy,* Ministry of Water Resources, New Delhi, India.

— (2005): *Report on 3rd Census of Minor Irrigation Schemes*, Ministry of Water Resources, Minor Irrigation Division, New Delhi.

Government of Madhya Pradesh (1992): *Action Plan of R&R of Oustees of MP (SSP), Parts I and II,* Narmada Valley Development Authority.

Guggenheim, Scott (1990): "Development and Dynamics of Displacement" in A.P. Fernandez (ed.) *Workshop on Rehabilitation of Persons Displaced by Development Projects: A Report,* Institute for Social and Economic Change and MYRADA. Bangalore.

Gujarat Infrastructure Development Board (2009): *BIG 2020 – Blueprint for Infrastructure in Gujarat*, Gandhinagar.

Gujarat PUCL (1992): Human Rights Issues at Manibeli: Fictions and Facts *(Mimeo),* Report by Gujarat PUCL on behalf of Human Rights Group of Narmada Abhiyan.

— (1993), "Drowning of Manibeli," *PUCL Bulletin*, Vol. 12(7), Report by M. Mohanty, C.C. Kalara on behalf of Human Rights Campaign on Manibeli.

Gulati, Ashok; Tushaar Shah and Ganga Shreedhar (2009): *Agriculture Performance in Gujarat since 2000: Can it be a* Divadandi *(Lighthouse) for other States?* International Water Management Institute, Anand; and International Food Policy Research Institute, New Delhi.

Gupta, Rajiv K (2001): "Human Rights Dimension of Regional Water Transfer: Experience of the Sardar Sarovar Project," *International Journal of Water Resources Development*, Vol. 17 (1).

GWRDC (n.d.): *Ground Water Resource Estimation.* Retrieved from http://www.gwrdc.gujarat.gov.in/Gwremaps/GWRE_B.K..htm

GWSSB (1983): *Water Demand for Non-agriculture Use from Narmada Project.* Gandhinagar.

GWSSB (1985-2005): *Master Plans*, Gujarat Water Supply and Sewerage Board, Gandhinagar.

— (2003): *Sardar Sarovar Canal Based Drinking Water Supply Master Plan*, Gujarat Water Supply & Sewerage Board, Gandhinagar.

Hague Deceleration (2000): "Ministerial Declaration of The Hague on Water Security in the 21st Century," *Second World Water Forum*, Hague. March.

Hanke, S.H. (ed.) (1987): *Privatization and Development*, Institute for Contemporary Studies Press, San Francisco.

Haskoning (1999): *Netherlands Support to the Water Sector in India: Impacts on Institutional and Development and Change*, Haskoning, Netherlands.

Hazare, Anna (1997): *Ralegaon Siddhi: A Veritable Transformation translated by Pendse*, S.B. from original Marathi book, published by Ralegaon Siddhi Pariwar.

Heath, John (1989): "Privatization: Modalities and Strategy," in V.V. Ramanadham (ed.) *Privatization in Developing Countries*, Routledge, London, pp. 118-124.

Hirway, I (2005): "Ensuring Drinking Water to All: A Study in Gujarat." *4th IWMI-TATA Annual Partners Research Meet*, February 24-26.

— (1999), *Integrating Environmental & Economic Accounting: The Case of Gujarat – Phase 1*, Gujarat Ecology Commission, Vadodara, Prepared by Centre for Development Alternatives, Ahmedabad.

— and Shital Lodhia (2004): *Status of Drinking Water in Gujarat: Towards a Sustainable Approach*, Unpublished Report submitted to World Health Organisation, New Delhi.

— and P.P. Patel (1994): *Dynamics of Drinking Water in Rural Gujarat*, Action Aid India, Bangalore.

— and Subhrangsu Goswami (2008): "Functioning of the Drinking Water Component of the Narmada Pipeline Project in Gujarat," Special Article, *Economic and Political Weekly*, March 2008.

— *et al.* (2002): *Quick Valuation of Depletion and Degradation of Environmental Resources in Gujarat*, sponsored by Gujarat Ecology Commission, Vadodara, Prepared by Centre For Development Alternatives, Ahmedabad.

— and Subhrangsu Goswami (2005): "Performance of Narmada Pipeline Project in Gujarat", *Research Highlights No. 10*, International Water Management Institute, Anand.

— (2006 a): *Concurrent Monitoring of Worlds Largest Pipeline Drinking Water Project: Study of Narmada Pipeline Project in Gujarat*, Unpublished Draft Report submitted to International Water Management Institute, Anand and PRAVAH, Anand.

— (2006 b): *Status of Drinking Water in Coastal Saurashtra: Need for Area Specific Approach*, Unpublished Draft Report submitted to Coastal Salinity Cell, Rajkot.

— (2008): "Functioning of the Drinking Water Component of the Narmada Pipeline Project in Gujarat", Special Article, *Economic and Political Weekly*, March.

Hooja, Rakesh; Ganesh Pangare; and K.V. Raju (2002): *Users in Water Management: The Andhra Model and Its Replicability in India,* Rawat Publications. New Delhi, India.

Howe, Charles W, Dennis R. Schurmeier and W. Douglas Shaw, Jr. (1986) "Innovative Approaches to Water Allocation: The Potential for Water Markets," *Water Resources Research*, 22 (4), pp. 439-445, April.

Independent Review Mission (1991): *Independent Review, Sardar Sarovar.* World Bank.

Indian Express (2009): "Government orders removal of diesel pumps installed along Narmada canal.", Ahmedabad edition, June 20.

Indu, Rajnarayan and Alka Rawal (2007): "Impact of High Incidence of Kidney Stone: A Study of Coastal Villages in Junagadh, Gujarat", Paper presented in workshop on *Crisis in Drinking Water in Coastal India*, EDI, Ahmedabad.

Institute of Rural Management, Anand (IRMA) (2001): *White Paper on Water in Gujarat*, report prepared for Government of Gujarat, IRMA and UNICEF, Gandhinagar.

International Council of Scientific Unions (1972): *Man Made Lakes as Modified Ecosystem.* Scope Report 2. Paris.

International Water Power and Dam Construction (1988): *Reservior and the Environment—Experience in Managment and Monitoring.* Volume 40, Issue 9.

IPCC (Intergovernmental Panel on Climate Change) (2001): *Impacts, Adaptation, and Vulnerability*. Cambridge, England, Cambridge University Press.

IRMA – Institute of Rural Management (2001): *White Paper on Water in Gujarat*, prepared for the Department of Narmada, Water Resources and Water Supply, Government of Gujarat, Gandhinagar.

Iyengar, Sudarshan (1997): *Resettlement and Rehabilitation in SSP: A Review of Policy and Implementation.* GIDR. Ahmedabad.

—, and Vinayak Dave (1998): *Status of Project-affected Families from Madhya Pradesh Resettled in Gujarat.* Reconnaissance Report. GIDR. Ahmedabad.

Iyengar, Sudarshan (1998): *Gujarat* in V.M. Rao (ed.) *The Poor in a Hostile Society: Glimpses of Changing Poverty Scenario in India*, Vikas Publishing House Pvt. Ltd., New Delhi.

Jayaratne, R.S. (1990): "Involuntary Resettlement Consequent to the Construction of Three Large Dams and Reservoirs under the Mahaweli Development Project, Sri Lanka" in A.P. Fernandez (ed.) *Workshop on Rehabilitation of Persons Displaced by Development Projects: A Report*, Institute for Social and Economic Change and MYRADA. Bangalore.

Jodha, N.S. (1981): "Role of Credit in Farmers' Adjustment against Risk in Arid and Semiarid Tropical Areas of India," *Economic and Political Weekly*, Vol. 26 (42-43).

Johnson, III Sam H. (1997): "Irrigation Management Transfer: Decentrlizing Public Irrigation in Mexico," in *Water International*, 22(3), September, 159-67 and in Maria Saleth (Eds), 2002; *Water Resource and Economic Development*, Edward Elgar Publishing Ltd., UK.

Joshi, Archana (2006): *Groundwater Conservation: A Perspective Plan for Ahmedabad Urban Complex Area, Master of Planning* Dissertation Report (unpublished), School of Planning, CEPT University, Ahmedabad.

Joshi, M.B. (2001): "Operation of Sardar Sarovar Conveyance System", *International Journal of Water Resources Development*, Vol. 17, 1 March.

— and Kapadia, Vivek (2010): "Sharing Water in the 21st Century: Rethinking the Rationale", *Irrigation and Drainage*, 59: 92-101.

Joshi, Vidyut (1987): *Submerging Villages – Problems and Prospects*, Ajanta Publications (India). Delhi.

— (1991): *Rehabilitation – A Promise to Keep (A Case Study of SSP)*, The Tax Publications. Ahmedabad.

— (1991): *Rehabilitation at Sardar Sarovar Narmada Project* in Pathak, Mahesh T. (ed.): *Sardar Sarovar Project: A Promise for Plenty*. Oxford and IBH Publishing. New Delhi.

Kalpavriksh (1995): "Cry Our Beloved Narmada," *The India Magazine*, New Delhi. February.

Kay, Melvyn; Tom Franks and Laurence Smith (1997): *Water: Economics, Management and Demand.* London: E&FN Spon, London.

Khosla, A.N. (1964): Report submitted to the Central Water Commission. New Delhi.

Kothari, A. and S. Singh (1988): *The Narmada Valley Project: A Critique*, Kalpavriksh. New Delhi.

Kothari, Smitu (1996): "Whose Nation? Displaced as Victims of Development," *Economic and Political Weekly*, 31 (24).

Krishiworld.com http://www.krishiworld.com/html/water_resources2. html (accessed on 28th April, 2010)

Krishna Jai, Chaturvedr R.S., Srivastava L.S. (1976): *Geological and Seismological Investigations of the Environs of Narmada Valley around Navagam Dam Site in Gujarat State.*

—(1989): "Dams and Seismicity.".

Krishna Kumar (1996): "Narmada Debate: A Test of Modernity," *Indian Express*, July, 11.

Kumar, M. Dinesh (2002): Reconciling Water Use and Environment: Water Resource Management in Gujarat, Resource, Problems, Issues, Strategies and Framework for Action, *Report prepared for State Environmental Action Project* supported by the World Bank. Gujarat Ecology Commission.

— (2005): "Impact of Electricity Prices and Volumetric Water Allocation on Energy and Groundwater Demand Management: Analysis from Western India", *Energy Policy*, 33 (1).

— (2007): *Groundwater Management in India: Physical, Institutional and Policy Alternatives*, New Delhi: Sage Publications.

— (2009): *Water Management in India: What Works, What Doesn't*, New Delhi: Gyan Books.

— (2010): *Managing Water in River Basins: Hydrology, Economics, and Institutions*, New Delhi: Oxford University Press.

— and Jos van Dam (2009): "Improving Water Productivity in Agriculture in India: Beyond 'More Crop per Drop'", in Kumar, M. Dinesh and Upali Amarasinghe (Eds) Water Productivity Improvements in Indian Agriculture: Potentials, Constraints and Prospects, *Strategic Analysis of National River Linking Project of India Series 4*. Colombo: International Water Management Institute.

— and O.P. Singh (2001): "Market Instruments for Demand Management in the Face of Scarcity and Overuse of Water in Gujarat," *Water Policy*, 3 (5).

— (2005): "Virtual Water in Global Food and Water Policy Making: Is There a Need for Rethinking?" *Water Resources Management*, 19: 759-789.

— and Upali Amarasinghe (Eds) (2009): "Water Productivity Improvements in Indian Agriculture: Potentials, Constraints and Prospects", *Strategic Analysis of National River Linking Project of India Series 4*. Colombo: International Water Management Institute, Colombo.

— A. Narayanamoorthy, O.P. Singh, M.V.K. Sivamohan, Manoj Sharma and Nitin Bassi (2010), "Gujarat's Agricultural Growth Story: Exploding Some Myths", *Occasional Paper 2*, Institute for Resource Analysis and Policy, Hyderabad.

— Ajaya Kumar Malla and Sushanta Tripathy (2008): "Economic Value of Water in Agriculture: Comparative Analysis of a Water-Scarce and a Water-Rich Region in India", *Water International*, 33 (2): 214-230.

Kumar, M., A. Dinesh, A. Narayanamoorthy and M.V.K. Sivamohan (2010), Gujarat's *Agricultural Growth Story: Exploding Some Myths*. Occasional Paper 2, Institute for Resource Analysis and Policy, Hyderabad.

Kumar, M.D. (1995): "Dynamics of drinking water in rural Gujarat: physical, social and institutional dimensions", paper presented in the workshop of *Drinking Water Problems in Rural Gujarat: Issues and Perspectives*, GIDR, January 25.

Levenhagen, D.W. (1987): *Overview Environmental Impact Report: Sardar Sarovar and Narmada Sagar Complex Project Area, Narmada River Basin, India*. Consultant's Report. World Bank. Washington D.C.

Lobo, L., S. Kumar (2009): *Land Acquisition, Displacement and Resettlement in Gujarat,1947-2004,* SAGE Publications, New Delhi.

Lobo, Lancy and Shashikant Kumar (2009): *Land Acquisition, Displacement and Resettlement in Gujarat: 1947-2004*, Sage India, New Delhi.

Luders, Rolf J. (1991), "Massive Divestiture and Privatization: The Case of Chile," *Contemporary Policy Issues*, Vol. 9, No. 4, pp. 1-19.

M.S. University, Baroda (1993): *Studies on Ecology and Environment*, Department of Geology, Project Report submitted to SSNNL, Government of Gujarat.

Mahapatra, L.K. (1990): "Rehabilitation of Tribals Affected by Major Dams and Other Projects in Orissa," in A.P. Fernandez (ed.) *Workshop on Rehabilitation of Persons Displaced by Development Projects: A Report*, Institute for Social and Economic Change and MYRADA. Bangalore.

Mallik, Pradeep (2010): "Dams! Sarovar Project A Mess," *Ahmedabad Mirror*, Ahmedabad. March.

Manthan Adhyayan Kendra (2005): *Unravelling Bhakra*. (A Report), Badwani, Madhya Pradesh, India. April.

Marie-Helene, Z. (2006): *Urban Water and Waste Water.* New Delhi: Oxford University Press.

Mathew, A. (1989): "Industries, Mines and Disposition of Indigenous People," in W. Fernandes and Enakshi Ganguly Thukral (eds.) *Development, Displacement and Rehabilitation*, Indian Social Institute. New Delhi.

McKinsey Global Institute (2010): *India's Urban Awakening: Building Inclusive Cities, Sustaining Economic Growth*. April.

Meade, Ronald B. (1991): *Reservoirs and Earthquakes. Engineering Geology*. Volume 30, Issue 3-4.

Mehta, S.S. (2010): "Access to Urban Water and Sanitation to the Poor in Gujarat and Maharashtra". *A Research Report Prepared for PAS Project*, (draft, mimeo), CEPT University.

Meinzen-Dick Ruth S., L. Brown, H.S. Feldstein and A.R. Quisumbing (1997): "Gender, Property Rights and Natural Resources," in *World Development*, 25 (8), pp. 1303-1315.

Merrey, D.J. (1996): "Institutional Design Principles for Accountability in Large Irrigation Systems", *Research Report 8*. Colombo: International Water Management Institute.

MHUPA and UNDP (2009): *India: Urban Poverty Report 2009,* Oxford University Press, New Delhi.

Milgroms, Paul and Roberts, John (1992): *Economics, Organization & Management*, Prentice Hall Publication, pp 543.

Ministry of Agriculture and Cooperation (1985): *Report of the Inter-departmental Committee on Soil Conservation and Afforestation, Sardar Sarovar and Narmada Catchment.* Government of India. New Delhi.

Ministry of Environment & Forests (1986): *Environmental Aspects of Narmada Sagar and Sardar Sarovar* – A Study for Narmada Planning Group.

— (1986): *Cost Benefit Analysis*. Circular No.11-61/85-FRY (Cons). Government of India, January, 13.

— (1987a): Office Memorandum No. 3-87/80-1A. June, 24.

— (1987b): Memorandum No. 8-372/83-FC. September, 8.

— (1987c): Memorandum No 8-646//84-FC. October, 7.

— (1990): Memorandum No.8- 29/89-FC. March, 15.

Ministry of Irrigation and Power (1972): *Report of the Irrigation Commission.* Government of India. Volume III, Part I. New Delhi.

Ministry of Urban Development (2008): *Handbook of Service Level Benchmarks*, Government of India.

Mitra, Amit (1995): "Development Distress," *Down to Earth*, Vol. 3, No. 21.

Modi, N. (2011): *Convenient Action: Gujarat's Responses to Challenges of Climate Change*, Macmillan Publishers India, New Delhi.

Morse, Bradford and Thomas Berger (1992): *Sardar Sarovar: Report of the Independent Review*, Resource Futures International Inc. Ottawa.

Musgrave, Richard A. and Peggy B. Musgrave (1989): *Public Finance in Theory and Practice*, 5th Edition. McGraw-Hill Book Company. New York.

Muthayya, B.C. and J.S. Mathur (1975): *Rehabilitation of Displaced Villagers*, National Institute of Community Development. Hyerabad.

Nagi Reddy, C.P. *et al*. (1981): *Relocation and Resettlement Plan for Project Affected Persons Under Singoor Project*, Centre for Economic and Social Studies. Hyderabad.

Naik, P.J., R.S. Patil and R.S. Maradi (1990): "Reservoir Submergence and Rehabilitation in Six Villages in the Upper Krishna Project," in A.P. Fernandez (ed.) *Workshop on Rehabilitation of Persons Displaced by Development Projects: A Report*, Institute for Social and Economic Change and MYRADA. Bangalore.

Nankani, Helen B. (1990): "Lessons in Privatization in Developing Countries," *Finance and Development*, Vol. 27, No. 1, pp. 43-45.

Narmada Bachao Andolan (1990): *Narmada – A Campaign Newsletter*, New Delhi. December.

— (1992): "Response to ARCH-Vahini's Critique of Narmada Bachao Andolan and Anti-dam Arguments," *(Mimeo)*.

— (1990): *Sardar Sarovar Project: An Economic, Environmental and Human Disaster*. Bombay.

Narmada Control Authority (1991): *Minutes of Meeting: Review of Activities of the Environmental Subgroup*, Item No XXXVIII 6 (417).

— (1992): *Silting of Sardar Sarovar.* Publication Number 2/92

—(1993): *Seismicity and Sardar Sarovar Project.*

—(2007): *Status Report on Sardar Sarovar Project.*

Narmada Planning Group (1981): *Projectised Studies: Sardar Sarovar (Narmada) Project.* Government of Gujarat.

—(1989): *Planning for Prosperity: Sardar Sarovar Development Plan.* Gandhinagar.

—(1996): *Report of the Expert Committee on financial Resources for Sardar Sarovar Project.*

Narmada Control Board (1989): *Minutes of Meeting: Rehabilitation Subgroup*, Narmada Valley Development Authority. December.

Narmada Planning Group (n.d.): "Cropping pattern and Water demand study in Narmada Command Area", Volume-I *Base Studies*, Government of Gujarat, Gandhinagar.

Narmada Project Dam Design Circle (1982 and 1983): *Sedimentation Studies and Life of Reservoir: Sardar Sarovar (Narmada) Project (FRL 455 ft)*. Vadodara.

Narmada Valley Development Authority (1992): *Action Plan of Rehabilitation and Resettlement of Oustees of Madhya Pradesh (Sardar Sarovar Project)*, Parts I and II. Government of Madhya Pradesh. Bhopal. January.

— (1993): *The Bhil Track: A Study of Displaced Tribals*, Government of Madhya Pradesh. Bhopal.

Narmada Water Dispute Tribunal (1978): *Report of the Narmada Water Dispute Tribunal*. Government of India. New Delhi.

—(1978): *Further Report on the Narmada Water Dispute Tribunal with it's Decision*. Government of India. New Delhi.

Narmada Water Disputes Tribunal Award (1979): *Sub-Clause IV(1) – Provision for Rehabilitation*, December.

Narmada Water Disputes Tribunal. (1979): *The Report of the Narmada Water Disputes Tribunal,* Volumes 1 and 2. New Delhi.

National Commission on Farmers (2005): *Serving Farmers and Saving Farming – Jai Kisan: A Draft National Policy for Farmers*, Draft Fourth Report, Ministry of Agriculture, Government of India.

National Remote Sensing Centre (2009): *Assessment of Irrigation Potential created in ABIP funded Irrigation Projects in India using Cartosat Satellite data*, Hyederabad.

Nitin Desai Expert Committee (1983): *Report of the Committee to Review the Existing Criteria for Working out Benefit Cost Ratio for Irrigation Projects*. Government of India.

Padhiari, H., (2005): "Water Service Markets in Surface Irrigation Systems: Institutions and Socio-Economic Impacts." IWMI-Tata Water Policy Programme, paper presented in *3rd ITP Annual Partners' Meet*, (unpublished research) Anand.

Paranjpye, V. (1989): *The Narmada Valley Project: A Holistic Evaluation of the Sardar Sarovar and Indira (Narmada) Sagar Dams,* Indian National Trust for Art and Cultural Heritage (INTACH). New Delhi.

— (1990): *High Dams on the Narmada*, INTACH. New Delhi.

Paranjpye, Vijay (1990): *High Dams on the Narmada: A Holistic Analysis of the River Valley Projects*, Indian National Trust for Art and Cultural Heritage. New Delhi.

Parasuraman, S. (1993): "The Anti-Narmada Movement in India: Can the Resettlement and Rehabilitation Policy Gains be Translated into a national Policy," *Working Paper Series, No.161*, Institute of Social Studies, The Hague.

— (1996): "Outcomes for Displaced: Two Case Studies," *Economic and Political Weekly*, 31 (24).

Parthasarathy, R. (2000): "Participatory irrigation Management Programme in Gujarat: Institutional and Financial Issues," *Economic and Political Weekly*, Vol. 35 (35 and 36), August, 26 to September 2.

— (2002): "Reforms in Irrigation Management and Peoples' Participation," in Rakesh Hooja, Ganesh Pangare and K.V. Raju (ed.) *Users in Water Management: The Andhra Model and its Replicability in India,* Rawat Publications, Jaipur and New Delhi.

— (2008): "Irrigation Water Pricing: Analytic of Competing Sources," in Vishwa Ballabh (ed.) *Governance of Water: Institutional Alternatives and Political Economy,* Sage, New Delhi.

— and Jharna Pathak (2008a): "The Guiding Visible Hand of Participatory Approaches to Irrigation Management," in V Ratna Reddy and S Mahendra Dev (ed.) *Managing Water Resources: Policies, Institutions, and Technologies,* Oxford University Press, New Delhi.

— and P.K. Vishwanathan (2008b): "Are Water Policies A Case of Reverse Engineering in India?" *Proceedings of the 7th Annual Partners Meet Volume - 2*, IWMI-TATA Water Policy Research Programme, Hyderabad, April.

— (2010): "The Role of Irrigation in the Growth Story of Gujarat," in Ravindra Dholakia and Samar K. Datta (ed.) *High Growth Trajectory and Structural Changes in Gujarat Agriculture*, Macmillan Publishers, New Delhi.

Patel, A.S. and H.F. Patel (1992): *Some Policy Issues for Canal Water Irrigation in Gujarat.* Department of Economics, Sardar Patel University.

Patel, Anil (2001), "Resettlement in the Sardar Sarovar Project: A Cause Vitiated". *International Journal of Water Resources Development*, 17 (3): 315-328.

— *et al.* (1990): "The Tragic Story of the Displaced: A Case Study of the Tribals Affected by Sardar Sarovar," in A.P. Fernandez (ed.) *Workshop on Rehabilitation of Persons Displaced by Development Projects: A Report*, Institute for Social and Economic Change and MYRADA. Bangalore.

— (1991): "Rehabilitation and Resettlement in Sardar Sarovar Project: Are the Critics Right?" *Social Action.* New Delhi. July-September.

— (1994): "What do the Tribals Want in the Valley of Narmada*?"* Paper presented at an *International Conference on Indigenous People and Water Reservoir Development Projects*, Montreal, Canada, April.

— (1995): "Reforms or Derailment: Activists' Dilemma Revisited," Paper presented at Refugee Studies Programme (RSP) Conference on *Development induced Displacement and Improvement*, Oxford, U.K. January.

— (1997): "What do the Narmada Valley Tribals Want?" in W.F. Fisher (ed.) *Towards Sustainable Development*, Rawat Publications. Jaipur.

— (2001): "Resettlement in Sardar Sarovar Project," *Water Resource Development*, Vol. 7(3).

—; A. Mehta; S. Iyengar; T. Parekh and R. Mishra (1992): *Morse-Berger Review: A Search for Truth,* ARCH-Vahini. Mangrol.

Patel, H.G. (1990): *Financial Working of Irrigation Projects—A Case Study of Four Projects in Gujarat.* Department of Economics, Sardar Patel University.

Patel, V.M. (1991): "Sardar Sarovar Project : Benefit-Cost Analysis" in M.T. Pathak (ed.) *Sardar Sarovar Project: A Promise for Plenty*, Oxford and IBH Publishing Co. Pvt. Ltd., New Delhi.

Pathak, M., V.D. Shah (2010): "Five Decades of Gujarat Agriculture: Some Reflections" in R.K. Dholakia, S.K. Datta (Eds): *High Growth Trajectory and Structural Changes in Gujarat Agriculture*. Macmillan Publishers India Ltd.

Pathak, Mahesh (1989): *Environmental Planning for Sardar Sarovar project.* SSNNL. Gandhinagar.

—(ed.) (1991): *Sardar Sarovar Project: A Promise for Plenty.* Oxford and IBH Publishing. New Delhi.

—(ed.) (1995): *Development Imperatives and Environmental Concerns, Sardar Sarovar Project.* Agro Economic Research Centre. Vallabh Vidyanagar.

Patkar, Medha (1988): *Sardar Sarovar: Vikas Ki Vinash* (in Hindi), Parisar, Pune.

Perry, Chris J. (2001a): "World Commission on Dams: Implications for Food and Irrigation", *Irrigation and Drainage*, 50: 101-107.

— (2001b), Water at any Price? Issues and Options in Charging for Irrigation Water, *Irrigation and Drainage*, 50 (1): 1-7.

Plusquellec, H. (2002): *How design, management and policy affect the performance of irrigation projects: Emerging modernisation procedures and design standards*. Food and Agriculture Organisation of the United Nations, Bangkok ftp://ftp.fao.org/docrep/fao/004/ac799e/ac799e00.pdf (accessed on 3rd February, 2010) 20.

Raj, P. A. (1990): *Facts: Sardar Sarovar Project*, Sardar Sarovar Narmada Nigam, Gandhinagar.

Raj, P.A. (1991): Erroneous Zones, *All About Narmada*, Government of Gujarat, Gandhinagar.

Ram, Rahul (1993): *Muddy Waters,* Kalpavriksh. New Delhi.

Ranade, Rahul and M. Dinesh Kumar (2004), "Narmada Water for Groundwater Recharge in North Gujarat: Conjunctive Management in Large Irrigation Projects", *Economic and Political Weekly*, XXXIX (31): pp. 3498-3503.

Rao, S.V.L.N. (1992): *Pricing of a Public Utility: A Survey.* Department of Economics, South Gujarat University.

Raskin, P., P. Gleick, P. Kirshen, G. Pontius and K. Strzepek (1997): "Water Futures: Assessment of Long-Range Patterns and Problems", Stockholm Environment Institute, Stockholm.

Ravindra H. Dholakia and Samar K. Dutta (Eds) (2010): *High Growth Trajectory and Structural Changes in Gujarat Agriculture*, Macmillan Publishers, New Delhi.

Reddy, I.U.B. (1992): *Displacement and Rehabilitation*, Mittal Publications. New Delhi.

— (1994): *Industrial Development and Problems of the Uprooted*, Rawat Publications. Jaipur.

Rich, Bruce (1988): *Statement before the Subcommittee on Foreign Operations Committee on Appropriations*, US Senate. June, 14.

Rondinelli, Dennis A. and John D. Kasarda (1991): Privatizing Public Services in Developing Countries: What Do We Know?", *Business & the Contemporary World*, Vol. 3, No. 2, Winter, pp. 102-113

Rosegrant, Mark W. and Hans P. Binswanger (1994): "Markets in Tradable Water Rights: Potential for Efficiency Gains in Developing Country Water Resource Allocation", *World Development*, 22 (11): pp. 1613-1625.

— and Renato Gazmuri Schleyer (1994), "Reforming Water Allocation Policy through Markets in Tradable Water Rights: Lessons from Chile, Mexico, and California", EPTD Discussion paper 2, International Food Policy Research Institute, Washington.

— (1996) Establishing Tradable Water Rights: Implementation of the Mexican Water Law, *Irrigation and Drainage Systems*, 00: 263-279.

RTI India (2008): Gujarat used only 7% Narmada dam water. As reported in Yahoo News 29th May, 2008. http://www.rtiindia.org/forum/4391-gujarat-used-only-7-narmada-dam-water.html (accessed on 3rd January, 2010).

Sah, D.C. (1993): *Monitoring and Evaluation of Resettlement and Rehabilitation Programme for Sardar Sarovar (Narmada) Project*, Report 16. Centre for Social Studies. Surat.

— (1994a): *Monitoring and Evaluation of Resettlement and Rehabilitation Programme for Sardar Sarovar (Narmada) Project*, Report 19. Centre for Social Studies. Surat.

— (1994b): "Avoiding Impoverishment Risks: Towards a Policy Frame of Dam Affected People, " Paper presented for the Seminar on *Development, Displacement and Rehabilitation* at Centre for Social Studies. Surat. 11-13 May. Revised version of the paper was accepted for publication in *Review of Development and Change* [Vol. 3 (1) 1998].

— (1994c): *Commercialisation in Agriculture and Living Condition,* (M&E Report 21), Centre for Social Studies. Surat.

— (1995a): *Monitoring and Evaluation of Resettlement and Rehabilitation Programme for Sardar Sarovar (Narmada) Project*, Report 20. Centre for Social Studies. Surat.

— (1995b): "Development and Displacement: National Rehabilitation Policy." *Economic and Political Weekly*, Vol. 30 (48). December 2.

— (1996a): *Monitoring and Evaluation of Resettlement and Rehabilitation Programme for Sardar Sarovar (Narmada) Project*, Report 21. Centre for Social Studies. Surat.

— (1996b): *Monitoring and Evaluation of Resettlement and Rehabilitation Programme for Sardar Sarovar (Narmada) Project*, Draft Report 23. Centre for Social Studies. Surat.

— (1996c): "Resettlement and Rehabilitation: Sardar Sarovar Project on the Narmada" *(Mimeo)*, Report No. 22. Centre for Social Studies. Surat.

— (1996d): *Living Condition in 31 New Sites* (M & E Report 24), Centre for Social Studies. Surat.

— (1997): "Second Generation Problems of Involuntary Migration," *Indian Journal of Agricultural Economic,* Vol. 52(4).

— (1998): "Avoiding Impoverishment Risks: Towards a Policy Frame of Dam Affected People," *Review of Development and Change*, Vol. 3(1).

— (1999): "Pressures on Land, Employment and Migrations," *Review of Development and Change,* Vol. 4(2).

— (2003): *Involuntary Migration: Evidence from Sardar Sarovar Project,* Rawat Publications, Jaipur.

— and J.N. Soni (1993): *Monitoring and Evaluation of Resettlement and Rehabilitation Programme for Sardar Sarovar (Narmada) Project*, Report 15. Centre for Social Studies. Surat.

Saleth, R.M. (1999): "Irrigation Privatization in India: Options, Framework and Experience", *Economic and Political Weekly*, 34(26): A86-A92.

Sardar Sarovar Narmada Nigam Limited (SSNNL) (1988): *Planning for Prosperity. Sardar Sarovar Development Plan*, Narmada Planning Group, Gandhinagar.

Scudder, Thayer (1991): "A Sociological Framework for the Analysis of New Land Settlements" in Michael M. Cernea (ed.) *Putting People First: Sociological Variables in Rural Development*, Oxford University Press, Second Edition. New York.

— (2003): *India's Sardar Sarovar Project,* (Unpublished Manuscript).

Sen, A.K. (1981): *Poverty and Famines,* Oxford University Press. Oxford.

Shah, G. and Biswaroop Das (1986): "Voluntary Organisation Development," *(Mimeo)*. Centre for Social Studies. Surat.

— (1988): "Voluntary Organisation in Gujarat," *(Mimeo)*. Centre for Social Studies. Surat.

Shah, H. (2009): *Narmada Yojana: Nehro ne badle pipeline ketli vyajabi?* (Gujarati) Arthbodh.

Shah, T. (1993): "Groundwater Markets and Irrigation Development", *Political Economy and Practical Policy*. Oxford University Press, Bombay.

— (2009): *Taming the Anarchy. Groundwater Governance in South Asia*, Washington DC: RFF Press.

—, S. Bhatt, R.K. Shah, J. Talati (2008): "Groundwater Governance through Electricity Supply Management: Assessing an Innovative Intervention in Gujarat, Western India". *Agriculture Water Managegment,* 95(11), 1233-1242.

—, S. Bhattacharya (1993): *Farmer Organisations for Lift Irrigation: Irrigation Companies and Tubewell Cooperatives of Gujarat*, Network Paper 26. Overseas Development Institute, London.

—, S. Verma (2008): "Co-management of Electricity and Groundwater: An Assessment of Gujarat's Jyotirgram Scheme". *Economic and Political Weekly,* Vol.43(7), 59-66.

Shah, T., A. Gulati, P. Hemant, G. Shreedhar, R.C. Jain (2009): "Secret of Gujarat's Agrarian Miracle after 2000". Review of Agriculture. *Economic and Political Weekly*, Vol. XlIV, no. 52.

Shah, Tushaar (1991): "Water Markets and Irrigation Development in India," *Indian Journal of Agricultural Economics*, Vol. 46, No. 3, July-September.

— (1993): "*Ground Water Markets and Irrigation Development in India: Political Economy and Practical Policy,"* Oxford University Press. Bombay.

Shah, V.I., Desai, N.V. and S.V. Joshi (1997): *Control of Brackishness in Drinking Water*, Central Salt and Marine Chemicals Research Institute (CSMCRI), Bhavnagar.

Shah, Zankhana and M. Dinesh Kumar (2008), "In the Midst of the Large Dam Controversy: Objectives, Criteria for Assessing Large Water Storages in the Developing World", *Water Resources Management*, 22 (12): 1799-1824.

Sharan, Girja (2007): *Harvesting Dew to Supplement Drinking Water Supply in Arid Coastal Villages of Gujarat, Paper presented in workshop on "Crisis in Drinking Water in Coastal India*, EDI, Ahmedabad.

Sharma, U.S. (1996): "Some Aspects of Drinking Water Supply in Rural Gujarat (Adequacy and Operation – Maintenance)", Proceedings of National Seminar On *Rural Water Supply Sanitation*, Organised by Centre For Development Studies, Thiruvananthapuram.

Sharma, Usha (1995): *A Study on Some Aspects of Rural Drinking Water Supply in Gujarat: Adequacy, Quality and Operation-Maintenance*, Sardar Patel Institute of Economic and Social Research, Ahmedabad.

Sheth, Pravin (1991): *Sardar Sarovar Project: Dynamics of Development*, Vikas Bharati Institute of Policy Studies, Research and Futurology. Ahmedabad.

— (1995): "Politics of Narmada Project," *(Mimeo)*. Madhya Pradesh Institute of Social Science Research. Ujjain.

Shingi, Prakash M. and V.N. Asopa (2002): *Independent Evaluation of Check dams in Gujarat Strategies and Impacts*, Centre for Management in Agriculture, Indian Institute of Management, Ahmedabad.

Shroff, Manu (1995): *Economic Viability of SSP* in Pathak, Mahesh (ed.) *Development Imperatives and Environmental Concerns*. Agro Economic Research Centre. Vallabh Vidyanagar.

Shunglu, V.K., G.K. Chadha and Jaiprakash Narayan (2006): *Report on the Status of Rehabilitation of Project Affected Families in Madhya Pradesh*, Sardar Sarovar Project Relief & Rehabilitation Oversight Group.

Singh, O.P. (2004): "Water-intensity of North Gujarat's Dairy Industry: Why Dairy Industry Should Take a Serious Look at Irrigation?", Paper presented at the Second International Conference of the *Asia Pacific Association of Hydrology and Water Resources* (APHW 2004), Singapore, June 5-9.

Singh S, (2003): "Large Dams and Environment" in Prasad K (ed.) *Water Resource and Sustainable Development*, Shipra Publications, Delhi.

— (2006): "Credit, Indebtedness and Farmer Suicides in Punjab: Some Missing Links." *Economic and Political Weekly*. 41(3), 3330–31.

Singhal, N., V. Patwari, V. (2009): "Evolving arrangements for local water diversion-delivery in SSP." *Management Traineeship Segment Report*, Institute of Rural Management, Anand.

Sinha, B.K. (1996): "National Policy for Rehabilitation: Objectives and Principles," *Economic and Political Weekly*, 31 (24).

Sneh Setu Foundation (2008): *Impact of Sardar Sarovar Project Water on Selected Villages of Gujarat*, A study report, September, Vadodara.

Solanki, Rupa (1994): *Condition of Tribal Women in Relocation Sites*, A report submitted towards partial fulfilment of coursework in Master of Social Work, Gujarat Vidyapith. Ahmedabad.

Soni, Trupti and Gautam Appa. (1993): "Sardar Sarovar Project Downstream Impact," Paper presented in *Workshop at Narmada Forum*, Delhi School of Economics and Institute of Economic Growth.

SSNL (1988): *Planning for Prosperity. Sardar Sarovar Development Plan*, Narmada Planning Group, Gandhinagar.

— (2003): *Environment & Amelioration Programme (Fisheries)*, Gandhinagar, January. No. 1," in *Texas Canal Lining and Pipeline Installation Project*", http://www.cocef.org/aproyectos/ExComHidalgo2003_05ing.htm (accessed on 28th December, 2009)

— (2009): "Benefits-Irrigation", Sardar Sarovar Narmada Nigam Limited. <http://www.sardarsarovardam.org/benefits.htm> (accessed on 12 October, 2009).

— (2009): *Water Use Plan—SSP*. Gandhinagar.

—(2010): *Report of the Expert Group for Strategy for Accelerated Development of Sardar Sarovar Project command Area*. Gandhinagar.

— (2010): Personal discussions with Executive Director, Agriculture, Sardar Sarovar Narmada Nigam Limited, Gandhinagar.

SSNNL's letter of 24/04/98 to Executive Member, NCA.

Sudarsen, V. and M.A. Kalam (eds.) (1990): *The Uprooted: Development, Displacement and Resettlement,* Gian Publishing House. Delhi.

Supplementary Affidavit to the Writ Petition (Civil) 319. 319 (Supreme Court of India, July 1997).

Supreme Court of India (2002): *Arundhati Roy.... Contemner, Contempt Petition (crl.) 10 of 2001*, Judgment, March 6, 2002, New Delhi.

SWRC (1988): "Towards People's Power in Jhabua," *People's Action.* May.

TAHAL Consulting Engineers Limited, Israel (1987): *Water Resources Planning for the State of Gujarat,* A report submitted to SSNNL, Gandhinagar, Gujarat.

Talati, J., J. Liebrand (2003): "Evolving institutions for irrigation management in Sardar Sarovar Project Command", *IWMI-Tata Water Policy Program*,(unpublished research) Anand.

—, T. Shah (2004): "Institutional Vacuum in Sardar Sarovar Project: Framing 'Rules-of-the-Game'". *Economic and Political Weekly*, 39(31), 3504–509.

Tata Economic Consultancy Services (1981): *Methodological Framework for Economic Appraisal of Narmada Project.*

Tata Institute of Social Sciences (1993): *The Sardar Sarovar Project: Experiences with Resettlement and Rehabilitation*, Tata Institute of Social Sciences. Bombay.

Thakkar, Himanshu (2005): "Who Pays for Our Giant Follies?" *Info-change News & Features,* October.

The Hindu (2006): *GoM's confidential report on A Brief Note on the Assessment of Resettlement and Rehabilitation (R&R) Sites and Submergence of Villages of the Sardar Sarovar Project.* Note of the Group of Ministers Dated 9 April 2006. Chennai. April 17.

Thobani, Mateen (1997), "Formal Water Markets: Why, When and How to Introduce Tradable Water Rights", *The World Bank Research Observer*, 12 (2): 161-179.

Thukral, Enakshi Ganguly (1989): "Dams for Whose Development?" in W. Fernandes and E. Ganguly Thukral (eds.) *Development, Displacement and Rehabilitation: Issues for a National Debate*, Indian Social Institute. New Delhi.

— (ed.) (1992): *Big Dams, Displaced People: Rivers of Sorrow Rivers of Change,* Sage Publications. New Delhi.

Times of India (2009a): "Shelve Narmada pipeline plan: Expert." Ahmedabad edition, September 12.

— (2009b): "Narmada pipeline project scaled down." Ahmedabad edition, November 13.

Tolba, M.K. (1988): *Sustainable Water Development: Opportunities and Constraints*, Vol. 13 (4).

TRENDSNIFF. (2008, September): Retrieved December 2010, from www.trendsniff.com: http://trendsniff.com/2008/09/16/targeting-top-20-growth-cities-in-india/

United States Department of the Interior, Bureau of Reclamation (1976): *Design of Gravity Dams.* Denver.

— (1983): *Economic Appraisal of Sardar Sarovar Project*. Bombay.

Unknown (2009): "State to Press for a Taller Narmada Dam," *Indian Express*. October 23.

US Bureau of Reclamation (1973): *Water System Automation*, Current Information on Automation of Reclamation Water Systems, United States Department of the Interior, Denver, Colorado.

Vaidyanathan, A. (2006): *India's Water Resources – Contemporary Issues on Irrigation*, Oxford University Press, New Delhi.

Varghese, B.G. (1994): *Winning the Future: From Bhakra to Narmada, Tehri, Rajasthan Canal*, Konark Publishers Pvt. Ltd., New Delhi.

Varma, S.C. (1985): *Human Settlement in Lower Narmada Basin*, Narmada Valley Development Authority. Bhopal.

Vermillion, Douglas (1997): *Impacts of Irrigation Management Transfer*, IIMI, Research Report 11, April.

Vice Chairman, NVDA's letter No. NVDA/D (Fish)/88/3867 of 1/11/88 addressed to the Chairman, NCA.

Vyas, J. (2001), "Water and energy for development in Gujarat with special focus on the Sardar Sarovar project", *International Journal Water Resources Development*, 17(1):37–54.

Vyas, Jay Narayan (2001): "Water and Energy for Development in Gujarat with Special Focus on the Sardar Sarovar Project," *International Journal of Water Resources Development,* Vol. 17, 1, March.

Wallingford, H.R. (1993): *Environmental Changes Downstream of Sardar Sarovar Dam.* Ex 2750.

—(1993): *Sardar Sarovar Project Hydrology and Water Resources.*

—(1993): *Sediment and Backwater Aspects of SSP Report.* Ex 2761.

—(1993): *SSP: Command Area Environmental Impact Assessment Progress Report.*

—(1995): *Potential for Agro Chemical Pollution from the Sardar Sarovar Command Area, India.* Ex 3177.

Wallingford, H.R. (1995): *Environmental Changes Downstream of Sardar Sarovar*. Oxfordshire, UK. Report No. EX 2750. of World Bank.

Wamana Consultants (1994): *Study of Preparation of Detailed Command Area Development Plan: SSP.*

2030 Water Resources Group (2009): *Charting Our Water Future: Economic Frameworks to Inform Decision-Making*, 2030WaterResourcesGroup @mckinsey.com

Watt, Jacqueline (2008), "The Effect of Irrigation on Surface-Ground Water Interactions: Quantifying Time Dependent Spatial Dynamics in Irrigation Systems," Thesis submitted to Charles Sturt University for the Degree of Doctor of Philosophy, School of Environmental Sciences, Faculty of Sciences, Charles Sturt University, March.

Wood, John R. (1993): "India's Narmada River Dam: Sardar Sarovar under Siege," *Asian Survey*, Vol. 33, No. 10.

World Bank (1985): *Narmada River Development, Gujarat—Sardar Sarovar Dam and Power Project*, Staff Appraisal Report 5107-IN; and *Water Delivery and Drainage Project*, Staff Appraisal Report 5108-IN.

—(1990): *Review of Sardar Sarovar (Narmada) Projects Economic Analysis*. India Agriculture Operations Division.

—(1993): *Project Completion Report: Narmada River Development, Gujarat.*

—(1995): *Project Completion Report-India-Narmada River Development-Gujarat Sardar Sarovar Dam and Power Project*, Report No. 14159.

— (1995): *Project Completion Report (Sardar Sarovar).* World Bank.

— (2005): *India's Water Economy: Bracing for a Turbulent Future*, World Bank, Washington.

Zoology Department (1994): *Environmental Impact Assessment Studies on inland and Marine Fisheries Relevant to the Command Area of Sardar Sarovar Project.* M.S. University. Baroda.

Index